The treatment
of industrial wastes

second edition

Fully automatic package deionization system.
(*Illinois Water Treatment Company.*)

The treatment of industrial wastes

EDMUND B. BESSELIEVRE, P.E.
Late Fellow, American Society of Civil Engineers
Consultant, Forrest & Cotton, Inc., Dallas, Texas

and

MAX SCHWARTZ, C.E./M.E.
Member, American Institute of Plant Engineers
President, Max Schwartz/Consulting Engineers, Inc.
Los Angeles, California

second edition

McGRAW-HILL BOOK COMPANY
New York   St. Louis   San Francisco   Auckland   Düsseldorf   Johannesburg
Kuala Lumpur   London   Mexico   Montreal   New Delhi
Panama   Paris   São Paulo   Singapore   Sydney   Tokyo   Toronto

Library of Congress Cataloging in Publication Data:

Besselievre, Edmund Bulkley, date.
  The treatment of industrial wastes.

  Bibliography: p.
  Includes index.
  1. Factory and trade waste. 2. Sewage—Purification. 3. Water—Pollution. I. Schwartz, Max,
1922-   joint author. II. Title.
TD897.B38 1976   628.5′4   75-9835
ISBN 0-07-005047-3

1 2 3 4 5 6 7 8 9 0   KPKP   7 8 5 4 3 2 1 0 9 8 7 6

The editors for this book were Jeremy Robinson and Patricia A. Allen, the
designer was Elliott Epstein, and the production supervisor was George E.
Oechsner. It was set in Linotype by Bi-Comp.

It was printed and bound by Kingsport Press.

Contents

Preface to the second edition

Five years have passed since the Preface to the first edition was written. Much has changed during those years and public attitude toward industrial wastes is foremost among these. Industry is no longer given the option to treat its wastes or allow the local municipalities to assume the burden. Every plant manager is being bombarded with official notices, surcharge forms, permit applications, and threats of heavy fines. The mails bring a flood of environmental protection publications, pollution control equipment, and claims of recycling miracles.

The federal regulations, while quite explicit, are just now filtering down to the local communities. Interpretation of the exact wishes of every governing agency is quite difficult. But pollution control is now the problem of every industry, large and small, with organic and inorganic wastes.

The second edition of "The Treatment of Industrial Wastes" brings the book up to date in philosophy and content. Treatment units once called "unconventional" are now installed as regularly as clarifiers. I have included material based on the needs of small industries and have tried to guide plant managers through the phases of any small pollution control installation, be it manual or fully automatic. This includes interpretation of the laws, discussion of pollution control equipment, etc. The book also contains information on control manholes and meters, now so crucial for wastewater discharge permits.

Unfortunately, with skyrocketing prices, it has been impossible to provide meaningful figures. When cost figures are stated, the corresponding date is generally mentioned. The most realistic method of obtaining estimates is to contact the manufacturer, state exactly what is required, and obtain his present cost figures.

The second edition, as the first, deals with implementation of theory. To prevent the destruction of our natural resources, the treatment facilities must actually be constructed, not just discussed in technical journals. I have tried to provide a means for determining the most economical method of installation, thus expediting its construction.

Portions of this revision required extensive research, compilation, and organization of materials. In particular, I was aided in the area of inorganic waste treatment systems by Nina Klein, B.S., a graduate in chemistry from UCLA. In the areas of federal, state, and local regulations and guidelines, and governmental incentives to industry, I was aided by Stephen Klein, B.A., UCLA, J.D., Southwestern University School of Law.

Their efforts augmented my own background in civil and mechanical engineering, making this book a complete presentation of the treatment of industrial waste.

Max Schwartz

The treatment
of industrial wastes

second edition

Chapter One

Introduction

Today, all parts of Earth are linked by supersonic jets, satellite TV, and the common problem of pollution. The same factor that served to shrink Earth is threatening to destroy her. Technology has been hailed as the savior and the executioner of mankind. The industrialized nations have comforts and luxuries never before dreamed possible. But they have paid for these technological advances with the very Earth on which we live. Resources are the stuff with which technology is built. The industrial nations possessed natural resources within their borders or obtained access to resources in distant colonies. It was by virtue of this that they became industrialized. Industry was supplied with unlimited raw materials. No thought was given to the limitation of the resources. If Earth spontaneously generated these materials once, she could replenish the supply again. But the old nemesis Time came into play. It has taken Earth 4 billion years to evolve to her present state. Technology will cause the depletion of her resources in a few lifetimes. Natural replenishment of resources proceeds at an agonizingly slow snail's pace while consumption races along by geometric leaps and bounds.

Of what natural resources are we speaking? Coal, oil, iron, copper, air, and of course water, to name a few. It is easy to understand how processes can "use up" coal, oil, etc. But how can air and water be depleted? The answer is simple. Air and water are not destroyed but rather rendered unfit for use. Unlike other resources, air and water are basic to life. But they must be used in their natural, unaltered state. It is true that Earth has capacity for cleansing the air and water, but the pace at which we are polluting these resources makes natural processes impossible.

It is now becoming obvious that the resources of Earth are indeed limited, as is Earth's capacity to cleanse herself. In her present state Earth may be likened to a spaceship. The philosophy on board a spaceship bound for a distant point is reuse. Everything must be reused again and again. This must become the ultimate philosophy of Spaceship Earth.

On a more concrete level, every individual resource user must eventually reuse his waste products. He may no longer render unfit the water he uses and expect the next user downstream to suffer the consequences. Since reuse has become a major theme in the United States, this book is devoted to helping industry reuse its water resources. Reuse does not necessarily mean each industry must recycle all its wastes. But the wastes must be rendered suitable for the next industry down the line. Even waters discharging into the oceans must be reusable, for aquatic life must not be harmed.

The ancient peoples were aware of the importance of water. Hippocrates, Hammurabi, and even the Caesars set down laws governing what could and could not be discharged into the stream waters. But population and technology were limited then, and compliance or noncompliance was not as crucial. As industry developed, more dictums were sent forth. In West Florida the following edict was proclaimed:

> In order to avoid the inconveniences that could injure the preservation of the herds, I command to all who make indigo, that they shall burn the weed as soon as they take it from the steeper or from the vats, and it shall not be lawful for anyone to convey the water from his vats into any stream whose waters are used by the inhabitants or their herds. And if there be any indigo factory erected without attention to this very important point, it shall be removed; or it shall be constructed in such a manner that the great injury in contaminating the water shall be removed.

Governor Manuel Luis Gayoso de Lemos y Amorin issued that statement in 1794.

That the intentions of the ancient leaders were not followed was obvious. Throughout the Middle Ages the streets of Europe were nothing more than open sewers. Colonists brought this practice to the New World. Countless deaths can be attributed to polluted water, the most catastrophic occurrence being the Black Death of 1348–1349.

But, you say, that was caused by insufficient sanitary waste treatment. All municipalities treat their sanitary wastes. This is not completely true. Dischargers into ocean waters are often lax about the degree of secondary treatment they deem necessary. "A little more 'fertilizer' for the ocean won't hurt," they claim. But scientific reports on such unpleasantries as red tides have linked the increased frequencies of occurrences with the high concentrations of organics in the coastal waters.

Biological contaminants are not the only source of major pollution problems. The growth of process industries has resulted in astronomical volumes of wastes, containing every conceivable inorganic compound and element. Mercury poisoning has been perhaps the most dramatically documented result, but cyanide, chromium, and copper along with oils and greases and plain detergents are just as toxic to fish and humans forced to consume the water downstream of the discharger.

What can be done? Not all wastes can be piped to the oceans for a quick burial. The oceans are limited, as has been stated. Aside from that, the volume of water consumed by industry is too great. The water must be available for reuse. The more equitable solution sees each industry reusing its own discharge. This is in fact the ultimate goal of the most recent legislation: zero discharge by 1985. But in the meantime, what has been done and can be done?

Industry, it must be said, has been treating its wastes. But treatment is an expensive business, and as production increases, old treatment facilities become obsolete. To help industry cope with the costs of treatment plants, the federal government began by allotting $305 million for research and development under the Clean Waters Act of 1966. The 1972 amendment to the Federal Water Pollution Prevention and Control Act provided for additional grants.

With the 1972 act, a new era of environmental concern was launched. The act established a system of national pollution limitations and standards. The objective of this program was to restore the quality of the nation's waters with an ultimate target to obtain zero discharge of pollutants by 1985. To obtain this objective, various stages of compliance were set with target dates. This was known as the National Pollutant Discharge Elimination System. It involved a gradual, progressive, and enforceable set of regulations requiring both local government and private industry to comply with the long-range program.

The field of inorganic waste treatment is relatively new when compared to the treatment of organic waste. Municipal sewage treatment plants have been in existence for a long time, and much has been learned by science and experience. But the treatment of industrial waste containing inorganic contaminants is in its early stages of development. Much of the equipment being utilized is borrowed from standard water purification systems used on process feedwater. Often this equipment is very expensive and designed to handle only small amounts of flow compared to the quantities involved in the discharge of industrial wastewater. Demineralizers, reverse osmosis units, activated charcoal filters, and reduction and oxidation units are in this category of equipment.

Treating inorganic wastes requires large quantities of expensive chemical additives used for neutralization, oxidation, and regeneration. The major obstacle for industrial treatment of its wastes is therefore economic. Because of high costs, industries favor construction of "building block" treatment facilities. That is, they would first treat the waste for the most troublesome contaminant, and then proceed to treat for other less noxious pollutants. But the regulations which govern discharge of contaminants are difficult to understand. Each local authority has a different rate of compliance with national guidelines. New industrial communities are likely to be more restrictive than old factory towns which rely on their local industry for their livelihood. Industry is naturally disillusioned with legislative acts that allow one industry to discharge its wastes virtually untreated while demanding that another only a few miles away construct expensive facilities.

In addition, the cost of waste treatment is not necessarily in proportion to the size of the company. A small metal-finishing shop making drawer hardware may have a much higher treatment cost than a giant furniture company that assembles the furniture. The chemicals used in production may vary from shop to shop. They will have a great effect on the cost of treatment. The concentration of contaminants, rather than the total weight of these contaminants, is often the criterion used by local authorities. Here a large plant may have a less stringent problem because of the dilution of contaminants.

Pollution control is the wave of the future. If we are to survive on this Earth, the basic compound water must be fit for human and animal consumption. So industries are facing up to the fact that the most economical way to meet federal guidelines is to recycle the water and reclaim the waste chemicals. It does not make economic sense to purify one's wastewater so that the next industry may use it.

There then emerge five basic objectives in the control of industrial waste. Each of these must be met at the lowest short- and long-term costs. The objectives are removal of all pollutants; recovery of any valuable by-products; recycling of recovered material including water; providing for a simple operation; and low capital investment.

In order to design an economical treatment plant, the importance of accurate fact gathering must not be minimized. It provides the engineering basis for the design of effluent control and the operability of the treatment system. Once the plant is built, maintenance must be as strictly followed as fire prevention regulations. Indiscriminate dumping of oils, paints, and chemicals into floor drains may nullify all the effort of a treatment system. Without good housekeeping, waste control investments are of little value.

The following fourteen chapters present the author's approach to solving the various pollution problems, with an emphasis on the economics of the situation. The time is now to begin to conserve our most important resource, pure water.

Chapter Two

Selection and coordination with the consulting engineer

When an industrial facility is confronted with a waste disposal problem, it frequently seeks help from an unbiased party. This is the first and most important step in obtaining a final, economical, and ecologically satisfying answer to the problem.

Management, in the person of the plant manager, has the task of selecting the engineer or engineering firm to design the treatment plant. Most industries do not have experienced industrial waste engineers on their staff. They are therefore forced to look outside for help, keeping in mind the desire not to become obligated to a particular purveyor of equipment, apparatus, or process system.

A consulting engineer is a good choice to make. He has experience with a variety of similar work, a large and broad background to bring to bear on specific design tasks, a large and broad team of specialists to draw upon, and a knowledge and evaluation of other specialists.

The consultant is also more flexible than a member of the plant manager's staff. He is free to consider bold, progressive ideas whereas an in-house engineer suffers many constraints. Not affected by internal politics, the consultant is free to exchange ideas and design concepts with anyone.

RESPONSIBILITIES

The plant manager may select the engineering firm on the basis of reputation in the field of industrial waste control, the recommendations of management, or the professional quality of the proposal. The plant manager must be satisfied with the firm with which he is about to contract. A reputable firm will recognize the following responsibilities to his client.

5

1. To regard himself and his staff as the agent or agents of the client in all aspects. To regard any information obtained as to secret or classified methods or processes as entirely confidential.

2. To have and to keep an open mind as to the method of treatment that he will employ until he has explored all aspects of the problem.

3. Not to rely on outside influences or sources in designing the plants or facilities simply because an idea has been successful in a seemingly parallel case.

4. To have personal acquaintance and experience with all new and recognized methods and processes of liquid waste treatment. To evaluate them individually with respect to the requirements of the case. To decide upon the particular method or process, or combination of methods or processes, only after careful consideration and personal inspection.

5. To design a plant or facility that will employ the equipment or apparatus best suited to the needs of the project. Whether this results in the use of equipment of one manufacturer or a combination of units made by several is unimportant if such a combination will produce the desired result—a practical, economical plant.

6. To study the problem carefully and objectively to determine:

    a. Whether the client really needs a separate waste treatment plant or can discharge the wastes, untreated or partially treated, into the sewerage system of the local municipality.

    b. Whether it is possible to combine the wastes with those of an adjacent plant or plants to obtain the advantages of reactions between the chemical or solid elements of the various wastes at low cost.

    c. Whether, by some internal rearrangements in factory or mill operations, by better process control, the concentration and volume of the wastes may be so reduced as to reduce the size, cost, and complexity of the treatment plant design.

7. To discuss the problem frankly with all the regulatory authorities having jurisdiction over the stream or body of water into which the final effluent is to be discharged, to establish a maximum degree of treatment actually needed and to resolve any differences or conflicts that may exist.

8. To base his final selection of chemical aids on laboratory tests, not superficial field tests. Such tests should be performed in a recognized laboratory facility under the direction and supervision of the engineering firm.

9. To use the best facilities available to ascertain the proper method of treatment in a specific case. Derive information not solely from literature, books, pamphlets, reports, and articles in the technical press but by inspection trips to existing plants in use on wastes with similar characteristics. Industrial waste treatment is changing so rapidly that the responsible engineer will not select or discard a process, old or new, on the basis of hearsay only.

SERVICES

Once the engineering firm has been selected, all aspects of the new relationship must be discussed. The project involved may be simply a study of existing treatment facilities. The project could entail the design of a million-dollar reclamation-recycling-treatment center. Whatever the project, the exact scope of the services to be provided must be established.

Suggested Checklist

> Preliminary report
>> Wastewater constituents
>> Permits required
>> Flow calculations
>> Contact with regulatory agencies
>
> Cost estimate
>
> Design services
>> Industrial engineering
>> Architecture
>> Civil engineering
>> Mechanical engineering
>> Chemical and process engineering
>> Structural engineering
>> Electrical engineering
>
> Construction services
>> Occasional inspection
>> On call for problems
>> Continuous inspection

With the scope of services established, both parties must take care not to meander. Meetings must begin immediately, especially when the problem has reached crisis proportions. At the early meetings the schedule should be established. It is also important to determine the method of control to be exercised during the project.

SCHEDULING

In order to manage a project economically and efficiently, the plant manager must arrange a detailed schedule agreeable to all participants. Occasionally, the engineer includes a rough schedule of his work with his proposals. This may include the dates of preliminary and final drawing submittals. But this type of schedule is not adequate when a very tight construction schedule is required. Long-term delivery items must be given priority. If the new work interrupts production, accurate scheduling is essential.

Mark	Description of activity	Aug.	Sept.	Oct.
A	Preliminary layout plan			
B	Preliminary equipment specifications			
C	Review and approval			
D	Design of holding tanks			
E	Design of treatment tanks			
F	Design of clarifiers			
G	Design of filter stations			
H	Design of piping system			
I	Design of site facilities			
J	Design of utilities			
K	Shop drawings of tanks			
L	Shop drawings of piping			
M	Requisitions			
N	Purchasing			
O	Fabrication			
P	Installation			
Q	Site construction			
R	Utility installation			
S	Start-up and testing			

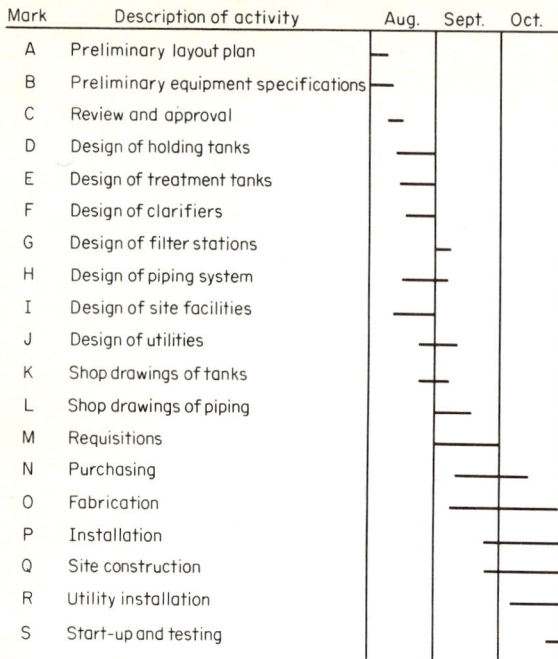

Fig. 2-1 Typical bar chart.

There are two common methods of graphically describing a design and construction schedule. The simpler one is a bar chart. Figure 2-1 shows a typical bar chart for a sample project. The other, the critical path method (CPM), is shown in Figure 2-2.

The bar chart indicates the date of the beginning and completion of each unit of operation. These units may be "preliminary design of treatment area," "shop drawings of pipe supports," or "requisition of pumps."

The encircled letter on the critical path method diagram, Figure 2-2, indicates the unit of operation of decision listed on the bar chart. The number between circled tasks indicates the estimated time in days or weeks to complete

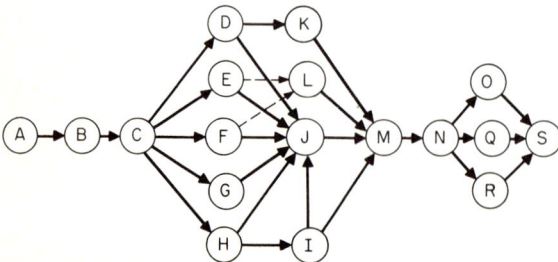

Fig. 2-2 Typical critical path network.

each unit of operation. By adding the total elapsed time along each path from *A* to *S*, the longest path can be determined. This longest path is called the critical path. Any delay in this specific path would delay the entire project. On complex, long-duration projects, the data of the critical path system are processed by computer. In this way, final completion dates are continuously corrected by actual elapsed time.

Whatever system is used, the plant manager should keep in mind that the chart should be in keeping with the size and complexity of the job. It should not be too complicated, for maintenance of the chart can be a full-time job in itself. There will be many changes on each job before it is finalized.

Some project engineers use cards attached to a 2- by 3-ft board with a piece of drafting tape. Noted on the upper left corner is the start and on the right upper corner is the completion. As the work on the project progresses, more steps, decision points, and alternatives become apparent. Shifting the cards on the board is easier than erasing and rewriting. After the network diagram is established, the critical path can be easily established.

The project engineer and the plant manager should discuss and together determine all aspects of the schedule. Some of the critical and limiting schedule requirements might include:

1. Agency deadlines for pollution abatement

2. Disruption of production

3. New utilities

4. Stockpiling of products before construction

From the engineer's point of view, there are some additional limiting aspects:

Results of testing—wastewater analysis, soil

Results of treatment testing

Availability of data

Availability of personnel on staff

Establishing a schedule at the beginning of the project allows the plant manager to plan his production, determine allocation of funds, assign plant and administration personnel, and review the progress of the engineer and contractor. The engineer can also determine his staff assignments, emphasize critical items, and coordinate other projects in the office. The schedule also tends to keep both the plant manager and the engineer from delaying or procrastinating decisions or actions. After reviewing the scheduling, management can determine whether it will do part of the construction, fabrication, and installation with its own in-house force or contract the work out. On time and material engineering contracts, the plant manager has a good tool to continuously judge whether the cost of the engineer's services is consistent to the contract.

CONTROL

As the representative of management, the plant manager should demand periodic review of all plans, specifications, and reports. In fact, many engineering firms will insist upon such reviews, with approvals or corrections recorded. An efficient method of review requires that the engineer submit a preliminary study report. Later, during the design phase, he is expected to submit three sets of plans, specifications, and cost estimates, at different intervals—10, 50, and 90 percent completion.

LIAISON

To avoid confusion, one representative from plant engineering and one from the engineering firm should be selected to guide the project. All communications should be through these two individuals. Each accumulates information, requirements, questions, and other pertinent data from the various divisions in his own company. The plant manager is in close contact with the representatives of management, production, maintenance, quality control, and purchasing. He must be sure that the requirements of each department are integrated into the whole of the project. The engineering representative must supply correct data and answers to the group leaders of the process, civil, industrial, mechanical, structural, and electrical engineering groups.

The project engineer should encourage frequent dialogues, conferences, and communications between management and the consultant, but all such activities should be directed through the project engineers themselves.

COMMUNICATION

Once the groundwork has been laid for communication, how do the participants ensure "meaningful" exchange? How can the two parties best exchange ideas? The medium of communication most generally accepted in these circumstances is graphic representation. It takes the form of flow diagrams, plans, details, sketches, bar charts, and tabulations, to name a few. But here, as in any language, a dictionary or glossary is necessary because a common understanding of symbols is critical. Each discipline has its own recognized symbols and nomenclature. It would behoove the plant manager to have included in his library the standard symbols for each field of engineering, since it goes without saying that a single symbol on an engineering drawing can mean more than a paragraph of written words. See Chapter 6 for recommended piping symbols.

The need to transmit drawings, specifications, cost estimates, etc., between management and the engineer will arise frequently. Two types of forms, shown in Figures 2-3 and 2-4, are available to expedite the transactions.

MAX SCHWARTZ
CONSULTING ENGINEERS, INC.
6464 Sunset Boulevard
LOS ANGELES, CALIFORNIA 90028

Phone 462-2369

LETTER OF TRANSMITTAL

DATE Aug 2, 1974 JOB NO. 1303
ATTENTION Mr. Ozzie Grisat - Process Engr.
RE: Rinse water overflow disposal
Automatic Plating Line

TO AJAX HARDWARE CORP.
825 So AJAX AVE
CITY OF INDUSTRY, CA 91747

GENTLEMEN:

WE ARE SENDING YOU ☑ Attached ☐ Under separate cover via ________ the following items:

☐ Shop drawings ☑ Prints ☐ Plans ☐ Samples ☑ Specifications

☐ Copy of letter ☐ Change order ☐ ________

COPIES	DATE	NO.	DESCRIPTION
3	7/31/74	P-2	Preliminary Layout - Overflow system
3	7/31/74	SP-2	Preliminary Specifications - overflow system

THESE ARE TRANSMITTED as checked below:

☐ For approval ☐ Approved as submitted ☐ Resubmit ____ copies for approval

☐ For your use ☐ Approved as noted ☐ Submit ____ copies for distribution

☐ As requested ☐ Returned for corrections ☐ Return ____ corrected prints

☑ For review and comment ☐ ________

☐ FOR BIDS DUE ________ 19____ ☐ PRINTS RETURNED AFTER LOAN TO US

REMARKS Please advise if this meets with your requirements for peak flow.

COPY TO E. Marks - Plant Manager

SIGNED Max Schwartz

If enclosures are not as noted, kindly notify us at once.

Fig. 2-3 Typical letter of transmittal form. (*New England Business Service, Incorporated*)

The letter of transmittal, Figure 2-3, can make three copies using NCR paper. The original and one copy are generally sent to the recipient, and one copy is kept in the sender's file. The recipient signs one copy with any comments noted and returns it to the sender.

The message reply form, Figure 2-4, is usually used for simple memos which do not require formal typed letters. These forms also make three copies, using NCR paper. Two copies are sent to the recipient and one remains with the sender. Here too the recipient notes his comments and returns one copy to the sender. In this manner a complete history of the exchange of thoughts is

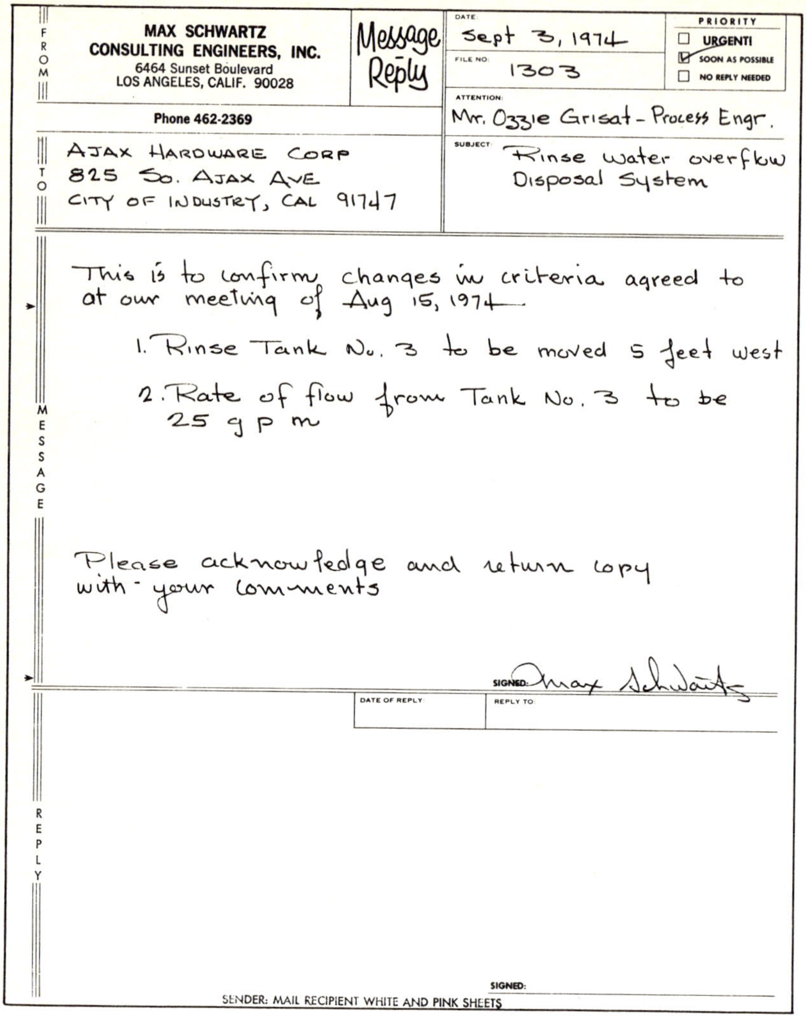

FROM
MAX SCHWARTZ
CONSULTING ENGINEERS, INC.
6464 Sunset Boulevard
LOS ANGELES, CALIF. 90028

Phone 462-2369

Message Reply

DATE: Sept 3, 1974
FILE NO: 1303

PRIORITY
☐ URGENT!
☑ SOON AS POSSIBLE
☐ NO REPLY NEEDED

ATTENTION: Mr. Ozzie Grisat - Process Engr.

TO
AJAX HARDWARE CORP
825 So. AJAX AVE
CITY OF INDUSTRY, CAL 91747

SUBJECT: Rinse water overflow Disposal System

MESSAGE

This is to confirm changes in criteria agreed to at our meeting of Aug 15, 1974

1. Rinse Tank No. 3 to be moved 5 feet west
2. Rate of flow from Tank No. 3 to be 25 gpm

Please acknowledge and return copy with your comments

SIGNED: Max Schwartz

REPLY

DATE OF REPLY: | REPLY TO:

SIGNED:

SENDER: MAIL RECIPIENT WHITE AND PINK SHEETS

Fig. 2-4 Typical message reply form. (*New England Business Service, Incorporated*)

clearly recorded. It is difficult to have too many of the memos during any project. They serve countless purposes, from confirming telephone conversations, recording questions and answers, and serving as a tickler file to keeping a chronological history of the project. These memos also carry legal weight in any litigation. They can establish who broadened the scope of work or who first requested a change.

There are unwritten laws governing the coordination of any project. The plant manager should be aware of these laws for proper coordination. One must always confirm instructions and engineering commitments in writing. It is risky to assume that the job will be done because of a verbal agreement. Many busy people have poor memories, and almost everyone will take a matter a great deal more seriously if he sees it in writing. On critical items it is wise to include a third party for a copy of the memo, as a witness.

EXPEDITING ACTION

In water pollution problems time is of the essence. Remember, in carrying out a project, not to wait for manufacturer representatives, vendors, and others to furnish critical data. Go after them and keep after them. Many projects progress in a direct proportion to the amount of follow-up and expediting that is applied to them. Expediting includes planning, investigating, promoting, and facilitating every step of a project. Develop a technique of looking immediately for an alternative way around each obstacle encountered, some other recourse or expedient to maintain progress without losing momentum. On the other hand, be careful about excessive demands of the engineer for action. Too much insistence and agitation may result in antagonism instead of cooperation. Before committing an opinion or promoting an undertaking, wait until you have had a reasonable opportunity to obtain and study the facts. Then be careful of the accuracy of your statements.

Strive for conciseness and clarity in oral and written reports and instructions. Try to keep each memo or letter confined to a single subject.

CONFERENCES

Project conferences are important for exchange of ideas. They should not be too large or too small. Large meetings frequently dissipate the subject over a number of conflicting or irrelevant points of view, in a generally superficial manner. A considerable amount of skill is required to manage a sizable meeting so as to keep it on the proper subject.

Small meetings, three or four persons, can usually hammer out a program or dispose of knotty problems much more effectively, but there is the possibility that not all interested parties are represented.

COORDINATION

As coordinator, the plant manager has a great deal of responsibility. The following is a suggested checklist to keep the project in line:

Define your objectives

Plan the job by outlining the steps to be accomplished

Prepare a definite schedule

Assign definite responsibilities for each item

Make sure that each man has sufficient help and facilities

Follow up; check up on progress of work

Revise your schedule as required

Watch for "bottlenecks," "log jams," and "missing links"

Hit lagging items hard

Drive to finish on time

FINANCIAL ARRANGEMENTS

The plant manager is usually not involved with the financial aspect of selecting an engineer. Nevertheless, it is wise to be aware of the particular compensation arrangement. On projects involving design of new facilities, where these facilities are neither too complicated nor repetitious, a payment of a percentage of the overall construction cost is usual.

Other projects that do not involve extensive construction, or the risk of major complications, lend themselves to a fixed lump sum. For projects requiring extensive studies, without a defined scope, a salary cost times a multiplier, plus direct nonsalary expenses defined in the contract agreement, is the most equitable form of compensation.

It goes without saying that the plant manager should not cause, especially when working under the first two contracts, any meandering, detours, or extensive studies that would justifiably cause demands of "extra compensation" or hard feelings, at the least. If the scope of the work has been changed, an addendum to the contract must be made immediately. This is in the interest of fairness and strengthens mutual confidence.

PURCHASE ORDERS

The contract determines the method of compensation. How important then are formal purchase orders? To what extent should formal purchase orders be used when there are changes, additions, or reductions in the original scope of the project? These are vital questions. Normally a purchase order requires a substantial amount of time. Often a requisition, cost estimate, justification, and approval by top management are required before a purchase order for change of scope is approved. This is often too time-consuming for an emergency change. The plant manager and the engineer must make nonverbal and verbal agreements to enlarge the scope of the project. After the heat of the emergency has long passed, there may be a disagreement as to the amount of extra work. It is essential that such changes be documented by records as soon as possible.

EXTENT OF SERVICES

Another financial problem is determining exactly when the obligated services of the engineer are ended. Long after the project is completed, should the engineer be expected to determine the changes to the system caused by a new process tank, or a change in process reagents? These nebulous areas of financial compensation are sources of ill feelings.

PROFESSIONAL

Every engineering firm regards itself as "professional." Professionalism carries with it many responsibilities. The extent of these responsibilities is directly related to the role assigned to the engineer by plant management. If the engineer is retained directly by management, he is considered a *prime professional*. As a prime professional he has the responsibility of overall planning and coordination of the design and construction, administration of all professional services needed, cost estimating, budgeting, and cost control. The prime professional is directly responsible to management.

When the engineer serves management indirectly through another professional, the engineer is the *professional-not-prime*. As a professional-not-prime, he must maintain control of work that he has professional responsibility over although he may have no concern for the overall aspects of the job. The plant manager in conjunction with management must decide the role he requires of his engineer.

The plant manager can expect the following *basic services,* regardless of the engineer's role. These, of course, should be mentioned in the contract. There are basic services in both the preliminary phase and the design phase of every project.

Preliminary Phase

In the preliminary phase these services include conferences with other engineers—management, building officials, and regulatory agencies—to establish the scope of the project and the requirements of alternate designs, soil investigation, economic studies of alternate construction systems, preparation of preliminary designs, and counseling on construction materials and methods of construction.

Design Phase

Basic services in the design phase include conferences; calculations for structural, mechanical, and electrical design; preparation of contract drawings; preparation of a reasonable number of prints of work in progress; and participation in establishing the material testing, fabrication, and construction inspection programs.

Construction Phase

If the engineer is retained as professional-not-prime, his responsibility ends when the design drawings and specifications have been turned over to the owner. From this point on, all further work involved with construction is handled by the project engineer or prime professional. When the services extend into the construction phase, basic services can include assistance in procurement of bids, consultation, and advice during construction; preparation of supplementary sketches and details needed to resolve field problems; review of shop and erection drawings; review of laboratory, shop, and mill tests; and observation of construction in progress. (*Note:* Such observation assists in quality control but does not guarantee contractor's performance.)

Predictable *special services* are also arranged by explicit notation in the contract. They often require compensation far in excess of the original engineer fee.

Special services can include land surveys and soil investigations. Technical inspection of construction can be provided by a full-time resident engineer or inspector. He can review and approve requests for all payments to contractors and issue certificates of completion to management. Other special services can include providing field measurements of existing construction and preparation of shop details.

Unpredictable special services are usually not included in contracts. The services are outside the scope of a lump sum contract and should be viewed as justifiable extras. Some services include excessive preliminary designs; preparation of alternate designs; design work abandoned before completion; design work completed for work not built; redesign for major changes after approval of preliminary plans; evaluation and recommendations on contractor's claims; assistance to management as an expert witness in litigation arising on a project; and observation of construction for a period prolonged beyond the period of services.

The role the plant manager assigns to the engineer must be given careful thought since the role carries implicit professional responsibilities. But where do any of these professional responsibilities end? This question brings up a very sensitive subject, the extent of liability on the part of the engineer.

LIABILITY

Many firms carry errors and omissions insurance policies, but there is a broad range of interpretation of what is meant by liability. An error can be anything from a north arrow incorrectly indicated on a plan, an underdesigned bolted connection of a critical member in a structure, to a treatment system that does not treat to the required degree. The cost of correction of any of these errors, once built, can be in the tens of thousands of dollars, far in excess of the professional fee paid to the engineer. It is these types of errors that are generally covered by an insurance policy.

Other factors which affect the engineer's liability include the question of inspection or supervision services during construction and changes made without the approval of the engineer.

Inspection

If management wishes to hold the engineer fully responsible for complete implementation of his services, he should be required to provide complete and continuous inspection. The engineer must be given authority to have the plans and specifications followed exactly. But proper continuous inspection is very costly. Construction inspection can include testing materials, compaction of soil, reinforcement, concrete, checking torsion on high-strength bolts, and so on through the scores of trades.

If the plant manager does not specifically require the engineer to take full responsibility for construction, most firms will provide liaison with the job. They then feel that "adequate" inspection has been provided. There is a great difference between liaison intended to ensure adequate inspection and actually supply inspection. Many engineers feel that "adequate inspection" is provided by visiting the job a few hours each week. Serious problems can usually be spotted, but complete, risk-free inspection is not being provided.

Because of the great time required to give detailed, continuous inspection, its cost should be carefully weighed by management. The entire question of inspection and supervision should be clarified in the negotiations or at least during the early phase of work. A practical solution for the dilemma can be found in assigning a project engineer from the plant engineering department to provide complete, continuous inspection. Such inspection would have the proper authority over the contractors. Any questions that arise would be referred to the engineer as part of the engineer's construction supervision, but responsibility for faithful reproduction of plans lies with management and management's representative.

ETHICS

Finally, in dealing with the engineering firm, the plant manager should recognize the code of ethics which governs the activities of these professionals. A conscientious engineer would consider it unprofessional and inconsistent with honorable and dignified bearing to act in any of the following manners:

1. To accept any remuneration other than his stated charges for services rendered to his client

2. To fail as a faithful agent or trustee in professional matters to his client

3. To attempt to injure, falsely or maliciously, directly or indirectly, the professional reputation, prospects, or business of another engineer

4. To review the work of another engineer for the same client except with the knowledge or consent of such engineer; or unless the connection of such engineer with the work has been terminated

Both parties in any contractual agreement and business relationship must realize that they are dealing with professionals who perform their duties and services in accordance with a code of ethics.

Chapter Three

The problem

Industry is sending out a loud collective cry for help. The cry becomes louder as the local governments begin citing for wastewater discharge violations and failure to renew industrial wastewater permits.

The plant manager of a local industrial plant relates the latest turn of events to his consulting engineer. The local sanitation district has threatened to shut down the plant. The plant manager, too busy coping with the daily crises of production, has no time to completely analyze the situation. He needs help.

In general, plant management will know the problem. The sanitation district has already complained of excessive copper, cyanide, or chromium in the wastewater. Now they want it eliminated. The problem is to identify the major source of pollution and economically control it.

To completely analyze and evaluate the problem, the engineer must know every aspect of the plant. Oftentimes this is difficult to do since many plants own secret processes and are wary of outsiders investigating the processes. Other plants, embarrassed about illegal waste disposal techniques, may conceal facts from the engineer. The engineer must be a detective as well as a waste control expert. A short checklist is helpful in evaluating the problem. A good rule of thumb to remember is that all the purchased and well water must be accounted for. A water management checklist should include some of the following.

CHECKLIST

1. Quantity of purchased water
    a. Flow rates per hour, day, month, year
2. Quantity of well-pumped water

3. Volume of sanitary waste discharged
   a. Number of employees per shift
   b. Number of shifts
   c. Showers
   d. Cafeteria

4. Sources of industrial water
   a. Overflow tanks
      i. Chemical composition
   b. Batch tanks
      i. Chemical composition
      ii. Frequency of discharge
   c. Scrubbers
      i. Size
      ii. Discharge point
   d. Boiler bleed-off
      i. Volume
      ii. Discharge point
   e. Cooling towers
      i. Volume
      ii. Discharge point
   f. Filter washdown
   g. Hydrostatic testing

5. Lagooning

6. Landscape water

7. Water consumed in process

8. Existing wastewater treatment facilities

9. Condition and discharge points of existing trenches

10. Total wastewater discharged

If the total water discharged is approximately equal to the purchased and pumped water, the engineer can be fairly certain that he has located all sources of discharge. At this time he may find it useful to meet with local officials to determine firsthand their specific requirements for the plant. The present laws are quite specific as to wastewater parameters, but a discussion may nevertheless be useful. Requirements are constantly changing. Moderate-flow industries were required to install control manholes, but this requirement has recently been modified. If a control manhole is not needed, industries with relatively low sanitary waste discharge may not want to separate the two systems. This eliminates the need for a second piping system. But each plant is different. To learn which modifications are necessary and which are not requires a careful analysis of the problems. Any industrial plant contemplating modifications to its industrial waste discharge system would be well advised to obtain a detailed appraisal of the problem from a qualified consulting engineer.

A less critical problem arises daily as the local sanitation districts send out requests for completion of the industrial wastewater permit renewals. Along with the renewal application is the instruction booklet. The plant manager calls his consulting engineer. He does not have time to decipher the code letters. He is too busy with the latest broken pump holding up production. The engineer is asked to look, as a favor, at the renewal form and the booklet and send in a brief summary of what new problems the plant has.

Being familiar with industrial classifications, the engineer can easily determine the type of chemical tests required on the wastewater.

If the yearly flow of wastewater is low, the engineer may simply ask for a list of chemicals used in the processes. Comparison with the restricted pollutants will alert the engineer to any impending problems. The brief summary may be returned to the plant manager with a list of plant modifications if they are required. These could include a control manhole or an interceptor. The report may include the kinds of chemical tests needed and their required frequency per year. For most large plants the report may present an outline of the ramifications of the industrial waste requirements. The plant may decide to request a complete analysis of the situation. If so, the engineer would begin with the checklist already described and be prepared to identify any potential problems.

With the consent of management he may go into the plant and consult with the plant engineering department, since they are the best informed about the problem. They know which operations are producing certain wastes, what chemicals are being used in these operations, how much of each waste is derived from each operation, and the volume of flow during an operating day or period of operation. This is important, since in some plants a process which produces a waste with a high concentration of one or more of the pollutants causing trouble may be discharged only a few times a day. The final handling of such wastes may play a very important part in the ultimate design. After discussing the processes and the elements that have been added and may constitute a pollutional item, the engineer or his representative should then tour the plant to see for himself the individual operations, take notes, and get a good overall picture. This tour will provide the engineer with valuable information which may enable him to introduce economies into the waste treatment phase by introducing changes in the operations within the factory or mill. Here is where diplomacy may be essential. Those who actually operate the productive units in the industrial plants are often jealous of their prerogatives and resist change. Therefore the engineer must have a well-documented case when he is ready to suggest changes and meet the objections of the personnel.

In a case in point, it was suggested to the plant engineering department of a large automobile manufacturing company that money could be saved in the final waste treatment if internal operational changes could be made to reduce the amount and concentration of certain wastes. The answer, loud and clear, was, "We make automobiles in this plant—we don't want the men bothered by someone making changes that will delay the work and maybe save a few cents a day. You fellows take the waste outside the wall of the plant and treat it as much as necessary." That was that. In another instance, in the manufacture of airplanes, where a lot of experimental work was done at odd places around the plant, some three-quarters of a mile away, it was suggested that economy would be served if all these small operations were concentrated at one point. Again the same outcry: it just could not be done. It was not done, but the small quantities of wastes from these isolated stations made necessary a number of collection systems and small pumping stations, and some unusual piping arrangements ensued. In such cases, if the engineer feels that important economies would result from action he proposes, he should not hesitate to take the matter to the higher echelons of management. For it is they who pay the bills, and economies are always welcome.

In cases where products are being manufactured by secret processes or patents, companies may be reluctant to let outsiders wander around the plant. Declaration of strictly ethical behavior by the engineer, that he will see not, hear not, and tell not, is sometimes taken skeptically. If such resistance is encountered in a plant, the management should be requested to appoint someone with a complete knowledge of the operations to work with the engineer and provide him with all the necessary information to enable him to design a practical and economical waste treatment facility.

What the exact problem consists of takes some time to discover, but since it is extremely important to a successful outcome, the engineer must not let himself be shunted off by internal objections. After he has satisfied himself as to the character, volume, and other factors pertinent to the wastes, he may find it advisable to have a conference with the officials of the city in which the factory or mill is located. At such a conference he can ascertain exactly what the city desires in the final discharge from the industry. In many cases the governmental regulatory authority determines what will be permitted in the final effluent from the municipal treatment plant. Usually the municipality then passes on these requirements to the industry. The ingredient or component of the municipal discharge which is considered most detrimental by the regulatory officials may not be of primary interest to the municipal officials. They have a well-operating sewage treatment plant, and they do not want it upset by certain chemicals contained in the industrial effluent. In one important case a plant was discharging copper cyanide wastes. The cyanide was of concern to the state officials. On questioning the chief operator of the municipal sewage treatment plant, it developed that he was most interested in the amount of copper that might be contained in the final effluent from the industry's waste. Copper would upset his sludge digestion process, and he did not want that. The cyanide was not directly harmful to the municipal sewage treatment works. Its presence was of no consequence to the chief operator. When assured that the copper content of the effluent from the industry would be at a minimum, he was willing to make temporary compromises.

Following the discussion with the city officials, the engineer may feel it is important to visit the headquarters of the regulatory authority and discuss the problem with them. In many cases the order for treatment of the wastes from an industry will simply say "you must treat your wastes to a certain degree by a certain date," or words to that effect. In most cases the authority will be interested primarily in one main component of the wastes from the industry in question. Knowing this, the engineer can in many instances sit down with the authority engineers and ascertain exactly what the immediate goal is.

In another case, in the manufacture of strip steel, the wastes discharged directly into a river. The usual order was given by the state health department: "Treat your wastes." There were five different types of waste, each of which required individual treatment. To design a plant to treat all the wastes would result in an estimated cost of almost $1 million. However, the engineer felt certain that the state authority was primarily interested in one component of the waste, the cyanide. A conference was arranged with the engineers of the health department, the designing engineer, and a representative of the industry, and after some discussion, the regulatory officials admitted that the cyanide was their most immediate concern. This being the case, they would permit the industry to construct the cyanide reduction phase of the treatment plant first, and follow with the other phases within the next 3 years. This made it possible for the industry to satisfy the authorities at a cost of $350,000 initially, with the balance to be spent in the future. This arrangement also made it possible for the engineer to design a coordinated complete plant, planned for building block construction.

In still another case, an airplane manufacturer discharged into a small stream that was the town's only source of drinking water. The requirements of the state health authorities for any waste discharge into this stream were very strict. Investigation by the engineer showed that the requirements of the city in which the industry was located were very much more lenient. On its own volition the regulatory authority advised the industry to prepare for treatment only to the extent of the municipal requirements and to discharge the effluent into the city sewerage system. The city effluent did not reenter the water supply system.

From the cases mentioned it can be seen that specific considerations can have material effect on the final outcome and cost of solving an industrial waste problem. Realizing this, it is the engineer's obligation to investigate all the alternatives in order to produce the most practical and economical design for his client.

THE REQUIREMENTS

As stated earlier in this chapter, a regulatory agency seldom elaborates on the extent of treatment of the wastes. Limits on concentration of different waste components have been established which are usually generalizations covering a wide area. Municipalities do the same for waste effluents to be discharged in the city sewerage system. These limits are therefore subject to variation or modification to fit the needs of specific cases. In many cases there are ambiguities in the regulations set up, which indicate a lack of careful study of their meaning. Ambiguities may also clue the engineer to the fact that the ordinance or requirements have simply been copied from those of some other municipality or regulatory agency whose circumstances may be different.

In one case the city requirements called for maxima of milligrams per liter of pollutants in industrial wastes, but at the same time added a limitation of "pounds per day of the individual elements." It was pointed out to the municipal authorities that the combination was possible of attainment only on one specific volume flow and a certain concentration of elements. Further discussion ascertained which pollutants were required to meet both restrictions and which were required to meet one. A decision was made as to which restriction had precedence.

INDUSTRY AND MUNICIPAL COOPERATION

There are numerous instances of cooperation between industries and the cities in which they are located. In some instances municipalities have been too anxious to induce an industry to come to a city. Incentives are often the money workers would distribute among the stores of the city, and the taxes that would be paid by these employees. Frequently, the local chamber of commerce and other business groups will promise concessions in taxes, low-cost land, etc., without fully comprehending the costs that ensue when a municipality is called upon to provide the additional public services that are required. Usually, a large industry will construct its plant on the outskirts of the town, an area to which the existing sewer and water lines of the city do not reach. If the industry has a large volume of liquid wastes which the city has agreed to handle, this may entail a heavy capital expense for new service lines, electric power lines, and street paving, not to speak of new schools and police and fire protection.

Industries are usually pleased to have municipalities approach them with offers of concessions for building within the corporate limits, or even outside, but the municipality must be careful not to promise benefits to the industry for which its citizens will be paying for generations in increased taxation.

On the other hand, in view of the increased demand for the treatment of industry's wastes to prevent stream pollution, the federal government and several of the states, as explained in Chapter 9, have made it seem beneficial to both industry and the municipalities to work out reasonable programs for industry to discharge its wastes to the local sewerage system. This is, of course, very beneficial to the industry, since it practically eliminates the heavy cost of providing waste treatment facilities at the plant, and also relieves the industry of the cost and trouble of operating them. In return, the industry should be willing to compensate the municipality for the privilege of discharging its wastes into the city system, in some form or other agreeable to both parties. Cooperation on both sides in such a program may redound to the financial benefit of each when properly and equitably agreed upon. It has recently been established that municipalities may obtain federal funds for construction of sanitary facilities when the program includes expenditures made necessary by the inclusion of industrial wastes. Ways in which industries in several cases have helped municipalities to carry the resultant operating load are explained in more detail in Chapter 9, as well as in the following discussion.

An outstanding example of industry and municipal cooperation for the handling and disposal of the combined wastes is that of the American Cyanamid Company at Bound Brook, New Jersey. In this case the industry built a large plant to handle its own wastes, large enough and of a proper type to handle the sanitary wastes of an adjacent group of municipalities, as well. This plant consists of the following phases and operations: Waste treatment at this plant has actually been in progress since the 1930s, when the initial treatment given was neutralization and lagooning. The effluent from this installation was suitable for stream conditions at that time, and the plant units provided then are still in service as integral parts of the primary treatment phase of the present greatly expanded treatment facility. When expansion of the plant and stream conditions indicated the need for additional treatment, extensive laboratory studies and pilot-plant operations were carried out over a period of 10 years to determine the development of a reliable and practical plan to meet the new conditions. These investigations alone cost about $1 million and culminated in the design and construction of an activated sludge biological-type treatment works, at an initial cost of $4.5 million.

This plant was put into operation in 1958. The annual operating cost is about $1 million. In this operation ten people are actively engaged, five in direct operation and five in quality control, plus a technical group engaged in research and development, consisting of a sanitary engineer and two scientists. When the new plant was designed, it was made capable not only of handling about 21 million gal/d of industrial wastes from the Cyanamid plant, but also of treating about 4 million gal/d of chlorinated domestic and industrial wastes from the Somerset–Raritan Valley Sewerage Authority, which embraces three local municipalities. The combined wastes are treated in six aeration tanks with a total capacity of 20 million gal. After about 19 h aeration, the effluent is settled in circular clarifiers, and the biologically active sludge is returned to the aeration tanks. The clarified effluent from the settling tanks is then discharged into a small brook which empties into the Raritan River. The wastes from the sewerage authority referred to are handled under a 20-year contract which provides for a specified quantity and quality of wastes to be accepted for the joint treatment, at a charge based on the cost of extra electrical power, chlorine, and laboratory costs. These charges do not include capital cost expense, routine operating costs of the activated sludge plant, or other minor expenses. This unique arrangement has been found to be not only workable, but flexible and beneficial to both parties. This cooperative plant also serves a number of other industrial plants in the three municipalities of the sewerage authority. The total charges made to the authority for the service rendered to it are less than $10,000 per year, a decided bargain, since three separate sewage treatment plants could not possibly be operated for this small sum. To comply with recently increased stream quality requirements, Cyanamid has new plans for additional biological treatment to operate at higher levels of efficiency than those presently employed.

Another example of industry and municipal cooperation was put into active operation in Longview, Texas, in 1966. Here the Joseph Schlitz Brewing Company has completed a large brewery, and since the sewage treatment plant of the city also needed enlargement and modernization, the designing engineers developed a plan for cooperative action on the part of both entities. Under this plan the city was to construct the plant and include the necessary additional capacity and units to handle the wastes from the brewery, estimated at about 1 million gal/d. For this service the industry was to pay a service charge sufficient to cover the extra expense of handling its wastes. In the actual operation, when the plant was put into service, the sewage flow from the city was estimated to be about 4.87 million gal/d, and the flow from the brewery about 1.13 million gal/d. The original design called for a separation of the two flows up to the final biological treatment on trickling filters. In practice, however, it was found that the low pH of the brewery wastes, about 4.5, was upsetting the biological action in the final stages. The flow pattern was then changed to mix the brewery wastes with the municipal sewage to produce a more normal pH of slightly less than 7.0.

Still another and much-publicized example of industry and municipal cooperation is that at Charleston, West Virginia, with the combined treatment of the wastes from the Union Carbide Chemical Company and the sewage of the city of Charleston. In this case, also, the city owns the treatment works. Union Carbide designed the plant and built it. They also made the preliminary and pilot-plant studies. The city, by its ownership of the plant, relieves Union Carbide of the responsibility of acting as a public utility, which eliminated possible legal difficulties that might have arisen if other communities applied to Union Carbide to treat their sewage. The inclusion of the municipal sewage was beneficial to this operation since it supplied nutrients required by the industrial wastes at no cost.

At Bayport, Texas, there is a $30 million treatment plant. This is an activated-sludge-type plant designed to handle all the wastes from a 7250-acre industrial park for heavy industries being developed by the Humble Oil & Refining Company. As a first step, three holding ponds, with a total capacity of 20 million gal, have been constructed, with the necessary system of collecting channels. The plant, designed to handle all the expected industrial wastes from this complex, offered improved control with lower costs for the individual industries sharing the service. The future estimated volume of flow to be handled by the installation is 40 million gal/d. The use of the joint plant will be coupled with in-plant changes and conservation by the industries. Only those wastes which responded to biological treatment will be accepted. Charges to the participating industries locating in this industrial park will be based on the strength and nature of the individual wastes.

In the case of the above-mentioned plant, the cost of the ultimate joint plant, estimated at about $30 million, and equal to about $4000 per acre of the land to be sold to the industries, was added to the price of the land in the park. This was less costly to the individual industries than building and operating individual treatment plants, and provided the advantage of skilled operation at all times, coupled with the benefits in treatment that result from mutual reactions between the various wastes, which it is believed will simplify the operation and lessen the cost of treatment.

In the future, the common advantages which are inherent in this cooperative endeavor will undoubtedly encourage its adoption. Properly worked, with due regard to the rights and equity of each party to the compact, it may be considered sound practice. As in any such agreement the advantages and disadvantages to each party must be given mature consideration. For its part the municipality has certain obligations to the local industries since they are usually payers of substantial taxes and provide the income of their employees, who may number thousands. These employees will have millions of dollars to spend in the local markets in the course of a year; they will use local supply industries; and they will support the local cultural and civic activities. They will be, in effect, citizens, entitled to the same benefits as all residents of the municipality. Consequently, they have the right to be heard when there is talk of restrictions on the industries' activities. On the other hand, the industries must respect the laws and do their part in protecting the public health and welfare by preventing nuisances that might arise from their operations, and air and stream pollution by their gaseous or liquid wastes.

The definite advantages to a municipality in permitting industries to discharge the proper types of wastes into its sewerage system are as follows:

1. If the sewage treatment plant is operating at less than its designed capacity, the service charges paid by industry will add revenue for the city.

2. If the sewage treatment plant is of the biological type, it costs no more to run it at full capacity than at a reduced volume.

3. If sludge digestion is practiced and the industrial wastes are of an organic type, the increased volume of gas from digestion will produce more power or heat, if power or gas is used for these purposes.

4. If the sewerage system has been constructed of a size to handle a large estimated future flow, the inclusion of industrial wastes will tend to keep deposits from forming in the sewer lines because of low velocities produced by the small sewage flow.

Some of the apparent benefits to industry in discharging their wastes into the local sewerage system are as follows:

1. Industry is relieved of financing and operating its own waste treatment facilities.

2. Dilution of its wastes into the municipal sewage reduces service charges for concentrations of waste ingredients over certain maximums.

3. Usually, low-cost handling of wastes is provided.

4. The mixture of sewage and industrial wastes eliminates the need for addition of costly nutrients in which the industrial wastes may be deficient.

5. Regulatory agency jurisdiction or inspection is reduced or entirely eliminated.

6. Industry property is freed for productive uses.

In any agreement by which a municipality agrees to accept an industrial waste flow into its sewerage system, the conditions to which the industry should be required to subscribe should be clearly stated, to avoid misinterpretation. These conditions are

1. A clear definition of the characteristics of each particular waste that will be acceptable for discharge into the system. This should include the maximum volume of waste permitted per unit of time and the maximum limits of toxic and other polluting elements, solids, oil, clogging elements, etc., shown to be present in the wastes.

2. Establishment of a proper and reasonable charge to the industry for the wastes admitted to the sewerage system, by total volume of flow per day and per unit of suspended solids, Biochemical Oxygen Demand (BOD), total solids, oils, toxic elements, etc., over and above established and clearly stated limits for each item. These charges must include, in fairness to the citizens of the municipality, a proportionate share of the financing costs on bonds or loans forced upon the city by additional facilities chargeable to the inclusion of industrial wastes.

3. Provision that in the event of the sewerage system becoming overloaded because of the normal increase in sewage flow from the growing population, the municipality retains the right to refuse further service to an industry unless the industry agrees to finance any additional facilities, plant units, equipment, and personnel that may be required to handle the retention of industrial wastes in the flow to the plant. The contract should also provide that in the event the wastes from an industry change in character or concentration of polluting elements to an extent that would materially affect the operation of the municipal sewage treatment, the city will have the right to exclude the industrial waste in question from furthe discharge to the municipal sewerage system unless and until the industry provides, at its own expense, any facilities necessary to return the wastes to the condition that existed when the original permission to discharge into the city system was granted. In other words, the municipality must not obligate itself to take whatever the industry chooses to discharge into its system, forever, without proper recompense for the extra costs that may be entailed by changes in composition or volume.

An example of this development occurred at Worcester, Massachusetts, in the early part of the century, when the city agreed to receive the wastes from a new steel plant. As the years passed, the wastes from the steel mill increased in volume to such an extent that the municipality incurred much heavy expense in adjusting the treatment works to keep up with it.

Another example was that of the Passaic Valley Sewerage System in northern New Jersey. This authority constructed a large collecting sewer along the Passaic River to eliminate pollution of the river from the wastes of the extensive textile industry which had developed in the cities along the stream. The pollution had become so bad that the river had been rendered useless for all normal human purposes, such as water supply, recreation, etc. A stipulation of the agreement made with the various industries tributary to the sewer was that they could discharge 10 percent of their total wastes untreated into the sewer and treat the balance as necessary before further discharge into the river. The consequence was, of course, that the industries discharged the most polluting and difficult of their wastes to the sewer, thus placing a very heavy and unexpected load on the treatment works established by the authority at the end of the sewer, necessitating unusual skill in operation of the plant.

An interesting example of cooperation between a municipality and local industries occurred in Kalamazoo, Michigan. The problem for the city in this case was primarily with the many paper industries. These industries had, for some years, maintained their own treatment facilities. Since these industries form the economic backbone of the city, they had a just claim for consideration on the part of the city when orders of the Michigan Water Resources Commission made it necessary for them to increase their treatment facilities. All these industries had been discharging their effluents into the badly polluted Kalamazoo River, which runs through the city. The order of the Commission was to the effect that the total amount of 5-d BOD that would be permitted was a maximum of 30,000 lb/d. This reduction of nearly two-thirds of the former requirement immediately raised the question whether the industries should enlarge their individual treatment facilities or utilize the sewage treatment plant maintained and operated by the city. An intensive study of all the many factors involved showed that this latter proposal was the most acceptable.

The existing plant of the city at that time provided primary treatment of the sewage only, strengthened by chemical precipitation when the composition of the sewage required it. An engineering analysis showed that the plant could be transformed into an activated sludge type of plant which would provide a satisfactory degree of treatment for the combined municipal and industrial wastes. The Michigan Water Resources Commission had not ordered the city to provide further treatment, but the engineering economic studies indicated that a joint venture with the industries would benefit the entire community. The changes, as advised by the engineering group, are designed to increase the capacity of the primary plant from 8 to 12 million gal/d, because the new secondary phase of treatment would require less detention time than was required in the primary phase. The secondary phase of the plant has been designed to handle an average flow of 34 million gal/d, which provides for some reserve capacity to accommodate future increases in population and industrial growth. The activated sludge phase used Eimco mechanical aerators. The sludge from the treatment will not be digested, but will be dried on open beds. The charges to industry for handling their wastes in the new system were

5-d BOD	40 cents per 100 lb
Suspended solids	20 cents per 100 lb
Flow	13 cents per 1000 ft³

These charges are to be based on automatic continuous sampling and a daily test of a composite sample.

PILOT PLANTS

Pilot plants—small installations of tanks and equipment—are justified when a search of the literature on industrial wastes fails to reveal any concrete evidence of work done by responsible researchers on a particular type of waste, or where the inherent character of the waste does not present a definite indication that it will respond to any normal, developed method or combination of methods. In cases where new types of wastes evolve from newly developed industrial processes, and where laboratory investigations indicate that certain established processes will have a beneficial effect on this new type, it may then be justifiable to consider the development of a pilot plant, to produce design factors that will be helpful in a full-scale project.

Pilot plants are not intended to be used, generally, when known factors have been proved by operating plants over a period of years and when reliable data are available in the literature on the type of waste under consideration. For instance, from experience over many years, factors for the development of plants for the treatment of wastes of an organic character by normal biological processes—trickling filters, and activated sludge—have been well established, and any competent sanitary engineer can prepare the basic design of a treatment plant that can be expected to operate satisfactorily.

In the past few years, however, with the development of new products, primarily those of chemical origin, the existing biological methods do not respond, and in these cases pilot plants, following initial laboratory investigation, are justified before design of full-scale units. Where it is felt that the expense of a pilot plant is justified, it should be placed under the supervision of a qualified operator, preferably the one who performed or supervised the previous laboratory demonstrations. A number of manufacturers of equipment and material now provide small, completely assembled units of their equipment for pilot-plant use, on a rental basis, or for purchase, if desired. Among these may be mentioned the manufacturers of the newly developed plastic media for trickling filters, namely, the Dow Chemical Company, Goodrich Tire & Rubber Company, the Ethyl Corporation, and Tex-Vit Company. Manufacturers of vacuum filters also have small units, completely assembled on skids: Dorr-Oliver, Inc., Eimco, Komline-Sanderson Company, Filtration Engineers, etc. The manufacturers of centrifuges also have pilot units: DeLaval Company, Bird Machine Company, and Dorr-Oliver, Inc.

It is extremely important that in order to design a practical, economical waste treatment facility, the foregoing factors be carefully discussed with all persons having control or jurisdiction over the specific plant discharge. It takes a little more time and effort on the part of the engineer, but the author has never found such cooperation to fail to produce satisfactory results. In fact, it is a shortsighted industry which does not insist that its engineer or adviser make these contacts and ascertain the immediate and long-range needs of a case before beginning any actual design, or even considering a method or the extent of treatment.

The information, data, and characteristics

To tackle any wastewater problem, information on the volume and characteristics of the wastes must be obtained. The data can be simply the volume, BOD, and suspended solids content. Or if the plant processes involve many reagents, the parameters required could include heavy metals, cyanide, chlorine, and others. Even renewing a discharge permit requires a comprehensive description of the wastes involved.

VOLUME OF WASTES

The measured volume of flow for each type of waste is the most important information to be determined in any study. It not only affects the concentration and total mass of the pollutants but is used in the determination of the surcharge levied against each plant owner. In Los Angeles County, the 1973–1974 annual surcharge of wastewater is approximately $104 per million gallons and an additional charge of $18.75/gal(min) for peak load measured in any 30-min period. If the factory does not have any record of flow volumes or the means of obtaining this information, the engineer must procure the data from meter readings or sample testing.

Methods

The oldest form of measuring liquid flow in an open channel is by the weir. See Figure 4-1. This method has been used for decades in irrigation, mining, and other operations in which water flows by gravity in a channel, pipe, or open drain. If plant wastewater flows through an open channel, a V-notch or

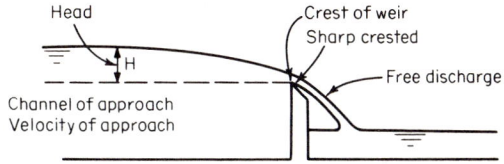

(a)

Fig. 4-1 Types of weirs. (*a*) A cross section of a weir showing the channel of approach, the weir, and the downstream channel. The head *H* is usually measured about 2.5 ft upstream from the crest of the weir. (*b*) A rectangular weir. (*c*) One type of trapezoidal weir, or a Cipolletti weir. (*d*) One type of V-notch weir. A 90° V-notch weir shown.

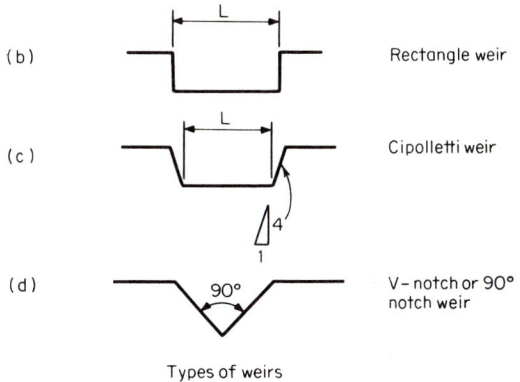

(b)

(c)

(d)

Rectangle weir

Cipolletti weir

V-notch or 90° notch weir

Types of weirs

other type of weir may be installed in the channel. The principle of flow measurement by a weir is as follows:

Weir. Basically the height may be a V notch of various angles, a trapezoidal notch, or other shapes. The main components of a typical V-notch weir are head (H), width of weir bottom nappe (submerged or drowned, L), free discharge, and crest of weir. The head is usually measured about 2.5 ft upstream of the weir. To avoid turbulence of fast-flowing water, a stilling basin may be installed adjacent to this point. Measurement of H is then made in the stilling basin. Weir formulas are as diverse as the shape of the notch. Some are as follows:

Francis formula for rectangular weirs

$$Q = 3.33H^{3/2}(L - 0.2H)$$

Francis formula for suppressed weirs

$$Q = 3.33 \times L \times H^{3/2}$$

Francis formula for Cipolletti weir

$$Q = 3.367 \times L \times H^{3/2}$$

Thompson formula for 90° triangular

$$Q = 2.54 \times H^{5/2}$$

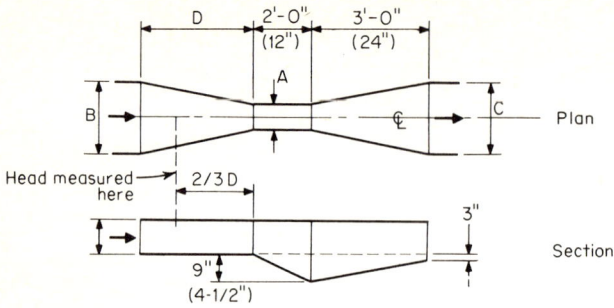

Figures in parenthesis refer to flume 1/2 A

Fig. 4-2   Sketch of a Parshall measuring flume.

Parshall Measuring Flume.   Another type of measuring device is the Parshall measuring flume. See Figure 4-2. This type of weir has a narrow specially designed throat with a tapered inlet and outlet. Two gauges are provided, one at the inlet and the other at the throat. For free flow the measurement of the head is made at the upper gauge only while for submerged flow the head is read at both upper and lower gauges.

With a proper recorder, the weir will record the deviation in flow at different hours of the day. It will also show the total flow for a given period. These units do not require constant attention.

Pipe Flow.   Flow in pipelines under pressure can be measured with a variety of flowmeters. One of the more popular types is the venturi tube as shown in Figure 4-3. This can be a constricted short length of pipe with a throat

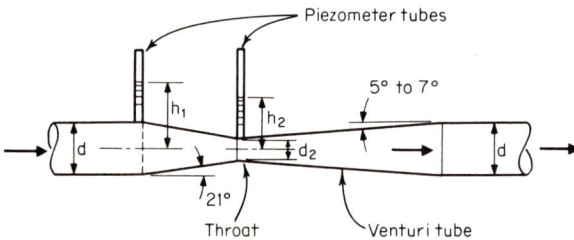

Fig. 4-3   A venturi tube. Flow is determined by the difference in static pressure at the throat and at a point upstream of the transition. Flow is determined by the following equation:

$$Q = C \frac{d^2 d_2^2}{4(d^4 - d_2^4)^{1/2}} (2gh)^{1/2}$$

in which d and d_2 are diameters of entrance and throat. C is a coefficient which reflects the construction features of the meter and the loss due to friction between the entrance and the throat. The value of C for standard makes of meters is quite close to unity.

diameter between one-half and one-quarter the main pipe diameter. The formula for determining flow is based on Bernoulli's theorem, neglecting friction:

$$Q = ckd_2{}^2 \sqrt{h_1 - h_2}$$

where c = coefficient between 0.97 and 0.99

k = a factor depending upon ratio d/d_2 (about 6.31 to 6.51) or as shown on Figure 4-3

Manual Measurements.   Wastes discharged into a pit which does not receive any other waste may be used to determine flow rate. The outlet from the pit may be closed, and the time required to raise the level of the liquid in the pit a foot or a fraction of a foot in a given time is observed. When this information is combined with the dimensions of the pit, the flow rate can be determined. If this type of system is employed, the data should be taken periodically through the normal operating period of one hour, shift, or day to determine any fluctuations in the flow.

When no open flume or pit is available, but the open end of the pipe may be reached, an operator can get a fairly accurate measure of the flow with a 5-gal bucket, or even a quart bottle, held over the end of the pipe, observing the time in which the container is filled. This again requires attention over one entire operating day.

Water Meter Determinations.   If the plant has a water meter which registers the total amount of water used in the factory during the day, the engineer has an alternative method of determining waste flow. If the water is not consumed in the process, purchased water readings can give a fairly accurate reading of the total waste discharge volume as well as periodic readings every 30 min. Purchased water may not be the plant's only source of water. Well water may also be used. Neglecting its volume could upset design calculations.

If the engineer is forced to base his discharge volume on water meter readings, there are several points he must take note of.

Water Discharged as Sanitary Wastes.   If the plant has several shifts and many employees, much of the water indicated on the meter could simply be sanitary waste. If the plant discharges into a local sewer system, this sanitary waste will pose no problems. The engineer can estimate the volume of domestic water after he has determined the number of employees per shift and the number of shifts per day. The Uniform Building Code allows 25 gal per capita for each 8-h shift. If the factory has a cafeteria, an additional 5 gal/capita (d) (shift) is counted. Some simple mathematical calculations will give the approximate amount of sanitary discharge.

Cooling Water.   Cooling water also needs attention. If equipment is water-cooled and the water is recycled, there is continuous bleed-off to consider. Bleed-off water frequently has corrosion inhibitors which may require treatment. Single-pass cooling water can pose quite a problem for the engineer. Industries frequently discharge this water directly into their sewer system. Many localities prohibit this practice since it overloads the sewage treatment plant with essentially clean water. Discharging single-pass water is also not economical. It forces the plant to purchase more water than it needs, to pay a surcharge on the excess water it discharges, and to treat a larger volume of water than necessary.

Landscape water may be another source of water loss that does not contribute to the waste flow. Neglect of this item could cause an overdesign of waste facilities.

PERIODICITY OF FLOW

This is the time period during which each type of waste is actually discharged. Some wastes may be discharged regularly every minute during the entire operating period of the factory. This period may be 8, 16, or 24 h. If, on the other hand, the wastes are discharged only at odd periods, such as batch discharge of treatment tanks, this must be accurately determined. The complete information on such sporadic discharges is important, particularly if the waste, or dump, contains a high concentration of a polluting or toxic ingredient, such as cyanide, chromium, or acids. The periodicity of flow frequently determines the advisability of providing holding tanks to facilitate the dribbling of concentrated wastes into other flows at a uniform rate over a period of hours for combination, neutralization with other wastes, or more efficient treatment. These facts must be individually determined for each type of waste.

Some municipal treatment plants encourage industry to dump their waste during off-hours. The facilities are usually running below capacity from midnight to sunrise and are better able to handle this extra load at that time.

SAMPLING FOR PARAMETERS

Most governmental regulations now require sampling of industrial waste discharge. Formerly, effluent streams were one of the most neglected areas of plant operation. The water has served its purpose and is ignored, forgotten, and avoided. Frequently, it is moving at high velocity in a pipe that is inaccessible.

Methods of Collection

The science of sampling is withdrawing a small portion of the whole in such a way that the small-volume sample represents the character and quality of the mass of the flow from which it is taken. Withdrawing a small portion can be done best when the entire lot is moving as a stream and increments of the stream are withdrawn, as shown in Figures 4-4 and 4-5.

Sampling of the stream should be at a point of free fall so that a sample cutter can pass at a uniform speed completely through the entire stream. The sample cutter should be designed so that it removes a small portion of the entire

Automatic timer synchronizes intermediate
operation of pump and sampler

Fig. 4-4 Suggested installation
for an automatic sampler by
Denver Equipment Division of
Joy Manufacturing Company.

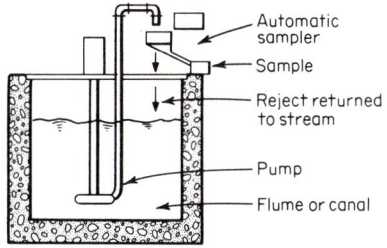

- Automatic sampler
- Sample
- Reject returned to stream
- Pump
- Flume or canal

stream without overflowing and delivers the sample so that it is not contaminated by foreign material or by previous or subsequent qualities of the stream.

Because most plant effluents contain materials of differing specific gravity which follow natural laws and tend to reside together, surface sampling should be avoided. Submerged vessels which overflow or in which heavier particles displace the lighter particles must also be recognized as inaccurate. A poor sample may report the presence of a particular constituent, but the concentration is false and unreliable owing to the displacement theory.

When the stream is small, pipe-contained, and under pressure, it is easy to provide a free-fall sampling condition. On many streams such as underground gravity lines, it is impractical to elevate the entire stream in order to provide free fall necessary for proper sampling. A method to handle this is called "sampling a sample."

For this procedure to work satisfactorily, it is necessary to make sure the contents of the entire stream are as uniform in quality as possible. This is accom-

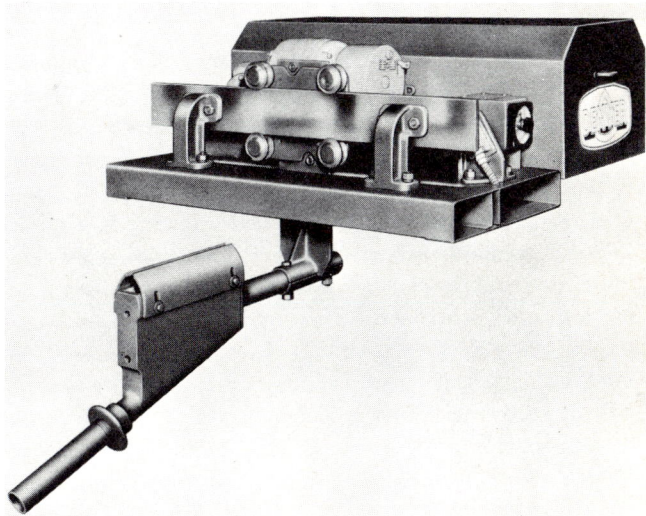

Fig. 4-5 An automatic sampler. (*Denver Equipment Division of Joy Manufacturing Company*)

plished through mixing, turbulence, baffles, etc. Then, when the entire stream is of one quality, a large amount of stream is withdrawn by a pump, and the pump discharge, as it returns to the stream, is cut by an automatic sampler. The large primary sample, withdrawn by the pump, is to compensate for possible imperfect mixing within the stream. The primary sample is then cut at a point of free fall to withdraw a small, representative final sample.

Certain sampling problems involve stream conditions that are affected by time and temperature. To overcome the possibility of change in a sample due to temperature conditions, it is recommended that the final sample be refrigerated. Small, low-cost refrigerators are used. The sample container is housed in the refrigerated case until the sample is removed for analysis.

Analysis

Once the samples have been collected, it remains to analyze them and determine the type and extent of the pollution. Primary determinations may be made in the industry laboratory at the time the samples are taken. Practically every modern industrial plant using chemicals or additives in its productive operations has a laboratory which is able to make many of the necessary tests.

Suspended Solids (SS). The simplest laboratory test is one to determine the amount of settleable solids present. This may be done by use of the familiar glass Imhoff cone, which holds 1 l of liquid and in the course of an hour or two will give a fairly accurate reading of the amount of unassisted settleable solids in the waste, by the accumulation of sediment at the bottom of the cone. This is readily ascertained in milligrams per liter by the graduations on the glass.

Graver Water Conditioning Company offers an automatic tape analyzer. This is claimed to be a concept in data collection for those interested in analysis of suspended solids in liquid streams. It detects insoluble impurities in the lower parts-per-million or parts-per-billion range. In operation the liquid to be analyzed is piped to the tape analyzer under 15 to 25 lb pressure. A preset volume of waste sample filters through the filter tape and is accumulated. When this volume fills the sample-receiver chamber, the tape advances and the machine is set for the next cycle. An override timer is so set that if an unusually large amount of suspended solids should plug the filter, preventing the desired volume from passing through the membrane filter, the tape will automatically move forward into the next cycle. Samples may be taken as frequently as every 10 to 15 min. The result of the sampling is a spot on the tape $\frac{1}{2}$ in in diameter which contains suspended solids. It is occasionally possible to calibrate the tape, on the basis of the intensity of color of the spot, so that just looking at the tape will give an idea of the concentration of impurities in the solution. At the time this information was received (June 1965) the only tape being used was a Millipore filter tape, available in 0.5- and 3-μm porosity. It is said that the instrument will pay for itself in a short time if the present methods of sampling and metering are being used. The unit is neatly housed and occupies a very small space.

Cary Instruments Division of the Applied Physics Corporation also offers another type of automatic sampler. It is designed to handle and collect up to 144 samples with its three containers, each holding forty-eight test tubes. The unit is designed to transfer samples automatically to and from the measurement cells of spectrophotometers, polarimeters, refractometers, and other analytical instruments, or to collect samples from a sample line, chromatograph, or fractionating column, etc. When operated in conjunction with an analytical instrument, it provides a fully automatic system for sample handling and analysis. It can operate unattended for hours. To protect the samples, the test tubes may be capped automatically with polyethylene balls, which are removed automatically whenever samples are withdrawn or placed in the test tubes.

Another laboratory item, recently developed by the Instrument Division of Warner-Chilcott Laboratories, is dubbed a "robot chemist." It is said to perform up to 120 wet chemical tests per hour, exactly duplicating the manual techniques of the laboratory analyst. It measures samples, adds reagents, and analyzes all automatically. It prints out test results in digital form, with identification of the sample. The operator programs the instrument for reagent volume, incubation-line temperature, and spectrophotometer wavelength; loads the instrument with up to 100 samples; and awaits the result. Different types of analyses can be performed in rapid succession without flushing, cleaning, or disassembly of the tubing. The unit is said to have high accuracy.

David Milne, Production Engineering Section, General Motors Corporation, Detroit, has written an article, A Rapid Method for Analyzing Wastes. This method is said to have been used in the laboratories of various plants of the corporation. Samples adjusted to a pH of 2 or 3 with hydrochloric acid are submitted in 300-ml plastic containers. Hundred-milliliter portions of each sample are poured into clean 50-ml beakers, along with 2 ml of nitric acid, 4 ml of hydrochloric acid, and 1 ml of 1 percent lithium nitrate solution. These solutions are then evaporated to dryness, without baking. The beakers containing the salts are then taken to the spectrographic laboratory, where 1 ml of 1:1 hydrochloric acid is mixed with the salts. One drop of this is placed in a shallow-cup $\frac{1}{4}$-in center-post high-purity electrode and excited for 40 s with a dc arc, using a conical-tipped counterelectrode. The spectrograph is a 2-m instrument with a 36,000-lines-per-inch grating. A 2-mm gap is used, and the slit setting is 50 μm. Standard methods are run, following the same procedure, so that working curves may be made, using the ratios of element-line intensities versus the internal-standard lithium line. The ratios obtained from the sample are then plotted on these curves to obtain element concentrations. For cyanide, oil, and pH values, other samples must be used because of the additions of acid to the first samples.

Many other automatic devices have been developed in recent years which relieve a lot of the work of laboratory personnel, such as automatic analyzers for cyanide and chromium, which perform analyses in a matter of minutes as compared with the normal time of about 2 h by standard laboratory technique. In avoiding possible penalties for imperfect treatment of wastes containing the toxic cyanides and chromes, such units would be a profitable investment for large plants. Autoanalyzers make accurate and complete analysis on a continuous basis. They require little manual attention, except for periodical replenishment of reagents. Units are available to monitor up to sixteen sources for some components at various levels of concentration, and up to four different components, at twenty, forty, and sixty analyses per hour, according to the complexity of the analysis. Thus, at a rate of twenty analyses per hour, a total of 160 analyses of a component may be had in an 8-h day compared with four analyses by normal manual laboratory procedure. The reluctance of operators to make the small amount of effort to supply the reagents seems a somewhat dubious hindrance in view of the benefits that may be derived from automatic analyzing. Types of wastes or components for which automatic analysis is available are

1. Ammonia
2. Antibiotics
3. Beer
4. Beverages
5. Chemicals
6. Chlorides
7. Copper
8. Carbon dioxide
9. Cyanides
10. Iron
11. Nitrogen oxide
12. Fruit concentrates
13. Paper pulp
14. Petroleum
15. Plating solutions
16. Plastics
17. Phenol
18. Glucose
19. Sulfates
20. Sulfites
21. Sulfur
22. Vegetable oils

ANALYSIS AND TESTING OF BOD AND COD

For many years the normal BOD test, requiring a time element of 5 d at regulated temperatures, has been a source of concern to sanitary engineers. Many things can happen, and have happened, while laboratory technicians wait for the 5-d period to end.

Dr. Robert Arthur, of Rose Polytechnic Institute, Terre Haute, Indiana, has developed a device which provides measurement of BOD in 1 to 2 h. According to Dr. Arthur, in this method an automatic recording respirometer could be used to provide additional treatment at waste treatment plants automatically. It is planned to install such devices at two locations on the Wabash River to test their effectiveness under actual conditions. If this idea proves practical, it will be a decided boon to the profession.

In another phase a rapid, inexpensive, and simple method for Chemical Oxygen Demand (COD) analysis has been used successfully on a wide variety of organic compounds. Results of extensive tests conducted at several analytical laboratories indicate the method to be sufficiently reproducible for use in wastewater analysis. The "rapid" method, as described below, seems to warrant serious consideration by treatment plant operators and laboratory personnel. It involves the addition of a sample of the waste to a flask containing mercuric sulfate. After mixing 25 ml of a dichromate acid, silver sulfate mixture is added, mixed, and placed on a preheated hot plate. A thermometer is inserted in the flask, and the temperature is checked as the mixture is heated (care must be employed in heating the sample because the temperature can rise very quickly). Water is then added cautiously, and after cooling, the solution is titrated with ferrous ammonium sulfate in the presence of a ferroin indicator. The rapidity of the method is the major advantage of the procedure. Generally, it is said that a trained laboratory technician can heat twenty samples in less than 30 min. The rapid method is also less costly than other methods and requires the use of only small-volume samples. It is warned that the method is not intended as a substitute for the standard methods of COD procedure, since in some cases it will not be as efficient in the oxidation of certain organic compounds.

Beckman Instruments Scientific and Process Instruments Division is marketing a chloride-ion analyzer for continuous automatic measurement in aqueous solutions. The range is from 0.1 to 10,000 ppm. It is recommended as being suitable for monitoring stream pollution analyses.

pH Determination. Acidity or alkalinity may be determined by a variety of methods. The simplest method is litmus paper, which will give whole-number values for the pH of the solution. pH can be determined by systems as sophisticated as probes with transmitter/recorder and controller units. See Figure 4-6. One simple type of unit is a hand-held battery-operated pH tester which gives

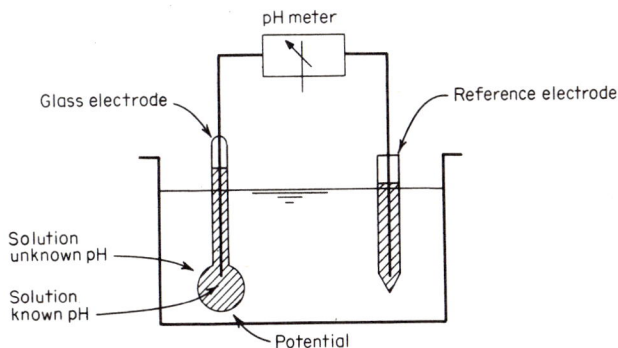

Fig. 4-6   Diagram of a pH meter.

CHEMICAL CONSULTANTS

INFLUENT & EFFLUENT WATER LOG FORM

SAMPLES PREPARED FOR: _____

_____

_____ zip_____

SAMPLERS NAME: _____ PHONE: ___ ___-____

CONTACT: _____

SAMPLE TIME: _____ TEMPERATURE: _____

LOCATION OF SAMPLING SITE: _____

_____

_____

SPECIAL INSTRUCTIONS: _____

_____

_____

SAMPLING CONDITIONS: COMPOSIT_____, ONE LOT_____,

ONE LOT MULTI_____, HYDRANT_____,

OTHER: _____.

- ☐ BOD$_5$ _____ mg/l
- ☐ COD _____ mg/l
- ☐ $\frac{BOD}{COD}$ RATIO _____ mg/l
- ☐ SUSPENDED SOLIDS TOTAL _____ mg/l
- ☐ DISSOLVED SOLIDS TOTAL _____ mg/l
- ☐ pH _____

- ☐ CYANIDE _____ mg/l
- ☐ FLOURIDE _____ mg/l
- ☐ SULFIDE _____ mg/l
- ☐ AMMONIA _____ mg/l
- ☐ ALUMINUM _____ mg/l
- ☐ ARSENIC _____ mg/l

Fig. 4-7 Influent and effluent water log form. (*Chemical Consultants, City of Industry, Calif.*)

direct readings from pH 1 to 13 with a 0.1 pH accuracy. It is easy to operate. Switch it on, immerse the sensing element in the solution to be tested, and read the pH value directly from the scale. It has the advantage of spot measurements and saves the time of taking samples to the laboratory for analysis. A typical unit with solid-state circuitry costs about $225 (1973). In the sophisticated model, the controller, high and low limits can be present at any point within the pH range. When these designated values are exceeded, a corresponding high and low power switch will be energized with sufficient 110 V ac power for a pump, motor, or other system. Most instruments have millivolt scales enabling the instrument to be used in any application where a plus or minus dc voltage

CHEMICAL CONSULTANTS

☐ ANTIMONY _____ mg/l		☐ TOTAL HARDNESS _____		
☐ BERYLLIUM _____ mg/l		_____		
☐ BORON _____ mg/l		----------------------------------		
☐ CADMIUM _____ mg/l		☐ OIL & GREASE (Hexane Extract) _____ mg/l		
☐ CHROMIUM - TOTAL _____ mg/l		☐ PHENOLS _____ mg/l		
☐ CHROMIUM HEXA.$^{6+}$ _____ mg/l		☐ SURFACTANTS (MBAS) _____ mg/l		
☐ COBALT _____ mg/l		☐ CHLORINATED HYDROCARBONS (Indicate Pesticide) _____ mg/l		
☐ COPPER _____ mg/l				
☐ IRON _____ mg/l				
☐ LEAD _____ mg/l		----------------------------------		
☐ MANGANESE _____ mg/l		☐ SODIUM _____ mg/l		
☐ MERCURY _____ mg/l		☐ CALCIUM _____ mg/l		
☐ MOLYBDENUM _____ mg/l		☐ MAGNESIUM _____ mg/l		
☐ NICKEL _____ mg/l		☐ POTASSIUM _____ mg/l		
☐ SELENIUM _____ mg/l		☐ LITHIUM _____ mg/l		
☐ SILVER _____ mg/l		----------------------------------		
☐ THALLIUM _____ mg/l		☐ BARIUM _____ mg/l		
☐ TIN _____ mg/l		☐ NITRATE _____ mg/l		
☐ TITANIUM _____ mg/l		☐ CHLORIDE _____ mg/l		
☐ ZINC _____ mg/l		☐ FREE CHLORINE _____ mg/l		
----------------------------------		☐ RESIDUAL CHLORINE _____ mg/l		
		☐ CHLORINE DEMAND _____ mg/l		
☐ COLOR _____ units		☐ BROMIDE _____ mg/l		
☐ DISSOLVED OXYGEN _____ mg/l		☐ SULFATE _____ mg/l		
☐ SPECIFIC CONDUCTANCE _____		☐ PHOSPHATE-ORTHO _____ mg/l		
		☐ PHOSPHATE-META _____ mg/l		
_____		☐ PHOSPHATE-ORGANIC _____ mg/l		
☐ TURBIDITY _____ J.T.U.				

Fig. 4-7   (continued)

is available for measurement. Millivolt and pH can be directly measured. This instrument is ideal for acid-base and redox titrations. Equipment used for pH determination in the sampling stages of the project can later be used as control equipment during the treatment stage.

Temperature.   It is important to ascertain the exact temperature of the wastes as they are produced. In treating wastes in large holding tanks, or basins, the incoming wastes frequently will have a higher temperature than the wastes in the tank. The mixture of these wastes may cause currents which will interfere with the settling of the light solids. If the raw wastes are at a higher than

```
                    CHEMICAL
                    CONSULTANTS

  □  PHOSPHATE TOTAL _____ mg/l    □  _____
  □  THIOSULFATE_____ mg/l     □  _____
  □  _____   □  _____
  □  _____   □  _____


  NOTE:
        All above determinations were made following the methods

        described in "Standard Methods for the Examination of Water

        and Waste Water", APHA, 13th Edition, 1971, unless otherwise

        specifically noted.
```

```
  STATEMENT OF ACCURACY OF DATA

        I hereby affirm that the above data comprise a true and correct

     representation of the water/waste water sample obtained or

     supplied from the above stated discharge point.

                                    Respectfully Submitted,

                                    CHEMICAL CONSULTANTS

                                    ----------------------
                                    Officer of Company--title
```

Fig. 4-7   (continued)

ambient temperature at the time of discharge, the samples too must be heated to the same temperature when tested.

CASE HISTORY—SAMPLING WITH MULTIPLE DISCHARGE POINTS

A sheet-metal fabricator produced aircraft and missile components. During its 35 years of existence the firm had expanded from a single building to five buildings covering an entire city block. Each building acquired had its own sewer connection. Ultimately the firm had seven different sewer discharge points that

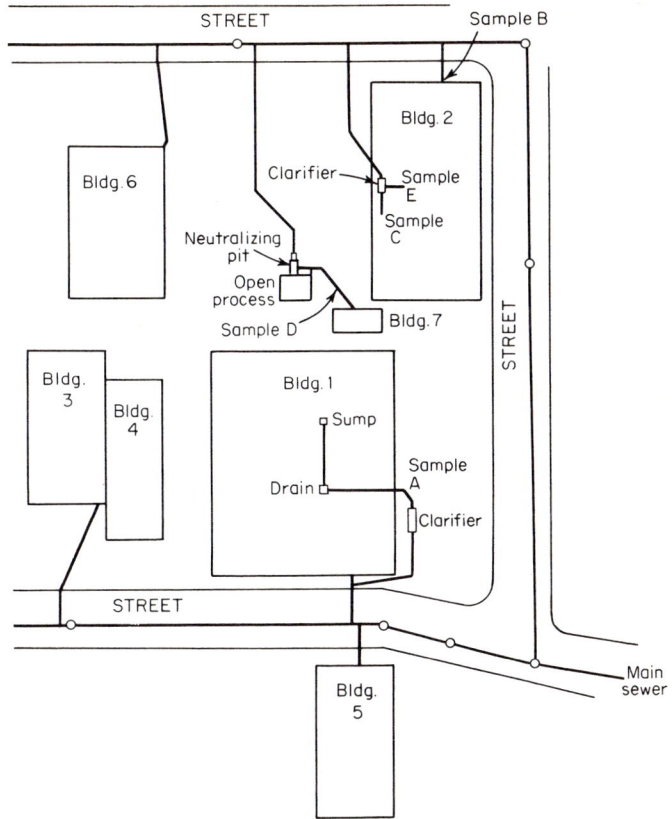

Fig. 4-8 Plan of an existing sewerage system of a manufacturing plant. (*Fansteel/Precision Sheetmetal, Los Angeles.*) This illustrates a typical plant that has grown from a single building with its building sewer to a group of buildings with many individual sewer connections. The sewer lines without clarifiers are for domestic waste only.

required seven different industrial waste discharge permits. New regulations by the local sanitation district based the number of permits required on the number of discharge points to the city's sewer system. Ownership was not considered. It was therefore possible that under present conditions five different sets of neutralizers, clarifiers, and monitoring stations could be required. The author was retained to make a preliminary study and cost estimate of the work needed to meet local discharge regulations and obtain a renewal of the discharge permits.

The first step was, of course, to obtain a description of the discharge. Since the information would be passed on to the local authorities, samples were taken and reported by a licensed chemical laboratory. A typical log form from a testing laboratory is shown in Figure 4-7. Figure 4-8 indicates sample locations. The report of the testing laboratory was as follows:

Sample A—Discharge from Building 1

pH	6.5
SS	13
COD	Not detected
Sulfides	0.3
F	5.0
Al	Less than 0.1
Cr	80.0
Cu	0.07
Fe	0.6
Na	111.0
Mg	0.4
Hg	Not detected
Ni	Not detected
Ti	Not detected
Zn	0.16
Oil	2.4

Sample B—Discharge from Building 2 (Domestic Water Only)

SS	34.5
COD	367.68

Sample C—Discharge from Film Processing of X-Ray Inspection Area

pH	7.4
SS	9.0
COD	19.64
CN	0.33
Ag	0.45
Oil	Not detected

Sample D—Discharge from Building 7

pH	2.1
SS	8.0
COD	100.16
F	162.5
Ti	Not detected
Nitrate	814
Oil	2.65

Sample E—Discharge from Building 2, Process Area

pH	6.4
SS	12
COD	112.84
CN	0.2
F	1.87

Al	Less than 0.1
Cd	Not detected—less than 0.01
Cu	0.5
Fe	1.15
Mg	2.5
Ni	0.1
Na	64.0
Ti	Not detected
Zn	0.78
Nitrate	57.2
Oil	2.8

Sample F—Discharge from Open Process Building

pH	11.9
SS	59
COD	62.72
CN	0.19
F	16.87
Al	8.5
Cr	9.14
Cu	0.11
Fe	1.05
Mg	0.31
Ni	0.1
Nitrate	1496
Oil	6.8

The quantity of discharge was determined by the engineer and estimated to be as follows:

Building 1	5 gal/min
Building 2	5 gal/min (1.5 gal/min from x-ray film process)
Outside process	10 gal/min (two 8-h shifts per day)
Building 7	Infrequent hosing with distilled water

Only the outside process had two 8-h shifts per day. The rest of the plant operated on the one 8-h shift.

In order to determine the total flow into a single outfall, the hourly and daily flow was calculated:

Building 1	5 gal/min $\times$ 60 = 300 gal/h
	300 $\times$ 8 = 2400 gal/d
Building 2	5 gal/min $\times$ 60 = 300 gal/h
	300 $\times$ 8 = 2400 gal/d
Open process	10 gal/min $\times$ 60 = 600 gal/h
	600 $\times$ 16 = 9600 gal/d

With all the information gathered, the engineer was able to design efficiently and economically a method of treatment that would eliminate the need for seven separate waste permits.

RIVER MONITORING

River water regulation falls within the controls of the federal government under the 1972 act amendments. The modern concept holds that the entire basin of a river, including the main stream and all its tributaries, must be considered as a unit. The regulations controlling the discharge of wastes into the stream must then be developed for the protection of the entire course, regardless of whether it is entirely an intrastate stream or an interstate one crossing other state boundaries.

This is a logical and forward-looking concept, and has already been put into effect by a number of river basin assemblies. Undoubtedly, the best-known, best-organized, and best-developed of these is the Ohio River Valley Water Sanitation Commission, familiarly known as ORSANCO. Its headquarters are in Cincinnati, Ohio, which has jurisdiction over the entire Ohio River and all its numerous tributaries. Some traverse some part of each of the eight member states of this compact, namely, Ohio, Pennsylvania, New York, Indiana, Illinois, Kentucky, Virginia, and West Virginia.

In existence since 1948, this organization has managed to clean up a very bad situation in water pollution. In its report for 1966, it is shown that almost 90 percent of the total of 1769 industries that discharge their wastes into the Ohio River system are provided with waste treatment facilities which satisfy at least the minimum requirements of the Commission. The monumental task accomplished by this dedicated work and coordinated action can best be realized when it is known that eight different governmental bodies, each with ideas of the importance of its own problems and responsibilities, were involved. To get them to agree to a single set of rules was indeed a notable accomplishment. Anyone interested in such coordination should obtain a copy of the comprehensive and extremely interesting book entitled "The ORSANCO Story," written by Mr. Cleary and published in July 1967, by The Johns Hopkins Press of Baltimore, Maryland.

The criteria that were set up vary for different sections of the river system, according to the use to which the water in a given section is to be put, that is, for potable water supply or for industrial use.

To assist in maintaining a balance between the desires of the officials of the several states in the compact and the rights of the industries discharging into the system, and for the purpose of setting up reasonable demands for treatment, as well as to benefit from the experience of those actively engaged in the industries in the area, the Commission has set up a number of advisory groups or committees, each representing different industry categories. These are

1. Aquatic-Life Advisory Committee
2. Chemical Industry Committee

3. Coal Industry Advisory Committee

4. Metal-Finishing Industry Action Committee

5. Petroleum Industry Committee

6. Power Industry Advisory Committee

7. Pulp and Paper Industry Action Committee

8. Steel Industry Action Committee

9. Water Users' Committee

These groups are composed of representatives of individual entities in each category, to give them a wide scope. Several of these committees have prepared comprehensive manuals relative to the special problems of their wastes, outlining methods of treatment, handling, sampling, etc.

The patroling of the total of 4499 mi of the streams which come under the jurisdiction of ORSANCO to determine the extent of observance of the regulations set up by the Commission having placed a great responsibility on the operating staff, the Executive Director conceived the idea of a robot monitoring system. His basic idea, translated into practical reality by the staff chemist, assisted by other members, resulted in the engineered design of the monitors. The original monitors were fabricated by the Engineering Specialties of Cincinnati, Ohio. The studies developed an automatic system employing electrode and transducer-sensing components, which seemed to offer the greatest promise for river monitoring. These devices were capable of providing continuous "output" signals, which would permit interrogation of the monitor whenever and as often as required. Photometric instruments were rejected as being able to provide only an intermittent signal, and there was also the probability of monitors being located in areas where it would be inconvenient to replenish the reagents in photometric instruments. This conclusion was fortified by an electronics consultant whose services were retained to supplement staff research. This specialist designed the circuitry and structural arrangements and then built what is called a "modular, multiple, automatic analyzer." The function of the unit is described by the name chosen for it, a "robot monitor." Coincident with its development, evaluation studies were made on commercially available telemeter equipment, data-logging devices, and telephone circuits that could be adapted for remote transmission of signals from the analyzer.

The ORSANCO monitor was designed to accommodate measurement of ten different water quality characteristics from as many as forty locations. At that time (1967), instruments had been devised to measure and transmit data on dissolved oxygen, chloride-ion concentration, hydrogen-ion concentration, specific conductance, oxidation-reduction potential (ORP), and temperature of water, as well as solar radiation in the area where the monitor is located. Feasibility studies have been made for a unit to measure gross beta radioactivity.

The robot monitor system consists of three separate but integrated components, a group of electronic analyzers combined with a transmitter, a telemeter interrogator and receiver, and a data logger and transcriber.

The analyzer component with its transmitter is a self-contained unit that can be located at any point along a stream. As river water circulates through the unit, the cells containing the electronic sensors respond to changes in quality. These changes are converted into electrical impulses for relay to the headquarters in Cincinnati.

The telemeter component in Cincinnati is connected with the analyzer units at river locations by means of a leased telephone-wire circuit. At a predetermined interval (every hour at the time of this writing), the telemeter interrogates each monitor station in sequence, and then stands by to receive a report on the quality characteristics being measured. Signals received through the telemeter circuit actuate data-logging devices. One of these is a transcriber that automatically types the information received on a tabulation sheet. This provides a visual record of quality data. At the same time a paper-tape punch is also activated; it codes the data in a form that can be fed into an electronic digital computer, programmed to perform a variety of evaluation processes.

In September 1960, the prototype analyzer, along with the telemeter and recording equipment, was placed "on stream." Its performance justified expectations, and ten additional analyzer units were constructed. By July 1965, a total of thirteen units were in operation. Meantime, a so-called "mobile" monitor was constructed for survey purposes and special investigations. It is a portable version of the robot analyzer, equipped with a strip-chart recorder. It will accommodate any two units of the standard modular analyzers for measuring quality characteristics.

On the basis of the successful operation of the system, the Commission anticipates that sensors will be added to measure other quality characteristics. Also, they are contemplating the possibility of circuitry arrangements in the existing robots for formulation of a "river quality index number." Such a number would represent a weighted composite of various measurements now performed by the monitor, and would be automatically computed at the monitor site and transmitted to central headquarters. Thus it would be possible to secure immediate and continual grading of river quality in such simple terms as "grade A, grade B," etc. The execution of this intriguing project is not so much a problem in electronics as one of selecting and weighting quality characteristics to arrive at a meaningful index number. Staff studies were being made (1967) on development of an index-number system.

In 1965 the monitor network consisted of four integrated components, as follows:

1. Seventeen Water Users' Committee manual stations operated on a voluntary basis by the managers of municipal and private waterworks

2. Eleven manual stations operated under contract arrangements with the U.S. Geological Survey

3. Thirteen robot monitoring units installed and maintained by ORSANCO, of

which ten relay data directly to headquarters and three record information only at the site

4. Fifteen sampling points for periodic assay of radioactivity in water, in river sediment and biota, and in fish, operated under contract with the Potamological Institute of the University of Louisville, Kentucky

The Cost of Monitoring

The collection and laboratory analysis of river samples by conventional methods is a costly procedure. For a single sample, the cost (1966) of measuring only five constituents is at least $10; if analyses are made for thirty constituents, the cost may go as high as $120. At these costs, control agencies are generally forced to limit the number and duration of their sampling programs for the assessment of stream conditions.

With respect to the cost of operating the manual components of the ORSANCO monitor network, the following tabulation provides an annual summary of data items received, the out-of-pocket expenses associated with their assembly, and the dollar value of the data. The "value" placed on the data represents what the actual cost would be if the Commission had to fund analytical services equivalent to those contributed by the Water Users' Committee at no charge and if ORSANCO had to pay the entire cost of services shared by the Geological Survey.

	No. of data items	Out-of-pocket expenses	"Value" of data	Value per item
Water users (17 stations)	31,570	$ 2,664	$71,376	$2.26
USGS (11 stations)	9,900	10,000	20,000	2.02
Total (28 stations)	41,470	$12,664	$91,376	$2.20

This accounting shows that the value attached to each item of data averages $2.20. When ORSANCO funds devoted to the project are matched against value received, it becomes apparent that the Commission acquires the equivalent of $7.22 of data for each dollar it contributes. The value of data received from each of the twenty-eight manual stations averages $3000 annually.

The cost of data acquisition using the robot monitor network may be judged from the following analysis based on actual experience with the thirteen-station system:

Amortization expense (assuming a 17-year life of equipment and a 4% interest rate on a total investment of $136,000)	$17,000
Operation and maintenance	14,000
Leased-circuit rental	6,000
Total annual expense	$37,000

The system produces about half a million items of data per year. Thus the cost per item averages $7\frac{1}{2}$ cents, as compared with $2.20 for the manually operated stations. In brief, for those quality characteristics that can be measured automatically, the cost on the average is only about $3\frac{1}{2}$ percent that of manual sampling and analysis.

In 1965 the Commissioner requested a staff appraisal of the robot monitor system with respect to the following questions:

1. Is the system serving the purpose for which it was developed?

2. What specific contributions to pollution-control practice have resulted or can be expected from operation of the system?

3. To what extent is expansion of the system merited?

4. Is it adaptable to the needs of other pollution control agencies?

The answers to these questions have been summarized as follows:

1. The purpose of the monitor system is to provide information on quality conditions on a continuous around-the-clock basis. This goal has been attained, and sections of the Ohio River and certain tributaries have been placed under 24-h vigilance.

2. One measure of the contribution made by automatic monitoring to pollution control management may be expressed in terms of economies realized in the collection of water quality data. Without the automatic system it would not be feasible to monitor streams in such a widespread region as the Ohio River system on a 24-h basis.

The robot monitor has attracted wide interest, and several local, state, and federal agencies have installed one or more units. A multiunit station installation was erected on the Potomac River in 1966 by the U.S. Department of Health, Education, and Welfare. Prior to this the federal government ordered units for surveillance studies in the Raritan Bay area of New Jersey and several places on the West Coast. The Interstate Sanitation Commission is operating a single unit in New York Harbor. The Municipality of Metropolitan Seattle, Washington, has installed four units in waterways adjacent to its sewage treatment plants. It should be pointed out, however, to those interested in applying this system to other places, that operating such a system is neither simple nor inexpensive. Acquiring the equipment is but a first step. And to justify the investment required by the installation and maintenance, the user must be prepared to engage in the continuing evaluation of data and diagnosis of river conditions. However, in view of the unusual opportunities for control of river pollution and the increasingly large number of river basin compacts that will undoubtedly be made, the author believes the extensive description given above is warranted.

As Paulson said at the Water Quality Symposium in Washington in 1965:

The fact of the matter is that water pollution must begin with the concept of an entire stream or river. To think in smaller terms is to lose sight of the complexity and magnitude of the problem. If we look for solutions through technology alone, we are failing to recognize that the water pollution control problem is as much a philosophical problem as a technical problem.

Prior to this precept, the federal government, in 1951, through the Federal Security Agency of the Public Health Service, Division of Water Pollution Control, published a series of fifteen Summary Reports, each covering a complete drainage basin of one of the major areas of the United States. These reports presented pertinent data on the basins and the extent of the pollution in each from municipal sources and industrial plants, and listed the municipalities and industries by name, their liquid discharges, and the extent of treatment they already had or what was needed to meet the requirements established at that time. The pollution was stated in terms of "population equivalent." Naturally, in the 16 years intervening, there has been considerable development in most of these basins, and much of the improvement in treatment suggested in the reports has been carried out. The reports proved to be a very valuable source of information.

Acting on the success of the ORSANCO compact, other similar river compacts have been set up, namely:

1. The Delaware River Basin Commission, with Pennsylvania, Delaware, New York, and New Jersey as participants

2. The Interstate Commission on the Potomac River Basin, which includes the cooperation of Maryland, Pennsylvania, Virginia, West Virginia, and the District of Columbia

3. The Klamath River Compact Commission, with California and Oregon as members

4. The New England Interstate Water Pollution Control Commission, including Massachusetts, Connecticut, Maine, New Hampshire, New York, Rhode Island, and Vermont

Other river control facilities and controlling organizations have been established for the Trinity, Neches, and Sabine Rivers in Texas, all of which flow through large industrial districts.

One of the largest waterways traversing a heavily industrialized district is the Houston Ship Channel, which is, in reality, an estuary of the Gulf of Mexico and reaches from Houston, Texas, to the Gulf of Mexico at Galveston, Texas. The shores of this estuary are lined with some of the largest industrial complexes in the United States, including oil refineries, steel mills, chemical plants, petrochemical industries, paper-products plants, fertilizer manufacturing, and many other establishments, all producing troublesome wastes. Since the waters of the Channel are saline in character, they are not available for potable purposes, but the state pollution authorities have set up strong requirements to eliminate pollution that would cause odors, unsightly nuisances, or damage to structures along the watercourse.

In Germany, an outstanding river control organization has been in existence for a number of years. This is for the River Ruhr, under control of a body known as the Ruhrverband, which controls the discharge of municipal and industrial waste into the stream and also provides treatment plants into which the industries may discharge their wastes for common treatment at a moderate cost.

There is no doubt that, with the program of the federal government being implemented, river basin control will be extended to more and more of the systems of the country and will be equipped with monitoring systems similar to the one described on the Ohio River, which has been so effective.

An example of the advantage to an industry of river monitoring is that of the Bowaters Carolina Corporation, at Catawba, South Carolina, which has installed a dissolved-oxygen continuous sensing and recording system for a distance of 9 mi downstream from the mill. This gives the mill management instant knowledge of the stream condition and eliminates the expense of an attendant to get samples three times daily, with travel of 20 mi for each trip. The attendance now is reduced to one trip each day for a check of the instruments. The waste discharge from the mill is regulated according to the volume of flow in the stream and the dissolved-oxygen content of the water.

In the setting up of stream standards, the significance of the factors must be thoroughly understood. Fecal and nonfecal coliform organisms must be evaluated. Pulp and paper mill wastes have a high total coliform content but a low content of fecal organisms. Studies show that less than 1 percent of the coliform bacteria in these wastes are of fecal origin. Most of them come from the wood-pulping operations. It is common for stream standards for coliform organisms to specify a limit of 5000 organisms per 100 ml. Pulp and paper wastes cannot meet this limit on the basis of total coliform organisms, but the level of fecal coliforms is well below the prescribed limit. Those who have studied the subject are of the opinion that the coliform standard for water is not a valid parameter for evaluating industrial waste pollution.

PUBLISHED DATA

Informed about the nature of the problem, the requirements of the regulatory authorities, and the data concerning the specific wastes, the engineer is ready to proceed with the determination of the type of treatment best suited to his particular project. Roughly, if the wastes contain a high percentage of organic matter, a biological oxidation method may be indicated, and that approach should be explored. If the wastes are mainly of a chemical character, chemical methods of oxidation, reduction, and treatment should be investigated. If the engineer does not have background experience in handling wastes of the type in hand, he must refer to the literature to find out how similar wastes have been handled by others. Better yet would be to make visits to operating treatment facilities at other plants.

The engineer should combine his reading of work performed with the information available on some of the experimental methods, for a new approach. Just because a new idea or development has not been tried in a full-scale plant is no reason why it should not be considered in a specific case if it appears to have possibilities. All new ideas have to start somewhere. Chapter 8 describes a large number of new ideas, developments, patented processes, etc.

Since the information on the character, volume, etc., of wastes is so important in successful treatment, a sample data questionnaire, which the author has developed and found very useful, is presented below.

INFORMATION QUESTIONNAIRE

General

1. Name of the plant _____

2. Address _____

3. Telephone _____ 4. Person interviewed _____

5. Official position of No. 4 _____

6. Plant manager (if other than No. 4) _____

7. Location of plant where problem exists (if not
   No. 2) _____

8. Type of plant (manufacture of) _____

9. Possible point of discharge of final effluent
   River _____ Lake _____
   Small stream _____ Open sea _____

10. Are other local industries discharging treated or untreated wastes into the same body of water? If yes, get data on types of wastes. _____

11. Is the potential body of water into which the final effluent will be discharged a source of potable water for a municipality? _____

12. Distance of point of ultimate discharge from the industrial plant producing the waste _____

13. Is it possible to discharge treated or untreated industrial waste into the local sewerage system? _____

14. If yes, name the municipality involved. _____

15. Has this municipality a sewage treatment plant? _____

16. If yes, give type and capacity (mg/d or m^3/d). _____
_____

17. Is this municipality willing to receive industrial wastes? If so, on what basis?

    a. To receive and treat raw wastes in the municipal plant on payment for volume and/or concentration of certain ingredients? Get data on charges.
    _____

    b. To build any plant additions made necessary to treat the wastes, at the expense of the municipality, the industry to pay the operating costs of the additional units? _____

    c. To permit the industry to pay for any additions needed, which then become the property of the municipality? _____

18. Has the municipality any regulations as to types, strength, or quality of wastes that may be discharged into its sewerage system? Maximum BOD _____ suspended solids, etc. If yes, obtain copies of such regulations. _____

19. Has municipality formulated a charge or charges or a formula for industrial wastes discharged into its system? _____ If yes, get full details. _____

20. Are the wastes of other industries in the area now being taken into this municipal system? _____ If yes, get as much information as possible on the types and volumes of these wastes. _____

21. What legally constituted regulatory agency or authority has demanded treatment of the wastes from this plant?

    a. City in which the plant is located _____

    b. State Board of Health or Pollution Control Board _____

    c. River Basin Compact Authority _____

    d. Interstate Water Pollution Control agency _____

    e. U.S. Public Health Service _____

    f. Federal Government Water Pollution Control Administration _____
    _____

22. Has the authority specified or implied the extent of treatment required? _____ If yes, give all possible details. Get copy of requirements. _____
_____

23. Is approval of plans and specifications required by authority before commencement of construction? _____ If yes, give details as to number of copies, where to be presented, etc. _____
_____

24. Has client been given a deadline for cessation of pollution? _____

25. Has client any waste treatment facilities now? _____ If yes, get full details and plans if possible. Get photographs of existing plant. _____

_____

26. Get names of other industries in vicinity discharging wastes. _____

_____

Get information on types of wastes. _____

27. Would client be interested in, or willing to join, a district plant to handle the combined wastes from other plants? _____

28. Is client interested in recovery of any component of the wastes? _____ If yes, determine feasibility of recovery. Get all information possible on amount of component in the wastes and the current market value. _____

_____

29. Is client amenable to suggestions for changes in productive processes to reduce the volume and concentration of wastes to reduce treatment costs? _____

30. Is client willing to construct the treatment works on a progressive basis if the authority will permit? _____

31. Is client interested in leasing equipment for the waste treatment facility, rather than outright purchase? _____

Details of the Wastes to Be Treated

32. Get same information for each type of waste.

    a. Acidity _____ pH _____

    Type of acidity _____

    b. Alkalinity _____ pH _____

    c. Volume of each waste flow _____
    Gallons per day or cubic meters _____
    Maximum hourly flow _____

    d. Periodicity of each waste flow _____
    Daily discharge over _____ hours
    Seasonal (per day, weekly, monthly, etc.)_____
    Types of suspended solids _____

    e. Suspended solids (mg/l) _____

    f. Total solids (mg/l) _____

    g. Organic solids (mg/l) _____

    h. Volatile matter (%) _____

    i. Precipitable solids (from test data) _____

    j. Biochemical oxygen demand (BOD, mg/l) _____

    k. Color of wastes _____ Concentration _____

    l. Toxic components of wastes
        Phenol _____ Cyanide _____ Chromium _____
        Other toxic components _____
        If client has an analysis of the wastes, obtain copy. _____

33. Allowable limits of wastewater constituents

    a. Chemical _____

    b. Source condition _____

    c. Allowable concentration _____

    d. Regulating body _____

    e. Remarks _____

34. Is ground area sufficient for waste treatment plant available in or close to the plant property? _____

35. If an area on plant property has been designated for waste treatment purposes, are there any immovable obstructions in the area? _____ If so, describe. _____

36. If, for any reason, client wishes the waste treatment facilities to be placed inside a plant structure, get data on area available, headroom, etc. _____
_____

37. Proximity of area designated for treatment facilities to residential or business districts _____

38. Are there any local zoning restrictions which would influence placement of waste treatment plant on company property? _____

39. Is client's plant the only industry in the municipality? _____

40. If yes, does the client's payroll represent a substantial income in the municipality? _____

41. Approximate amount of business, property, income, or other taxes paid by client to the municipality _____

42. Are client's relations with the municipality good? _____

43. Available electric current for equipment _____
    Phase _____ Cycle _____ Voltage _____

44. If power is purchased, give price paid by client (per kWh). _____ Is there a standby charge? _____ How much? _____

45. Is ample current available for the needs of the proposed waste treatment plant? _____

46. Does client purchase power from a local utility? _____ If yes, get name and cost of power. _____

47. Does client produce his own power? _____

48. Would client be interested in producing power from waste treatment solids if of proper type? _____

49. Would production of soil conditioner (fertilizer) from sludge be of interest to client? _____ If so, is there a local market for it? _____

50. Give client's source of water supply. _____

51. Cost of water from this source $_____ per year or per unit of water _____ Total annual water bill of client _____ Is this source able to supply additional amounts that may be needed for plant expansion? _____

52. Does client have his own water treatment facility? _____ If yes, describe.

_____

53. Type of water required by client:

    a. Potable (gal or m^3/d) _____

    b. For product use (gal or m^3/d) _____

    c. Softened water (gal or m^3/d) _____

    Degree of softening _____

    d. Clarified water (gal or m^3/d) _____

    e. Demineralized water (gal or m^3/d) _____

    f. Cooling water (gal or m^3/d) _____

    g. Washdown water (gal or m^3/d) _____

    h. Does any water enter into an edible product? _____ If so, give details.

_____

54. Would client be interested in reducing purchased water volume by treatment of wastes for recirculation and reuse? _____

Would there be any objection to reuse of treated effluent for plant purposes?

_____

55. Has client engaged a consultant to handle the waste problem? _____ If yes, get name and address. _____

56. What services does client want from the engineer?

    a. Study problem. _____

    b. Obtain data on characteristics of wastes. _____

    c. Make analyses and tests. _____

    d. Discuss problem with regulatory authorities. _____

e. Design plant or facility. _____

f. Obtain approval of authority. _____

g. Prepare complete plans and specifications. _____

h. Obtain bids for construction. _____

i. Supervise construction. _____

j. Instruct the operators. _____

k. Prepare an operating manual. _____

l. Supervise the initial operation of the plant. _____

m. Provide annual inspection. _____

For how long? _____

_____

(Name of person obtaining data)

Date _____

Chapter Five

The study

Engineers cannot ethically advertise their products in the same way that manufacturers of tangible objects can. They have to rely on future reports from the management of plants they have designed, testifying to their ability to fulfill their commitments to the satisfaction of their clients. Therefore, the time spent in carefully weighing and considering the many facets of design will be repaid in full by future engagement by other industries in similar projects.

Any industrial treatment project can be broken down into four phases: (1) preliminary, (2) preliminary design, (3) final design, and (4) construction.

PRELIMINARY PHASE

Most companies request a complete study of their industrial wastewater situation before contracting for the entire facility. The first phase must assure the client that he is in competent hands. The engineer outlines the existing situation and provides a framework from which to begin. The initial study generally includes a plot plan of the site with the existing drainage system indicated. The composition of the wastes, any existing wastewater segregation, location of manholes, and any crucial pipe sizes are also indicated. Also included in the report is a list of production capacities for various parts of the plant, as they relate to the flow rate and characteristics. Data should be tabulated for maximum production and minimum production, normal and future.

It is necessary that there be a clear and intimate knowledge of the entire plant drainage system. This can be shown by disregarding all parts of the plant except the wastewater conveyance system and points of discharge. Figure 5-1 is a schematic view of a plant drainage system. Both domestic and industrial waste sources are indicated. A sketch of this type helps greatly in identifying

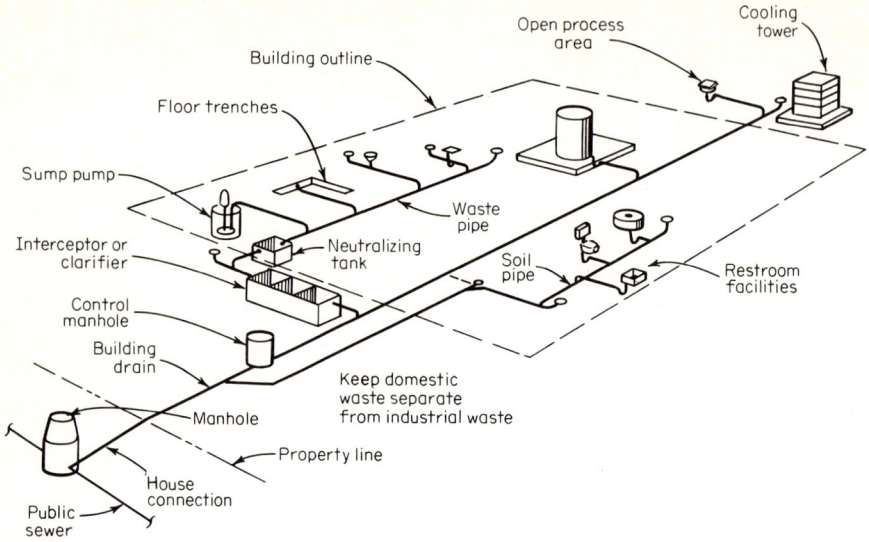

Fig. 5-1 Schematic of plant drain system.

sources of contamination, interconnection, and possible catastrophic spillage locations.

As soon as possible, the engineer should become personally familiar with all aspects of plant operations. The following suggestions for an investigative effort are made:

1. Begin by tracking all plant incoming water from its sources to its ultimate destination.

2. Do not overlook fume scrubbers, water-cooled rectifiers, heat exchangers, boilers, heating and cooling coils, air conditioners, and welders.

3. Ascertain that all inlet water lines to process solutions have antisiphon devices.

4. Inspect all heating and cooling coils for physical condition (conductivity meters for leak detection will be required when a waste treatment system is installed).

5. Thoroughly inspect all floors and foundations in the processing areas for possible leakage and consequent percolation to groundwaters.

6. Check all chemical storage areas for compliance with safety regulations and methods for handling "empty" and broken containers.

7. Make an inspection inside and outside of all process tanks and filters for physical condition. Check especially those tanks where process liquids are seldom if ever removed. All piping should also be examined.

8. Record all information acquired with the foregoing steps. It will prove very useful later.

Fig. 5-2 Trafalgar refinery waste treatment plant and aquarium, Cities Service Oil Company, Ltd., Bronte, Ontario, Canada. (*Cities Service Company.*)

ARCHITECTURE

Even in the preliminary study, architecture must not be ignored. In the past, structures built for industrial plants have been noted more by the absence of aesthetic appeal than by its presence. A new era has evolved in this respect. Today engineers who design industrial buildings take great care to provide structures that are pleasing to the eye, and many compete for prizes awarded for utility and beauty. In this perspective the engineer must give considerable thought to aesthetics. If he is not architecturally minded or equipped, he should engage an architect to contribute his ability.

When the city of Los Angeles was considering a very large sewage treatment facility, a committee of city officials visited a number of large plants in the country just to observe their architectural features. An example of a well-designed plant is given in Figure 5-2, showing the attractive buildings provided in the waste treatment plant for the oil refinery of the Cities Service Oil Company, Ltd., in Bronte, Ontario, Canada. Today, distinctive and pleasing architecture is recognized as an important factor in public relations. It earns the approval of the general public and eliminates the stigma formerly attached to an industrial plant in the neighborhood.

Past Practice

When the first mechanical equipment units were used in waste treatment plants in the United States, beginning about 1916, they were simply adaptations of units used in productive industries. Noise in industrial plants was taken for granted, and the plants were expected to be dirty. In some cases the operators of large mills objected to the elimination of noise in machinery units because they had come to know the familiar squeaks and clanks of individual machines. Unconsciously they could tell from many feet away if a certain unit was operating properly. If the noise was missing, something was wrong. Municipalities, however, were more interested in noise avoidance. Their engineers insisted that mechanical units installed in their plants be quiet in operation and good-looking.

Present Practice

The new outlook in design caused the manufacturers to redesign all units for sanitary engineering installations for quietness. The manufacturers began to cover them with attractive paint and provide railings and safeguards to protect the public. Men from industries, visiting the newer waste treatment plants, liked the quietness and appearance of the units, and they in turn required the same features in their plants. This attitude is well established today, so that attention must be paid to making plant structures compatible with other structures in the area. This is particularly true if the units will be close to public highways or visible from them.

SCOPE OF THE STUDY

The final design must encompass more than just the technical application of methods of treatment of industrial wastes. It must include aesthetic considerations, the relation of a particular installation to the local environment, and the economics of the industry. The study that the engineer will present to the owner, as an indication of his expertise, must indicate that the final design will take into consideration the following details:

1. Experience in the interpretation of the data

2. Knowledge of local environmental conditions

   a. Location of the plant

   b. Size and shape of the available site

   c. Possible objections by neighbors

   d. Zoning restrictions

3. Point of ultimate discharge of the effluent

   a. Stream, lake, ocean, etc.—nearby or remote

   b. Limits on pollutants—regulatory agencies

c. Piping to a distant point—if local regulatory requirements are too severe (i.e., drinking-water purity)

d. Piping into municipal sewerage system if city is willing to accept the wastes and if requirements are favorable when compared with cost of private plant

4. Sources of plant effluent

5. Chemical composition of effluent

6. Combining of wastes for single treatment versus individual treatment

7. Grouping of plant units

    a. Effect on economy and appearance of plant

    b. Combining new units with existing structures for reduction of cost and symmetry of appearance

8. Relation of laboratory tests or experiments to full-scale plant

9. Difference in results due to use of "pure" or "commercial" grade of chemicals

10. Need to transmit designer's ideas to manufacturers so as to get correct information and prices on proposed units

11. Approval of equipment used in plant by regulatory authorities

12. Planning necessary to prevent pollution catastrophes

Each point mentioned in the checklist is critical in the ultimate design of an economical, easy-to-manage treatment facility. Item 10 is important whenever the designer strays off the beaten track of facility design. In one case in the author's experience, mixer units were required for twenty-nine different sizes of tanks. These tanks were essentially holding tanks into which the waste flow for an entire day would be discharged. They would be held overnight and treated on the following day. Nothing was added to the contents of the tanks. Mixing units were required simply to homogenize the wastes during the holding period of from 10 to 12 h to prevent deposition of heavier solids and maintain uniformity in the wastes. Time was not of the essence. The first requests for quotations on these mixer units produced an astounding variety of prices, reaching in some cases to $40,000 for a single unit. The total horsepower load for all motors reached 565. This would have added an exorbitant sum to the initial and operating costs of the plant. The producers of the various units were consulted and the idea behind the use of the units was explained. Most of them insisted that the handling of so many gallons per minute was critical. Only one manufacturer seemed to understand; his actual quotation for the twenty-nine complete units, with drive units, supporting steel superstructures to span the tanks, all delivered at the plant site, was $97,000. The total power requirement was only 165 hp. Subsequently, information was received from the superintendent of this plant to the effect that, after 11 years, the company was completely satisfied with the operation.

Pollution Catastrophes

Item 12 of the checklist can have far-reaching importance. In upgrading metal-finishing facilities to reduce pollution, there is more to consider in the preliminary study and design than the flow chart of the pretreated and posttreated wastes. There is a great difference between the type of pollution which causes a nuisance to the public domain by gradually altering the environment to which it has been discharged, and a catastrophic pollution which not only causes a sudden and far-reaching change in the environment, but also poses a direct and immediate threat to the health of humans and aquatic life.

In the metal-finishing plant, nuisance pollution is characterized by the daily flow of waste rinse waters to the sewer, usually coupled with small-volume spillages incurred in workpiece transport between processes or during the addition of maintenance chemicals to the process itself. Spent process solutions gradually discharged along with waste rinse waters may also be considered nuisance pollution.

On the other hand, process solutions containing concentrated toxic materials, discharged suddenly and in large volume to the sewer, constitute potential catastrophic pollution. When these discharges are deliberate, they result from human carelessness and error, and may be eliminated only by stringent in-plant housekeeping measures. Electromechanical devices strategically located within the plumbing system may sense the passage of such discharges, but serve only to inform of the damaging event after its occurrence.

Accidental discharges of a catastrophic nature are far more insidious because they are not predictable. Fortunately they are very rare in the metal-finishing industry. Revision of a plant to prevent both nuisance and catastrophic pollution is within the province of the design engineer. He must make this an integral portion of his entire prospective treatment scheme. The control measures should be presented so as to allow evaluation by the plant manager prior to design discussion. Such evaluations may forestall delays and interruptions of work schedule which otherwise may be caused by the installation of treatment systems.

Piping Rinse Waters. As a temporary catastrophic pollution prevention technique, all rinse waters can be piped directly to the sewer. When the treatment facility is designed, the plans will show these waters piped to their individual treatment areas. Thus all floor inlets to the sewer may be plugged, preventing the escape of accidental spills of concentrated solutions. Spent process solutions in this system are pumped to holding tanks for removal by scavenger, or for treatment and gradual discharge. Floor sumps will be required for sump pumps. Obviously, where both cyanide and chromium solutions are in use, steps must be taken in floor segregation to assure that cyanide spent solutions are prevented from mixing with acid wastes. This precaution will avoid the possibility of generation of toxic hydrocyanic acid gases. Separate drainage sumps and holding tanks must be provided for this condition.

Holding Pits. A large holding pit or lined lagoon could be installed outside the process buildings. All flows exiting from the plant would pass through this pit or lagoon before entering the sewer. Electromechanical devices for the measurement of pH and conductivity would sense any irregularity and sound an alarm. All incoming water to the processing plant would be turned off. The retention time in the pit or lagoon would be sufficient to cushion and absorb the incoming concentrated solution or "slug" without adverse effect at the outfall. Thus, the slug would be retained in the holding facility for treatment or disposition.

Existing Floor Trenches. A third means of attack is based on the supposition that the plant in question has existing floor trenches through which all wastes are conveyed from the plant. These trenches must be large enough to be divided lengthwise into two subtrenches. By this dividing technique, one side of the trench would be used for rinse waters. The headers from the rinse water tanks could go directly to this portion of the trench, which would then be covered over to prevent entry of floor spillage. The other half of the trench would remain open and be used as the above paragraph describes. All wastes would end up in a sump of some type. Accordingly, this open portion of the trench would have no exit to the sewer and would catch all solution dumps and floor spills for pumping to a disposition point elsewhere.

MATERIAL AND EQUIPMENT SELECTION

The preliminary study ends with several proposals to the client, along with an overall appraisal of the situation. Once the owner has selected the approach he favors, the second step in the design begins. The engineer must select the size and shape of the individual units of the plant. Figure 5-3 illustrates a simplified version of industrial waste flow through a plant and the main treatment components involved.

The effluent may have to be pretreated to adjust the pH or to remove heavy solids or floating oils, in general, to make the cost of treatment more economical. There are two flows out of the treatment stage; one is the filtrate and the other is the separated solids. The filtrate may not be at this point acceptable for discharge into the sewer and therefore would require additional treatment or posttreatment. This may simply be the addition of chlorine and an adjustment of the pH. The solids discharging from the treatment equipment usually have a high percentage of water which must be removed. This water after separation is recycled to the effluent while the semidry solids are recycled or removed to a dump.

The engineer's selection of units requires an intimate knowledge of the available equipment units, apparatus items, controls, etc., that must be brought

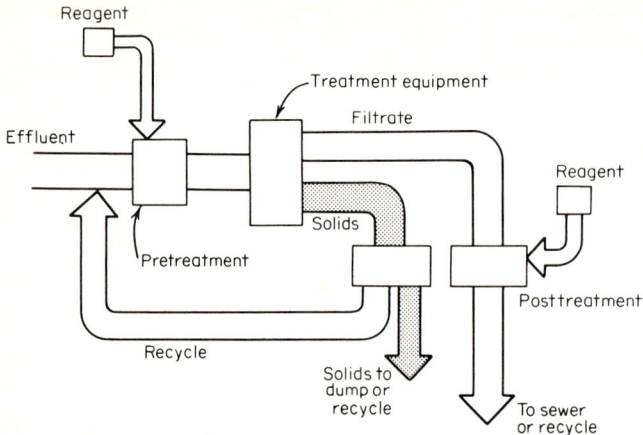

Fig. 5-3   Graphic diagram of flow of industrial wastewater.

together to make a concerted whole. Every engineer has his favorite type and shape of tanks, his preferred mechanisms and apparatus. There is nothing wrong with preferences for certain items provided that judgment is used in the selection and fitness for the case at hand. One must also be willing to admit the advantage of different brands of units in the particular case. There are many considerations in the selection of the size and shape of equipment for treatment, neutralization, sedimentation, flocculation, clarification, etc. Such considerations include:

1. The amount of land available for the proposed plant

2. The topography of the site—flat, hilly, rocky, etc.

3. The availability of overhead space

4. The possible attachment of new plant units to existing structures

5. The size of the plant

6. Preferences of plant management or personnel

7. Type of pumps for liquids, sludges, etc.

8. Metals required in pumps, pipelines, etc., for protection against corrosion or wear

9. Lining required for tanks to prevent damage by chemical components of the raw or treated wastes

10. Cost of tanks and other equipment

Tank Selection

The selection of the type of tank structure is of material importance in the final cost and design detail. In the design of municipal water and sewage treatment facilities it is common practice to use concrete for tank structures. There does not seem to be any particular reason for this, except that in the early days of construction of such plants, the tanks were normally set in the ground. Steel was therefore not considered suitable. The tanks were set in the ground to permit gravity flow of water and sewage. Also, the municipalities wanted a low profile for the plants. The operators disliked climbing ladders to inspect the operations. This was a concession to laziness, since the personnel wished to be able to see what was going on while walking on the ground. The general design of these concrete tanks has always been more or less unscientific, more for looks than practicality.

In one instance in the author's experience a large plant for a city had been designed by a well-qualified engineering firm. The city officials asked a local architect to review the plans. He did and called upon the author to explain why the walls of the sedimentation tanks were 1 ft thick. The architect said, "I could design those tanks with walls half as thick and save a lot of money." The reply was that those who inspected these plants after their construction seemed to think that unless the wall of a tank was 1 ft thick, it looked insecure. The plant operators could not walk safely on a thinner wall. There is, of course, no technical reason why walls of concrete tanks must be 1 ft thick. The author has seen tanks designed by foreign engineers with unsupported walls only 4 to 5 in thick which seem to stand up indefinitely. Nevertheless, the concrete tanks used in sanitary engineering constructions will probably continue to be 1 ft thick or more.

Steel Tanks. In industrial plants it has been common practice to use steel tanks for all purposes. They are usually completely above ground, where no special protection against corrosion is needed. Any leaks that may occur are detected immediately and their condition is readily observable. As in an oil refinery, tank-farm industries are accustomed to steel tanks above ground, and there is no aesthetic objection to them. In municipal water and sewage treatment plants the tendency is to camouflage them in some way that will conceal their purpose from the passerby.

If, in designing a waste treatment plant, the engineer has decided to use steel tanks, that is one step. However, other important considerations must then be taken into account. What will be the depth and diameter of these tanks? The relation of these two factors to the final cost of the plant is fully discussed in Chapter 6 and illustrated by data in tables.

If the engineer is a devotee of the circular design of tanks which are to be equipped with mechanisms for mixing, sedimentation, etc., he has a variety of different makes to call upon. If his preference is for the rectangular type of tank, that is, a tank with a length several times its width, he is restricted to only a few types of mechanical sludge and scum collection units. There are devotees of each of these shapes of tanks who will not use the other type under any circumstances. This is not a proper, ethical attitude for an engineer to take. There are good arguments on both sides, and no one has been, or ever will be, able to determine which of these types is measurably better than the other. The simple reason for this is that the only way to determine the superiority of one type is to have units of each type of appreciable size installed in the same plant. With each handling the same volume, concentration, operating schedule, and program of withdrawal of sludge, a comparison could be made. Since this is an impractical solution and the actual difference in the effluent from well-designed tanks would probably be small, it is unlikely that this question will be solved in the near future. The expense of duplication would not be justified.

There are many criteria besides personal preference that must be considered in selecting tank shapes and sizes. Paramount is the actual installed cost of the unit, ready for operating, including not only the actual cost of the tank structure, but also the cost of assembly and installation of the mechanical units and the cost of lining the tank, if that is needed. The expected allowance for maintenance and repair of mechanisms must also be included.

CONTROLS—
AUTOMATIC OR MANUAL

Another important aspect of the treatment facility is the control system. The best-designed treatment facility is worthless if it operates ineffectively or not at all.

To eliminate the imperfection of the human element and fight the ever-rising cost of labor, many industries have turned to automation. In wastewater, there is an even greater tendency to automate since automation saves treatment time and provides a continuous record of test results for future reference. In an industrial waste treatment plant it is essential to have clear records of what is happening; how much waste of each type is coming to the treatment plant, the content of polluting elements, the amount of chemicals being used; the character of the final effluent and its fitness for final discharge; the amount of power being used; as well as instant warnings of malfunctions of mechanical units and indications of proper operation. Also important is the need for knowledge of tank levels at any time and of the acidity and alkalinity of the wastes (pH control).

All wastewater treatment systems therefore require the use of a variety of control instruments. Although the field of instrumentation is a profession in itself, the basics should be known by anyone involved with wastewater treatment.

Regardless of the data required, most instruments include some or all of the following components, which may have several descriptions:

Probe	Receiver
Electrode	Controller
Sensor	Analyzer
Indicator	Integrator
Transmitter	Amplifier

The information obtained by these instruments includes:

pH	Conductivity
ORP (oxidation-reduction potential)	Liquid level
Temperature	Flow

These units tell the plant operators what is going on, sometimes hundreds of feet away. The dial or graphic chart for the unit is on the main control panel and writes a continuous record of the performance. In plants where toxic constituents, such as cyanides, chromium, etc., must be reduced to levels set by regulatory agencies, samples must be taken periodically and analyses made. Up until recently, the samples were obtained at set periods by an employee going to the sample points. He would fill a jar and return to the laboratory, where the standard analysis took up to 2 h to complete. In the interim there was no record of what might have happened at any particular point. A slug of improperly treated waste might have escaped. Now, automatic samplers are frequently required by law to be located downstream of any unit from which samples are important.

Automatic Samplers

There are many automatic samplers available. Some contain a standpipe assembly which can accurately measure the amount of sample being taken. The unit may come with or without metering pumps. The flow is maintained in a turbulent state to keep solids suspended. The sample through a unit continuously purges out the system at a rate of ½ gal/min at the end of each cycle, making each sample discrete. The sample frequency is every 30 s for a 2 gal/24 h composite.

The sampler is independent of flow and proportional to it. There are samplers with heaters to prevent freezing, others with a refrigerator unit to keep samples cold. The frequency may be adjustable from one sample per minute to one every 30 min. The sample size is adjustable and ranges from 60 to 1800 ml. The samples are put in polyethylene containers. The units have a maximum lift of 22 ft. Prices of the sampler units range from \$450 to \$1500.

The samples obtained are analyzed immediately, and the results are transmitted to the main panelboard. Thus any malfunction or faulty treatment is detected at once, and remedial action is begun. pH can be read at any number of points, and the reading can be transmitted to the main office and recorded. Liquid levels in the tanks may be determined and indicated on the control panel. Chemical usage can be measured by the automatic feeders, with the result indicated on the main panels. In other words, the operators do not have to travel around the plant, but can get an instant and true picture of what is happening anywhere. They can then control the process through remotely controlled valves, pumps and mixers, etc. The control valves are handled by air-operated piston actuators, air-operated diaphragm actuators, electric solenoids, float and motor operators. All control valves are available normally open or normally closed.

The services that can be performed by automatic units are fine, but they are still quite expensive. The engineer must use judgment in specifying them for a plant. He should avoid "overmation." The amount of automatic apparatus and control in any plant should be in proper ratio to the size and cost of the plant in which it is to be used. The size and operating staff and their other important duties should also be taken into account. For instance, if a plant is estimated to cost $50,000 and a man from the normal maintenance staff of the main factory is to care for it, he may have so many other duties that he fails to give the waste treatment plant the necessary attention. In this case, to avoid trouble with the regulatory agencies for faulty treatment, it will pay to provide much more automation than for a plant where there will be six or more operators always on duty. In the latter case the decision to install automation must also take into consideration that when men have nothing to do but sit around and read dials, they lose initiative. Operations that require them to visit certain parts of the plant at regular intervals will keep them alert and enable them to find certain defects that otherwise might go unnoticed.

Manual Samplers

The use of water test kits to determine the amount of contamination in effluent before and after treatment is an economical alternative to instrumentation. Most kits are available in durable plastic carrying cases which are easy to clean. Powder reagents are packaged in individual, premeasured polyethylene "powder pillows." Each pillow contains the exact amount of reagent for one test. Each test kit comes complete with step-by-step printed procedures. Some of the test kits which are suitable for wastewater testing are

> Chromate—diphenylcarbohydrazide method—range 0–1.5 mg/l
>
> Chromate, sodium—direct determination—range 0–1000 mg/l
>
> pH, wide range—range 4.0–10.0

One kit widely used in plating plants, wastewater plants, cooling towers, and other industrial areas is the hexavalent chromium test kit. A dry powder reagent, which is packaged in powder pillows, will produce a pink color if chromate is present. Measurement is made by using a chromate color disk. Range is from 0 to 1.5 mg/l hexavalent chromium.

Another test is for total chromium. This test determines both hexavalent and trivalent chromium. Hexavalent is determined as described above. Trivalent chromium is determined by the hypobromite oxidation of trivalent chromium to hexavalent state, followed by the conventional determination of hexavalent chromium by the diphenylcarbohydrazide method. Range is from 0 to 1.5 mg/l total chromium.

There is also a drop count titration test for hexavalent chromium. This test is useful for accurate chromium levels in water such as found in cooling towers, or where chromate levels are critical, as in wastewater and diesel engine coolants. The drop count titration develops a sharp end point, and the color changes from yellow to colorless.

CASE HISTORIES

It is often useful to follow the course of studies done for various industries of differing sizes. The following five cases cover metal-fabricating companies, hardware manufacturers, gasoline refineries, aircraft companies, and sheet-metal fabricators.

History 1

A small metal-fabricating company used a three-stage iron phosphate cycle operation to prepare its product for painting. The process, typical for a company of its size, consisted of four operational stages:

Stage 1   Clean and phosphate simultaneously

Stage 2   Water rinse

Stage 3   Chromic phosphoric acidified rinse

Stage 4   Dry off

The product was conveyed by chain conveyor through a spray tunnel enclosing the three treatment tanks. It was the third stage that was of critical importance with respect to industrial wastewater discharge regulations. The third stage contained chromic acid.

Chromic acid contains hexavalent chromium—classified as a dangerous contaminant. Investigation showed that local authorities permitted a maximum of 5 ppm hexavalent chromium in any discharge to the sanitary sewer or storm drain system. The hexavalent chromium content in stage 3 was between 120 and 150 ppm. The pH was also too low, being between 3.5 and 4.0. The chromium must be reduced to allowable levels or all chromium-bearing wastewater must be disposed of by a liquid waste transport truck.

Chromium Reduction

There are many package chromium reduction units which reduce hexavalent chromium to trivalent chromium. Chapter 8 describes these in greater detail. But this particular metal fabricator's production yielded only 1000 gal/week of contaminated wastes. It seemed economically infeasible to utilize sophisticated

Before treatment

After treatment

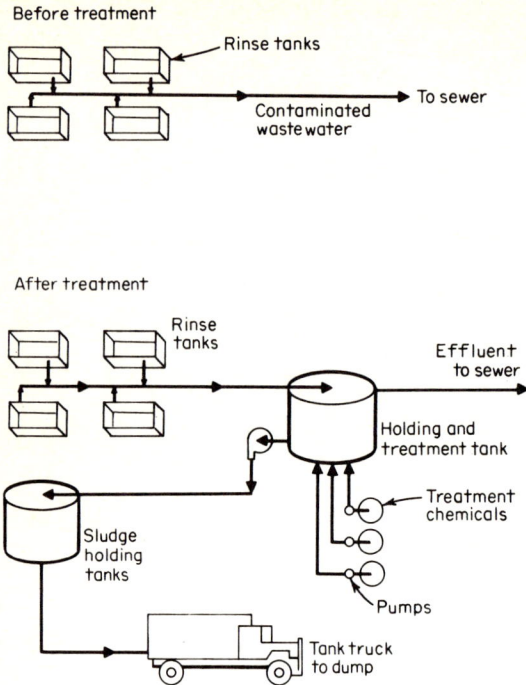

Fig. 5-4 Simple batch treatment system.

units in eliminating this wastewater problem. The study suggested that a more direct approach to the particular problem would be the custom design of a "batch treatment station." Figure 5-4 shows the wastewater pumped to the collection-treatment tank. There it is held until sufficient quantity is accumulated so as to maximize equipment utility. Treatment is accomplished in a manner chemically similar to the package units, but on a much smaller scale. A plant operator adds the required chemicals, eliminating much of the automatic equipment. Completion can be judged by use of a test kit. The batch system was deemed adequate to meet the regulations governing wastewater discharge. It cost $5000 or less.

History 2

A hardware manufacturing operation was in serious trouble with the local municipal treatment plant. Their many plating lines used copper cyanide as well as other toxic materials. Their operation discharged into a public sewer system which went directly into a newly built municipal treatment plant engaged in primary and secondary treatment.

The high content of copper, over 100 ppm from the plating lines, caused serious damage to the municipal treatment operation. Because the treatment plant discharged their effluent into the groundwater, any failure of the treatment plant presented a serious health hazard.

The influent to the plant from the area of the manufacturing operation temporarily bypassed the treatment plant to alleviate the health hazard. But this reduced the total amount of water to such a degree that the plant again did not function properly.

A study of the entire situation resulted in the following recommendations:

1. Improve housekeeping by reducing spillage onto floors

2. Revise rinsing procedure

3. Construct holding tanks for the dumping of process tanks

4. Separate toxic from nontoxic flow

5. Install a wastewater treatment station to precipitate copper and destroy cyanide

6. Install a recycling system to reuse treated water for rinsing operation

History 3

A small gasoline refinery produced wastewater in its production of gasoline and other petroleum products. For years oil-bearing wastewater was discharged into an adjacent sanitary sewer. New county ordinances required that this discharge cease immediately. The refinery met the problem by instituting several procedures suggested in a study of the situation.

1. Separate production water from rainwater

2. Install an oil-water separator to treat oil-bearing wastewater

3. Build a holding tank so as to discharge water into the sewer only during off-peak hours

History 4

A large aircraft company had a problem with the discharge of industrial waste into the adjacent storm drain system. A study of the facilities showed that there were two major sources of contamination. Seventy-three cooling towers and evaporative condensers located throughout the plant discharged their overflow water into open yard catch basins. As each of these units was periodically treated with chemicals, the wastewater contained contaminants which flowed to the storm drain channel.

Another source of contamination was the accidental spillage of chemicals and oils into the yard catch basins. This also contaminated the channel. Both acts were violations of the Water Quality Control Board regulations, which stipulate that only rainwater be permitted to enter the storm drain system.

The study detailed a twofold solution to the problem. All overflows from the cooling towers and condensers were to be piped to the nearest industrial waste piping system. In addition, automatic controls were installed in each overflow line to allow bleed-off only when the dissolved solid content of the cooling water reached a maximum point. This would greatly reduce the total amount of water discharged by each tower.

To reduce contamination by accidental spillage, a collection pit was to be installed at the outlet of the plant's storm drain system which would remove any oil which had inadvertently entered the system. During a rain this pit would be automatically bypassed, allowing the full flow to go directly to the public channel. The control board agreed that during a rain the concentration of oil in the storm water would be greatly reduced by dilution and would not contaminate the system.

History 5

A sheet-metal fabricator treated stainless steel and titanium steel parts. The treatment solutions contained sulfuric acid, nitric acid, and chromic acid. The sulfuric and nitric acids caused little trouble because these solutions required only neutralization. But there was a major problem with the chromic acid waste since the allowable concentration in the discharge to the public sewer was extremely small.

It was therefore proposed that the client follow this outline:

1. Separate chrome-bearing from the non-chrome-bearing wastewaters.

2. Install a chrome reduction package plant to convert hexavalent chromium to trivalent chromium.

3. Install a holding tank to collect the dump from chrome-bearing process water. The contents of this holding tank would be trickled into the feed of the treatment plant.

COMMUNICATING THE NEW DESIGN IDEAS

One can see that the "study" phase of an industrial wastewater treatment program is crucial to the entire "science" of the treatment design. Studies should be made available to other engineers in the field, to keep them abreast of the developments as related to particular industries and contaminants.

The engineer can help himself and others by preparing articles describing his works, to be published in the technical journals or simply for presentation at technical society meetings. The articles may be factual, describing a system installed or soon to be installed, but feasibility studies are also valuable.

The practicing engineer is often too busy doing things to write about what he hopes might come to pass in the future. But the profession and all others concerned would benefit if the engineers who are producing the designs for the plants that are actually preventing stream pollution would take time out to write about their works. In the last few years of increasing public interest in water pollution, the engineers have been accused of having been backward in the development of new methods for the better treatment of liquid wastes of industry. Actually, they have not been dilatory in this respect, but they have been laggard in telling about it. The good things they have done in waste treatment have gone unsung. One has only to look at the impressive accomplishments of the industries of this country, in spending hundreds of millions of dollars on the construction of waste treatment facilities and in providing the many millions annually to keep them running satisfactorily, to realize that the engineers of the country, particularly those operating in the industrial waste discipline, have been doing their part in full measure.

Chapter Six

The design

Having determined the major problems, the engineer must now get down to cases and actually "design." It is in design that the knowledge, experience, and judgment of the engineer are put to the test. His first move should be to ascertain whether his proposed approach will meet with the approval of the regulatory authorities. It is advisable, therefore, to present one's idea with supporting data in an outline form before the actual design starts. This may avoid costly changes if the agency should object to certain methods or aspects of the proposed plan.

TYPE OF OPERATION

Before actual design begins, the engineer should determine the period during which the waste treatment facility will operate. This may be concurrent with the normal operating period of the factory: 8 h, 16 h, or other periods deemed more suitable for the particular project. The type of operation—continuous, batch, or batch-continuous—must be decided. The selection of the best one for a particular case requires a number of considerations.

Continuous Operation

The wastes are treated as they leave the factory processes. The volume of flow and concentration of polluting elements may vary through the operating period. If the volume of concentrations varies considerably, constant attention will be required to effect the changes in treatment. This will entail a constant watch of the flowmeters and constant sampling and analyses. Such a system may require the use of pumps with a capacity equal to the waste flow from the individual processes. A pit or sump may be used to retain the flow for a given period

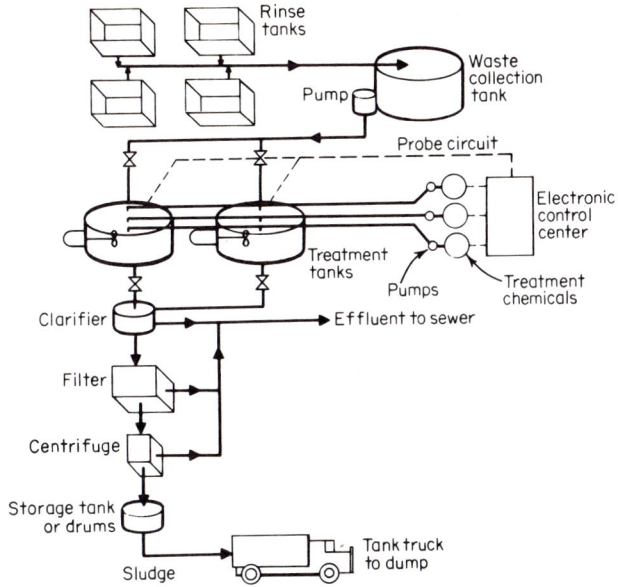

Fig. 6-1   Batch wastewater treatment system.

before pumping. In this case the pumps must be of a capacity greater than the maximum flow per minute of the wastes, or the pumps may never stop running. Holding the wastes in a wet well may permit sedimentation of solids and consequently require periodic cleaning. The treatment tanks for this system must also be of size and capacity to retain the wastes for the period of detention determined best for this particular waste. The influent and effluent arrangements must be carefully designed to prevent short-circuiting within the tanks. This method of operation is the cheapest in initial cost of tanks, but it calls for more manual attention and a correspondingly larger operating staff.

Batch Operation

Wastes are held for an entire operating period. During this period no treatment is given the wastes. They are treated en masse later over a given period of time. This method entails the use of two holding tanks for each waste line. The tanks must hold over half the period flow. The operation is simplified by the uniform flow of waste for treatment. See Figure 6-1.

Batch-Continuous Treatment

This treatment is a combination of the two preceding methods. It is best suited to large industrial waste projects. It involves pairs of holding or blending tanks equal in capacity to the entire flow from each waste line during the factory operating day. In this method each type of waste flows to one of the designated

Table 6-1   Relative Costs of Various Periods of Operation

	Daily operating period		
Item	8 h	16 h	24 h
Base plant cost*	$79,611	$68,010	$54,453
Cost of operation plus fixed and other charges	54,435	60,026	65,666

* All equipment, including holding tanks, treatment tanks, etc.
SOURCE: E. B. Besselievre, Use and Handling of Chemicals in Industrial Waste Treatment, *Sewage Ind. Wastes*, vol. 29, no. 11, November 1957.

holding tanks during an entire day. Treatment of the contents occurs on the following day. During the treatment period of the stored wastes, the flow is diverted to the second holding tank. This process requires the duplication of holding tanks, but provides a waste for treatment that is uniform in volume and concentration of components throughout the period. A great deal of manual attention is eliminated. All phases of treatment are assured close control.

Period of Operation.   If the batch-continuous method has been selected, the next step is determining the period of time that the waste treatment plant will be operated. If this period is set at 8 h, all the facilities in the waste treatment cycle must be sized to treat one-eighth of the total volume of waste in the holding tank per hour. The tank contains the wastes of the previous day. Eight-hour treatment periods require the largest treatment units to handle the flow. On the other hand, manual attention for the waste treatment will be reduced to one shift only. The principal advantage of this method is that there is ample built-in overload capacity to handle unexpected increases in flow. The treatment facilities can always be run for longer periods if necessary.

This method provides the best and least troublesome control. On the day when wastes are to be treated, the operator takes a sample of the homogenized wastes in the filled holding tank, analyzes it, and then sets the chemical feeder to run at a uniform rate of feed during the entire operating period.

The relative costs of the three time periods of operation discussed are shown in Table 6-1.

SELECTION OF
METHOD OF TREATMENT

If the wastes to be treated have a considerable concentration of organic matter and a high BOD content, the conventional, biological method of treatment should be selected. Due regard should be given to local considerations, such as space requirements and availability, possible objections to aerial nuisance or unsightly appearance, etc. With wastes of this character the engineer would have the choice of one of the two following biological methods:

1. Activated sludge process in one of its variant forms

2. Trickling filters

Each has its overall general design requirements.

If BOD is low but toxic contaminants are above legal limits, a nonbiological system will be needed. See Chapter 8 for general design of such systems.

Activated Sludge Process

The conventional process consists of primary sedimentation; aeration with compressed air or mechanical inspiration for a given period of hours, with return of activated sludge in selected proportions to the aeration tank; disposal of the excess sludge by consolidation with the sludge from the primary sedimentation step; discharge of the combined sludge to anaerobic sludge digestion; and final sedimentation of the effluent from the aeration phase.

Step Aeration. This is performed with or without primary sedimentation but provides for the raw or settled waste to be admitted to the aeration tank in descending volume as the waste progresses toward the discharge end of the tank, with final settling and return of a volume of sludge to the aeration tank.

Tapered Aeration. This is done with or without primary settling, but the total amount of compressed air is admitted to the mass of liquid in the aeration tank in descending volume in the length of the tank and with final sedimentation.

Contact Stabilization. This is done without primary sedimentation and with a long aeration period, perhaps 24 h or more, with reaeration of return sludge, but no extraction of sludge from the system.

Complete Mixing. This method is somewhat similar to contact stabilization. No sludge or solids are withdrawn from the system, and the time period of aeration is longer than normal.

This is a new and significant development in the activated sludge category. Rather than being a difficult process, it is actually a self-adjusting method requiring a minimum of skilled operation. It uses the entire aeration tank as a surge tank and equalizes the load over the entire tank and on the microorganisms in any one sector of the tank. In the event of sudden increase in the organic load, the mass in the aeration tank, being fully available, acts to dilute the organic load. This theory upsets the age-old idea that the activated sludge process is a very delicate method not readily adaptable to changing organic loads. The system provides for redistribution of the organic load uniformly throughout the aeration tank, instead of just at the head end. As one engineer has stated, it is "something for nothing." The method has been used to treat wastes from a cotton processing plant. These alkaline wastes did not need the usual preneutralization since the carbon dioxide produced by the metabolizing of the starch neutralized the alkalinity from the caustic and produced a pH suitable for the bacterial action. The results exceeded those normally obtained in the conventional type of activated sludge treatment. The raw-waste BOD averaged 420 mg/l, and the effluent BOD was 3 mg/l, a reduction of 99 percent or better.

Oxidation Ponds, Stabilization Ponds, Lagoons.   These units allow for natural oxidation using biological action or aeration by laying diffusion pipes at the bottom of the ponds. Mechanical aerators also can be adapted for use in ponds and lagoons. Some aerators draw air into the liquid. Other units, such as brush or rotor types, introduce oxygen by rotating a brush in the upper layer and spraying the liquid over the surface.

The selection of a particular type of unit and the aeration equipment is the result of a study of the efficiency of each type in the reduction of BOD per unit of aeration tank volume or the rate of application of the waste per unit of aeration tank volume, as well as the cost of one type of aeration equipment as compared with another. The power requirements per unit of volume treated are usually denoted in kilowatthours per million gallons. BOD efficiency is normally denoted in pounds per cubic yard (or cubic meter) of aeration tank volume per unit time.

Regulations.   Because most state water quality control laws severely restrict the quality of water contained or discharged upon surface of the ground, the pond or lagoon used in an activated sludge system must be provided with an impervious lining to prevent infiltration into the groundwater. (In general, the rule is that only water safe enough for drinking can be stored in unlined earthen basins.)

A synthetic rubber lining is an inexpensive method of lining an industrial waste pond. This material resists the effects of the sun and weather as well as physical abuse, oils, and acid-contaminated fluids. It is impermeable to a wide range of hydrocarbons and other inorganic chemicals and fulfills most state requirements. The liner is generally 30 ml in thickness and comes in sheets 5 ft wide. These can be seamed either in the field or at the factory. A recommended installation is shown in Figure 6-2. A sloping-sided pit is dug with a trench

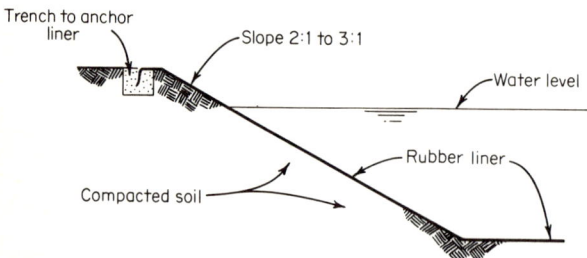

Fig. 6-2   Section through open reservoir showing a rubber lining.

around the perimeter to allow room to anchor the liner. The liner is then placed in the pit and anchored. An alternate to the synthetic liner is the concrete-lined pit. The concrete can be applied conventionally or pneumatically.

Whatever the choice, the lagoon-type body of water, used as a basin for treating wastes, must meet with local and national regulations. Chemical reactions between the liner chosen and the chemicals being used must be taken into consideration.

Trickling Filters

If the trickling filter method has been selected as best suited for a particular problem and type of waste, the engineer must make several determinations.

1. The rate of dosage, or application of the waste volume to the surface of the filter, in gallons per day per square foot or per square yard (cubic meters per square meter). This will determine the diameter of the filter beds.

2. The depth of the filter media. In low-rate filters the depth is usually 6 to 8 ft. In certain types of high-rate filters, a depth of 3 ft of media is common.

3. Type of media: conventional type with broken stone, slag, gravel, or other natural media or synthetic-plastic media.

4. Ratio of recirculation to be used if high-rate filters are adopted. Most of the various types of recirculation used are, or have been, patented, and their use and possible royalty fees should be investigated. Since most of the patented systems are owned by equipment manufacturers, any license fees are usually included in the price of the equipment.

5. Type of filter enclosure: concrete sidewalls, laced walls of concrete, wood or steel beam sections, or sloped media with low walls.

6. Type of cover, if environmental conditions make this desirable: domes of concrete or of plastic materials.

7. Type of underdrain system: vitrified tile, wood, or cement flats.

Filter Medium. The selection of the filter medium is of major importance in the design considerations. It is influenced by site conditions, area of site available, local sources of filter medium, cost of placement of medium, and cost of the underdrain system. Until recently the most common medium for trickling filters has been broken stone, usually of a granitic nature, resistant to dissolution by liquids or moisture. This medium is a natural product in many localities, and since the tonnage required in a large filter bed is considerable, hauling has been a significant item of expense, to be added to the cost of the stone itself and the placement in a filter bed structure. Because stone is heavy, weighing on an average around 90 lb/ft^3, the supporting structure, the bottom and sides of the filter, and the underdrain system must be able to stand the load. For instance, if broken stone is used, the load on 1 ft^2 of the underdrain for a stone depth of 6 ft is 543 lb. Another item of considerable expense in the use of broken stone is the actual placement of the stone in the filter structure. If the stone is dumped into the bed from an elevation of 8 ft above the underdrain system, the impact of this weight on the vitrified tile may cause disaster by crushing the tile. So until a complete layer of stone has covered the entire area above the underdrains, the first loads of stone must be carefully distributed by hand. This introduces an expensive cost item when it is considered that a filter bed 100 ft in diameter and with an internal area of 7854 ft^2, 291 yd^3 of stone will be required just to cover the tile to a depth of 1 ft.

In recent years this burden has been lightened considerably by the introduction of the high-rate filter, in which the filter effluent is recycled back to the filter a number of times. This is exemplified by the Biofilter method developed by the late Harry Jenks, and other systems which operate with different systems of recirculation. This high loading of filters has rates up to millions of gallons per acre per day, instead of the older ratings of 2 mg/d. It has resulted in reduction of the size of the filters and in the depth of filter media. Investigations have indicated that the bulk of the work done in the high-rate filters in the reduction of BOD takes place in the upper 3 to 4 ft of the medium. Filter depths may be reduced accordingly. These two factors alone make important cost reductions possible. Another item of heavy cost with low-rate filters is the provision that the enclosing walls and the bottom of the bed be made of concrete. The usual construction for low-rate filters is a wall 6 to 7 ft high around the entire periphery of the bed. The bottom is 1 ft thick. In a filter 100 ft in diameter the concrete in the wall and bottom will amount to 368 yd^3. At 1974 prices of $100 per cubic yard of concrete in place, this starter cost will be around $36,800. Other items such as the tile underdrain and the filter stone bring the cost of a 100-ft-diameter low-rate trickling filter to a significant amount. The savings to be had by using a high-rate unit and smaller-volume filter medium and structure would justify investigation and the time of the engineer in working out the details. An alternative underdrain system, using 2-ft-long, 2-in-thick strips of redwood would greatly reduce the cost.

Operation.   Today, the average trickling filter is wetted with the liquid by means of a revolving distributor. In the usual form this is composed of a central hollow column through which the liquid enters. From this column two or more distributing arms emerge and extend to the periphery of the filter bed. The waste liquid flows through these arms and out through openings in the back side of the arms. This initiates a jet action which is sufficient to cause the distributor to revolve and distribute the waste more or less evenly over the surface of the stone. These units revolve at speeds which vary with the volume of waste entering the unit and the head of the liquid in the center column. In order to compensate for variations in flow, some of these units have double-compartment arms, the lower compartment handling all liquid up to a certain point, after which the level of liquid builds up in the center column and the upper half of the arm comes into use. Other types of units have multiple arms designed to produce a raindrop type of distribution, and in others, where a definite speed of rotation of the distributor is considered important, the units are rotated by electric motors mounted on top of the center column. The type of unit to be used in a specific case is a matter of selection for the engineer. In cold climates it has now become quite common to top the filters with domed covers to prevent freezing of the wastes. This is practical, however, only with small units, because a domed roof to cover a filter 100 ft in diameter is very expensive. The modern examples of these rotary distributors give very little trouble, and since they have few wearing parts, they operate for long periods without repair or excessive maintenance. An occasional inspection and perhaps an adjustment of a spray plate are all that is required. Rotary distributors need very little head for their operation, 2.5 to 3 ft usually being ample to move units as large as 180 ft in diameter. This is in contrast to the original fixed-nozzle type of distribution, which required heads of 12 to 15 ft to obtain wide distribution.

Low Rate versus High Rate.   The difference in area between the older low-rate and the modern high-rate type of trickling filters is an important consideration in designing a waste treatment plant. Usually land use is important in an industrial plant, especially one located in an area where land is costly. The high-rate units require considerably less area than the low-rate, but because of the recirculation program, a pumping operation is introduced in one or two stages, which adds slight initial and operating costs. For example, in a plant to treat 1 million gal/d (3785 m^3) in a high-rate filter of the Biofilter type in which the recirculation rates are usually 1 to 1 in each of the two stages, there will be two sets of pumps with a capacity of 700 gal/min each.

One of the best surface-area savers, enjoying considerable application in the field of industrial waste treatment, is the plastic media filter. This is described in detail in Chapter 8. The great lightness of this material enables filters to be built with depths of 30 to 40 ft, thus distributing the load vertically instead of horizontally. Thus a filter to treat a flow of 1 million gal/d with a medium depth of 30 ft would have a diameter of only 30 ft. The supporting structures for the light media are much smaller in size than those for low-rate filters since the total load on a square foot with plastic media is only about 108 lb. Since the medium is self-supporting, heavy boundary walls are not required, and a suitable enclosure may be made of aluminum or wood. Items of cost when using the plastic media are for the assembly of the sheets into the cubic elements to be used in the filter and for cutting the sheets if a circular filter unit is necessary. Units using the plastic media are small in diameter, even for large flows. Therefore the distributing mechanisms are supported on the center column and do not require expensive supporting beams or structures.

In conventional, low-rate trickling filters, the enclosures have been made of material other than concrete, such as interlaced wooden beams. In one case old railroad rails, interlaced, were used. In some cases where no containing walls were used, the stone has been piled with sloping sides, but this is a waste of medium material.

TREATMENT EQUIPMENT

There are other types of treatment units that can be used in conjunction with the conventional biological treatment or as simple systems in and of themselves.

Interceptors and Clarifiers

When wastes contain significant amounts of solids, a simple method for their removal may be interceptors or clarifiers. Suspended matter in industrial wastewater tends to settle out when the flow is held quiescent. The tendency to settle is counteracted by currents and eddies, frictional resistance, and the viscosity of the water. In the case of smaller particles, viscosity can be the most important factor. The rate of downward travel is affected by the size, shape, and specific gravity of the particles suspended in the solution.

When wastewater is allowed to flow slowly through a tank, or to stand at rest in it, some of the suspended solids will settle to the bottom. The wastewater that has been partially clarified by the settling out of some of the solids is known as the effluent, and the material that has settled is termed the sludge. The settling velocities of particles of mineral sediment with a specific gravity of 2.65 at a

Diameter of particles (mm)	Settling rate (mm/s)
1.00	100
0.50	53
0.10	8
0.05	2.9
0.01	0.154
0.005	0.0388
0.001	0.00154
0.0001	0.0000154

temperature of 50°F are shown above. The settling rate will change with different values of specific gravity and temperature. For a specific gravity of 1.2, the ratio of velocity of particles of that density to that shown in the table is

$$\frac{(1.2 - 1.0)}{(2.65 - 1.0)} = 0.12$$

Types of Clarifiers. Clarification may be accomplished by either a horizontal-flow or a vertical-flow chamber. In a horizontal-flow tank, the wastewater flows horizontally from the inlet to the outlet. In a vertical-flow container, the water usually flows downward through a vertical pipe and then upward.

The common clarifier as shown in Figure 6-3 consists of a series of compartments designed to allow the waste to cool and become quiescent. This permits the heavy materials to settle and the gases as well as lighter materials to separate and collect on the surface of the water. The tank, which is normally made

All openings have 1/4" steel covers

Plan

Top 3" min. above grade

Provide vent and cleanout

Inlet

Water level

Outlet

4" pipe

4" pipe

3" clear

Sampling box sealed to tank with epoxy

12" clear

Section

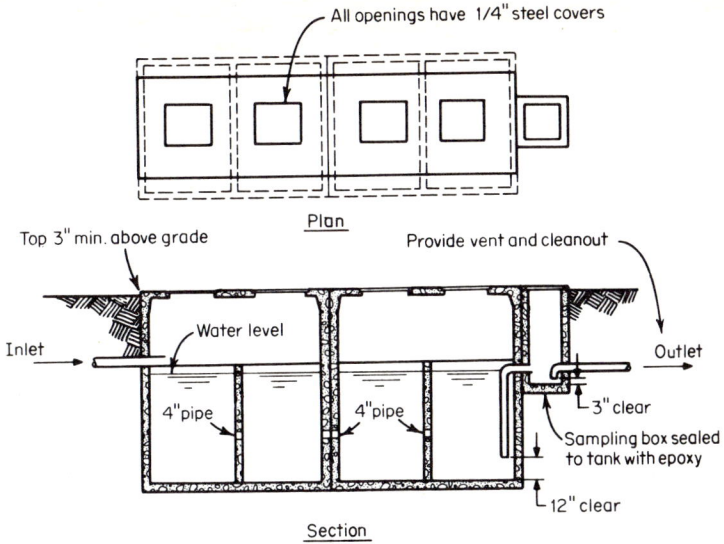

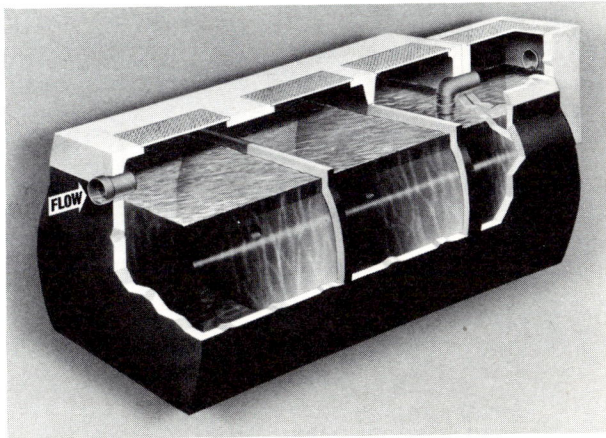

Fig. 6-3 Typical four-compartment concrete clarifier. (*M. C. Notting-ham Company of California.*)

of concrete, should be placed on natural soil or approved compacted fill. The bottom of the excavation should be level and well compacted. The excavation required for a precast concrete tank is usually about 2 ft longer and 2 ft wider than the tank dimension.

When installed outside of a building, concrete interceptors and clarifiers should have their sidewalls elevated above the surrounding ground surface to exclude storm water. When the tank is installed inside of a building, the top may be level with the floor.

Normally only the outside surfaces of the tanks are painted with bitumastic coating. Most governmental agencies prefer that the interior surface of the concrete walls be left uncoated and unprotected. They reason that if the wastewater is corrosive, the owner will be encouraged to neutralize the wastes before allowing them to flow into the clarifier.

Sand and other interceptors used for heavy solids should be designed and located so as to be readily accessible for cleaning. They should have a water seal of not less than 6 in. By installing properly located vents this equipment will not become air-bound if closed covers are used. The sludge collected at the bottom of the clarifier, interceptor, or sedimentation chamber is periodically cleaned by a vacuum hose. The hose is connected to a vacuum tank truck which is commercially available.

Grease Interception.   A common problem confronting most commercial kitchens and restaurants is the separation of oils and greases from their wastewater. This situation is a microcosm of large industrial plants such as wool scouring, laundry, textile, packing, and petroleum plants. Conventional precast concrete grease interceptors, shown in Figure 6-4, are generally used. Grease interceptors are available also as cast-in-place units. Standard units are made in 750-, 1000-, and 1250-gal capacity. If a greater capacity is required, multiple units are used in tandem. A low-capacity interceptor is usually a two-compartment design with the first compartment much larger than the second stage. Recently the Josam Manufacturing Company offered a series of grease interceptors with manual or automatic grease withdrawal systems. See Figure 6-5. These units are claimed to intercept

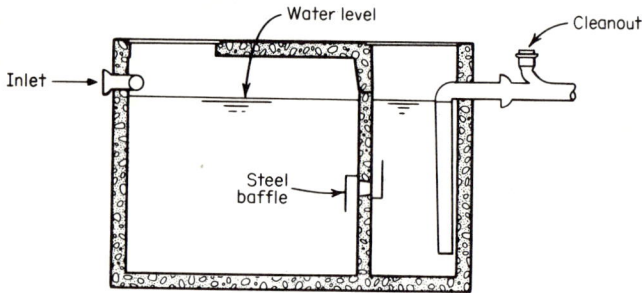

Fig. 6-4   Concrete grease interceptor. (*M. C. Nottingham Company of California.*)

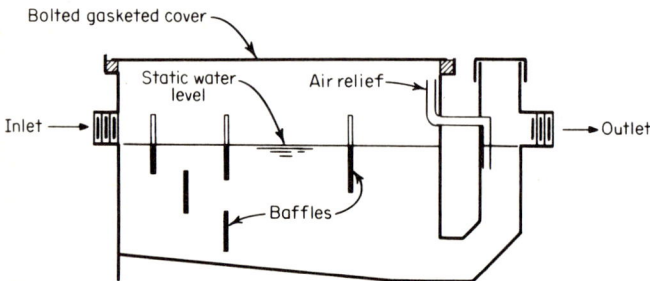

Fig. 6-5   Cast-iron grease interceptor. (*Josam Manufacturing Company.*)

up to 95 percent of the grease in the wastewater. In addition, the units evacuate heavier-than-water solids. The system's cascade design is based on the principle of the waterfall. Owing to the tumbling of grease-laden water over four levels, the grease is separated from the wastewater regardless of the wastewater temperature. Solids and sediment are evacuated through the channel on the bottom of the interceptor, reducing the possibility of decomposition of solids which cause foul odors.

The units are made of cast iron with a capacity of flow ranging from 7 to 75 gal/min. The grease-holding capacity is 14 to 150 lb. These grease interceptors automatically discharge the grease into a drum or other holding container. This eliminates the inconvenience of removing the cover of the interceptor and cleaning out the grease by hand. The units can be positioned in four ways: on the floor, partially recessed in the floor, flush with the floor, or in a pit.

Neutralizing. Neutralizing interceptors are similar to other types of interceptors but are usually coated internally with a material to prevent damage to the concrete from strong concentrations of chemicals. These units provide for chemical neutralization of the wastewater by either automatic or manual equipment. The interceptor usually has one compartment to serve as a sampling box, and may have any number of compartments properly sized for each type of waste. After neutralization, the wastes may be discharged into the sewer or discharged to a clarifier or other treatment unit.

STORAGE AND TREATMENT TANKS

Wastewater treatment systems usually require the use of one or more storage tanks. These tanks may be used to provide continuous flow through the treating phase, to store chemicals, or to hold the concentrated waste until it can be hauled away. Tanks can be ordered with specified data as illustrated in Figure 6-6. The following terms are used in a detailed specification of a tank:

Service. This describes the type of liquid to be stored.

Item Number. As in the case of every other piece of equipment, each tank should be identified by a mark number such as T1, T2, etc.

Quantity. The storage capacity of the tank should be given in gallons. Allowance should be provided for freeboard.

Diameter. The tank diameter should be given in feet and inches. Inside diameter (ID) or outside diameter (OD) should be indicated.

Code. This indicates whether the tank is to comply with an accepted industrial standard specification. Reference to ASME, Pressure Vessel Code, or another

Fig. 6-6 Tank data sheet.

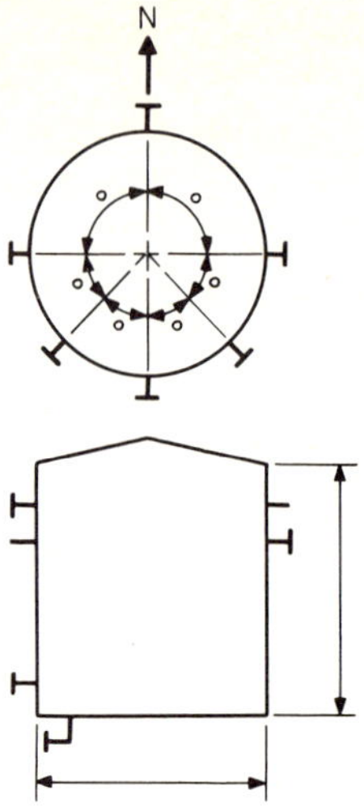

Tank Data Sheet

N

Service _____ Item No. _____

Quantity _____ Dia. _____ Height _____

Design conditions

Code: _____

Design pressure internal _____ psig

Design pressure external _____ psig

Design temperature _____ °F

Corrosion allowance in shell _____ inch
                      in nozzle _____ inch

Wind pressure _____ psf

Earthquake _____ g

Operating conditions

Pressure _____ psig

Temperature _____ °F

Material

Roof

Shell

Bottom

Support

Supplied by seller

Ladders and insulation supports
Paint
Surface preparation

Nozzles

Mk.	No.	Size	Rating	Facing	Service
N1					
N2					
N3					
N4					
N5					
N6					
N7					

code would automatically require that the fabrication of the tank follow the entire specification of that code.

Pressure.   To prevent tank failure from unusual pressures, maximum internal design pressures should be given in lb/in² gauge. The normal operating pressure should also be specified. Where the possibility of a vacuum condition exists within the tank, the differential pressure can be given as a design external pressure.

Temperature.   Temperature of the liquid, both operating and maximum, is important and must be specified.

Structural.   Tank thickness must be specified. The thickness is determined by the size of the tank and the corrosion allowance. One-eighth inch is the generally accepted corrosion allowance. This value is specified and added to the structural requirements. Large tanks should be designed for wind pressure and, in earthquake-prone areas, for seismic forces.

When the design engineer has determined the tank material, his data should include the plate thickness of the roof, shell, and bottom. The type of support should be described. Other tank appurtenances such as ladders, cages, insulation supports, platforms, and stairs should be described when necessary. The surface preparation and type of paint must also be specified.

Connections. A tank would serve very little purpose without the necessary pipe and other connections made to it. Types of fittings include flanged outlet, threaded nipples, half and full couplings, spigots, and needle-type spigots. It is very important to provide a nozzle schedule including a diagram showing the location of each fitting. The schedule should identify each type of nozzle by a mark number (N1, N2, etc.). The quantity of each type of nozzle should be listed. The pressure rating (150 lb, 300 lb, etc.) is also important and should be indicated. In the case of flanged nozzles the type of facing is next noted. When threaded pipe connections are to be used, the size and length of the threaded nipple or type of coupling should be shown.

The diagram should show an elevation of the tank with the location of each nozzle given in height above tank bottom. The plan view shows the orientation of each tank in degrees from the plant north direction. In the above manner all necessary information on any tank can be completely described on an 8½- by 11-in sheet of paper.

Concrete Tanks

If concrete tanks are decided upon and mechanical units, such as sludge thickeners, mixers, etc., are to be installed in them, the diameters of the tanks will be controlled to some extent by the regulations of the state or other regulatory authority. The officials learned, in the early days of the use of mechanically cleaned sedimentation tanks, that they must couple a detention period with an overflow rate in order to obtain units that would produce acceptable results. Omission of the overflow rate has had a disastrous effect in a number of cases in the author's experience.

At the present time practically all the states which prescribe certain factors of design, such as detention, now tie this in with a second factor, overflow rate, usually stated in gallons per square foot of tank area per day. In primary sedimentation tanks this may result in specifying a detention period of 2 to 4 h, while controlling the diameter of the tank by also specifying the overflow rate. If, in a given case, the overflow rate is the governing factor, this is a simple calculation:

$$\frac{\text{Average daily flow of wastes}}{\text{Overflow rate}} = \text{surface area of tank}$$

Of course, a contractor can build a concrete tank of any diameter the engineer chooses, but that is not the sole consideration. If a mechanical sludge collection unit is to be used, the manufacturer of the type selected should be consulted. Every manufacturer has standard sizes, usually progressing in increments of 2 ft in smaller sizes and 5 to 10 ft in larger sizes. If the engineer's calculations result in a tank with an odd-size diameter such as 23 or 57 ft, the manufacturer of the selected equipment can, of course, supply a mechanism to fit that tank. But the mechanism may cost more than the manufacturer's next largest size. This is because the manufacturers prepare the drawings for their shops with dimensions for each group of sizes. The shops may have templates or jigs for turning out all the units in a group. If an odd size is demanded, the extra cost is for the engineering department's work in preparing the new set of drawings for the change. The shop then must lay out the material and make cuts in regular shapes to fit the wanted size. All this entails labor, with time lost in the fabrication of the odd unit. In the selection of a standard-size unit, the order can be a simple instruction to prepare one unit of No. XYZ. Since the steel shapes and parts may already be in stock, shipment can be made promptly. Because oversize is not detrimental to the performance of the unit, a slightly larger tank to coincide with the equipment manufacturer's standards will save time and money. The equipment manufacturers are glad to provide information on their standards, and it is wise to use them.

The foregoing discussion is related to the circular type of tank. On the other hand, the devotee of the rectangular type of tank, using the chain-drag mechanisms, is also subject to control by the manufacturers of the components of those units. Their limits are usually in the width of one chain mechanism, or flight. The present limit of width is about 25 ft, so that if a rectangular tank is to be wider than that limit, the engineer must design the tank for two or more flights. If the desired width of tank is 40 ft, the usual practice is to provide for two flights, each 20 ft in width. In this case also the manufacturers will supply full details.

Depth of Tank

In sedimentation units where the area and diameter are frequently determined by overflow rates or some other factor, the depth is controlled by the volume of wastes to be handled in the detention period specified or selected, as follows:

$$\frac{\text{Total volume of flow, ft}^3}{\text{Surface area of tank, ft}^2} = \text{water depth of tank, ft}$$

However, if the circular or square tanks are to be equipped with mechanisms, the depth of the water in these tanks, regardless of detention, will be governed by the structural details of the manufacturer of the equipment to be used. Practically all the manufacturers of this type of equipment use truss arms extending

Fig. 6-7 A 68-ft-diameter holding tank with mixer and support, Pontiac Motor Division, General Motors Corporation, waste treatment plant. (*K & H Engineering Company.*)

outward from a central revolving cage. Thus, in all units, the length of the arm controls the height of the truss structure at its juncture with the center cage. In large units this depth of truss may be 10 ft or more. Since it is desirable to have the entire truss submerged below the surface water level in the tank to avoid interference with sedimentation, scum collection, etc., the arm depth determines the minimum water depth in the tank. Therefore, if the engineer's calculations call for a tank 100 ft in diameter and a water depth of 6 ft, this specification could not be met, since the minimum standard of the manufacturer for that size of tank would be 10 ft, in order to keep the complete arm structure below the water surface. These may seem like small details to be concerned about, but neglect of them, or insistence by the engineer on certain dimensions, could cause additional cost to the client and delay the shipment of the equipment.

If the tanks are simply for holding or storage of wastes, as in the batch-continuous program of operation, there is no need to worry about the equipment manufacturer and his standards. The tanks can be of any diameter and depth the engineer chooses. If mixers or agitators are used, as they should be, in any tank in which wastes are kept for more than a few hours, the structural-support members may be of any reasonable length. Figure 6-7 shows a 68-ft-diameter holding tank in a large industrial waste plant with the supporting structure for the homogenizing unit.

In industrial waste projects it is common practice to use steel tanks, entirely above ground and resting on concrete pads or ring foundations. In this phase important economies can be obtained by studying the relationship of tank diameter and depth in their effect on unit cost. In most industrial waste projects the installations are in the open, and so there is no reasonable limit on tank depth. The cost of steel tanks is definitely related to the amount of steel plate in the sides and bottom of the tank and the amount of concrete in the foundation, as illustrated in Table 6-2. Since a deep, small-diameter tank for a certain volume of liquid will have less square footage of steel than a shallow, large-diameter

Table 6-2 Effect of Tank Size and Depth on Cost Difference in Amount of Steel Plate and Concrete for Foundations in Tanks of Various Sizes and Depths

Tank volume, gal	Diam., ft	Depth, ft	Tank steel,* ft²	Concrete foundation,† yd³
100,000	58	6	3,735	40.5
	41	11	2,657	28.6
	29	20	2,482	20.2
200,000	72	6	5,428	51.0
	58	11	4,646	40.5
	41	20	3,896	28.5
300,000	94	6	8,711	65.8
	68	11	5,959	47.75
	51	20	5,427	35.75
400,000	107	6	11,009	75.25
	79	11	5,631	55.0
	59	20	6,441	40.75
500,000	119	6	13,174	83.50
	88	11	9,123	61.75
	65	20	7,402	45.0
600,000	131	6	15,947	92.0
	97	11	10,741	67.5
	72	20	8,595	50.5
700,000	142	6	18,514	99.50
	105	11	12,277	73.5
	78	20	9,678	54.5

* Tank steel is based on complete steel bottom and sides, with open top.

† Foundations based on ring foundations 3 ft wide by 2 ft thick.

SOURCE: E. B. Besselievre, How Does Tank Depth Affect Plant Costs? *Public Works*, April 1963.

unit, a saving is accomplished. If the tank requires a protective lining, there will also be important economies in this material. In a study made for a certain project the author developed the relative costs of steel tanks of a number of different capacities, diameters, and depths, and with and without protective linings. Table 6-3 shows the result of this study. In every instance the deeper tanks offered important savings over the shallower units. Where the use of deep tanks would require a pumping operation, there would be an extra cost for the power, but in consideration of the major savings to be made by the use of the deeper units, the slight additional power cost may be neglected. In the case of the study which produced the information in Table 6-3, the wastes were delivered to the tanks by gravity, so that no extra pumping was involved.

Table 6-3   Comparison of Cost of Steel Tanks of Various Diameters and Depths

Volume, gal	Tank size Diam., ft	Depth, ft	Cost tank*	Lining†	Total cost lined tank
100,000	58	6	$ 5,416	$ 7,058	$12,474
	41	11	4,916	5,020	9,936
	29	20	4,219	4,690	8,909
200,000	72	6	8,956	10,258	19,214
	58	11	6,736	8,730	15,516
	41	20	7,207	7,363	14,570
300,000	94	6	12,630	16,115	28,745
	68	11	11,026	11,264	22,290
	51	20	9,706	9,916	19,622
400,000	107	6	17,614	20,807	38,421
	79	11	12,209	14,422	26,631
	59	20	11,916	12,174	24,090
500,000	119	6	19,761	24,898	44,659
	88	11	13,228	17,242	30,470
	65	20	11,843	13,999	25,842
600,000	131	6	22,325	30,139	52,464
	97	11	17,185	20,300	37,485
	72	20	12,462	15,264	27,726
700,000	142	6	25,771	34,991	60,762
	105	11	19,643	23,203	42,846
	78	20	15,484	18,291	33,775

* Tank prices based on quotations made February 1956. Tanks are $\frac{1}{4}$-in steel plate, sides and bottom, open top, erected in place on concrete foundations by others. Prices include ladders.

† Linings based on Ceilcote material at $1.89 per square foot applied.

Fiber Glass or Filament-wound
Chemical Storage Tanks

Fiber glass tanks are made from a filament-wound material that provides a lasting protection against attack by almost any type of corrosive liquids. This type of material is effective throughout a wide range of service environments. The construction of the tank consists of a liner material which is thoroughly saturated with thermosetting resins to form the corrosion barrier. The liner is then overwound with resin and continuous high-strength fiber glass strands under tension. The structural wall is then built up to meet required specifications for thickness.

Filament-wound tanks are manufactured with a wide variety of openings and fittings including flanged and threaded nozzles, siphon drain nozzles, full bottom drains, and bottom elbow drains. Also available are gooseneck and V vents. Tanks are manufactured with access ways and manholes for top and sidewall locations.

Fitting into the Site

Before the engineer has finalized his tank sizes and depths, he must reconcile these with the area available for the treatment and the practical grouping of the several units. The same volume can be attained by an infinite combination of tank diameters and heights. Figure 6-8 illustrates the various capacities of different-size tanks. The manufacturer prefers to make tank sizes which utilize standard plate sizes. It is therefore imperative that the engineer check with the fabricator for his standard tank sizes before making final decisions.

In the case of the proposed industrial waste treatment plant for Pontiac Motor Division of General Motors Corporation at Pontiac, Michigan, the only area available on the factory property was an odd-shaped piece between two railroad spurs. This area was not straight in any dimension, but followed the curve of the spurs. The total available space was 2.71 acres. Immovable obstructions consisted of a high water tower, two high-tension power line towers, and innumerable underground water lines, sewer lines, drains, and conduits. None of these, according to the management, could be disturbed. With these limitations the designing of a very large industrial waste facility in the space allotted generated a number of important considerations. This was especially true when deciding upon the depth of the twenty-nine holding, homogenizing, and treatment tanks. The entire area was covered by the necessary units of the plant, leaving only clearance for the railroad spurs.

The only site adjacent to a motor parts manufacturer in Michigan was shaped like a segment of pie. This required a realignment of the usual order in treatment units. Some of the large secondary units were placed first in order to fit them into the area available.

A small steel mill in Ohio, with a pickling-liquor problem had a building on the site, alongside a railroad spur. This was an ideal location for housing

HEIGHT (FEET)	3'	3½'	4'	4½'	5'	5½'	6'	7'	8'	10'	12'
4	211	287	376	475	587	710	846	1,152	1,504	2,350	3,384
5	264	359	470	594	734	888	1,057	1,440	1,880	2,937	4,230
6	317	431	564	713	881	1,066	1,269	1,727	2,256	3,525	5,076
7	370	503	658	832	1,028	1,244	1,480	2,015	2,632	4,112	5,922
8	423	575	752	951	1,175	1,421	1,692	2,303	3,008	4,700	6,768
9	475	647	846	1,070	1,322	1,599	1,903	2,591	3,384	5,288	7,614
10	528	719	940	1,189	1,468	1,777	2,115	2,879	3,760	5,875	8,460
11		791	1,036	1,308	1,615	1,955	2,326	3,167	4,136	6,463	9,306
12		863	1,128	1,427	1,762	2,132	2,538	3,488	4,512	7,050	10,152
13		935	1,222	1,546	1,909	2,310	2,749	3,743	4,888	7,638	10,999
14				1,665	2,056	2,410	2,961	4,031	5,264	8,225	11,845
15				1,784	2,203	2,666	3,172	4,319	5,640	8,813	12,691
16				1,903	2,350	2,843	3,384	4,604	6,016	9,400	13,537
17				2,022	2,497	3,021	3,595	4,894	6,392	9,988	14,383
18				2,141	2,644	3,199	3,807	5,182	6,768	10,576	15,229
19				2,260	2,790	3,376	4,018	5,470	7,144	11,163	16,075
20				2,379	2,937	3,554	4,230	5,758	7,520	11,751	16,921
21									7,896	12,338	17,767
22									8,272	12,926	18,613
23										13,513	19,459
24										14,101	20,305
25										14,688	21,151
26										15,276	21,997
27										15,864	22,843
28										16,451	23,689
29										17,039	24,535
30										17,626	25,382
31											26,228
32											27,074
33											27,920
34											28,766
35											29,612
36											30,450
EST. GAL./FOOT:	53	72	94	119	147	178	212	288	376	588	846

FOR TANKS WITH 2:1 ELLIPSOIDAL HEAD (Height = ½R):

	3'	3½'	4'	4½'	5'	5½'	6'	7'	8'	10'	12'
EST. GAL.	26	42	63	89	122	163	211	335	501	979	1,692

Fig. 6-8 Storage tank capacities, gallons.

the entire installation. Considerable cost was saved by eliminating the need for construction.

An existing steel warehouse provided the housing for a chemical treatment installation in a custom dyeing establishment in Massachusetts.

GRAVITY SEWER SYSTEMS

In order for a plant to treat its wastewater, it must be able to convey the wastewater from the process units to the treatment facility. This usually is accomplished by pipe. The piping system may be designed for gravity flow, pressure flow, vacuum flow, or a combination of these.

Most sewer lines are designed for gravity flow. Such design eliminates the need for pumps. The pipes are sized to flow partially full. Sudden surges may cause a pipe to be supercharged, so that it may have a slight internal pressure. This pressure would be limited to a head of water determined by the elevation of the manhole cover.

Sewer lines should be designed to be "self-cleaning." This can be done by using a slope which will produce a minimum velocity of 2 ft/s for a small-diameter sewer. Where occasional peak flows which will flush out the sewer are expected, a minimum velocity of 1.5 ft/s is recommended.

To determine the hydraulic characteristics of the flow, most governing authorities generally accept the principles of the Manning formula:

$$Q = \frac{1.486}{n} Ar^{2/3}s^{1/2}$$

where Q = total discharge, ft³/s

n = coefficient of roughness of the pipe interior

A = area of cross section of water in pipe, ft²

r = mean hydraulic radius, a ratio which is equal to A/P where P is the wetted perimeter of the pipe interior

s = slope of the pipeline, read as a ratio of drop in feet per 100 ft of line

Figure 6-9 is a chart showing the hydraulic elements of flow in a pipeline. From this chart it can be seen that the maximum flow occurs when the depth of water is about 95 percent of the full depth. The maximum velocity is when water depth is approximately 85 percent.

To further simplify the designer's work, pipe size or flow in any particular pipe can be determined from a chart. The chart is based on Manning's formula. Figure 6-10 is a partial chart of the design capacity for circular sewers. It includes sewer sizes from 6 to 16 in in diameter. The slopes are between 0.0001 and 0.05 ft/ft.

One reason for designing a pipe to one-half to three-quarters full is to allow adequate air space for gases and sudden surges beyond design flow. The factors that affect velocity of flow are

1. The slope of the conduit between two points

2. The size of the conduit

3. The roughness of the interior surface

4. The depth of liquid at inlet and outlet

5. Nature of flow—steady or variable

6. Bends or obstructions

7. Viscosity of liquid

Hydraulic Elements

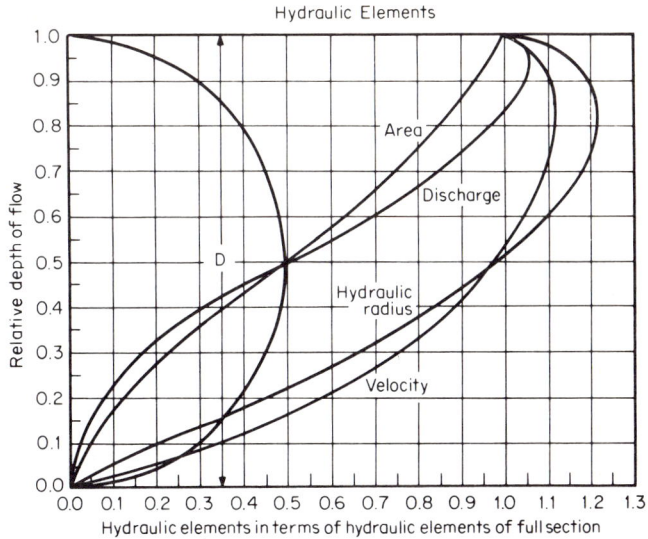

Fig. 6-9 Hydraulic elements of flow in a gravity pipe; discharge or flow in cubic feet per second; velocity of flow in feet per second; area of water flow in square feet; depth of water *D* in feet; relative depth of flow or depth of water flow/diameter of pipe.

The critical velocity is an important concept in gravity flow design. Generally speaking, it is the optimum flow velocity of the system. Beyond that velocity the flow turns turbulent. Because velocity also depends upon depth of liquid, the definition of critical velocity could be restated as the flow rate at which there is a maximum discharge for a given energy content of the water. The energy content of the water is proportional to the depth of the water and the square of its velocity.

General Requirements of Gravity Sewer Lines

When designing a gravity sewer line there are many factors to be taken into account. Pipes under 16 in in diameter should be sized to run half full. Pipes 16 in and over should be sized to run 75 percent of capacity. The slope should be adequate for maximum flow. The pipeline should be laid out free from obstructions which might accumulate debris and stop the flow. The pipelines must be properly vented to allow gases trapped in the line to escape. All inlets must be properly sealed. Underground pipes within buildings or other structures should be made of cast iron, lead, or approved plastic.

Changes of direction must be carefully designed so as not to obstruct the flow. Approved fittings of one-sixteenth, one-eighth, or one-quarter bends should

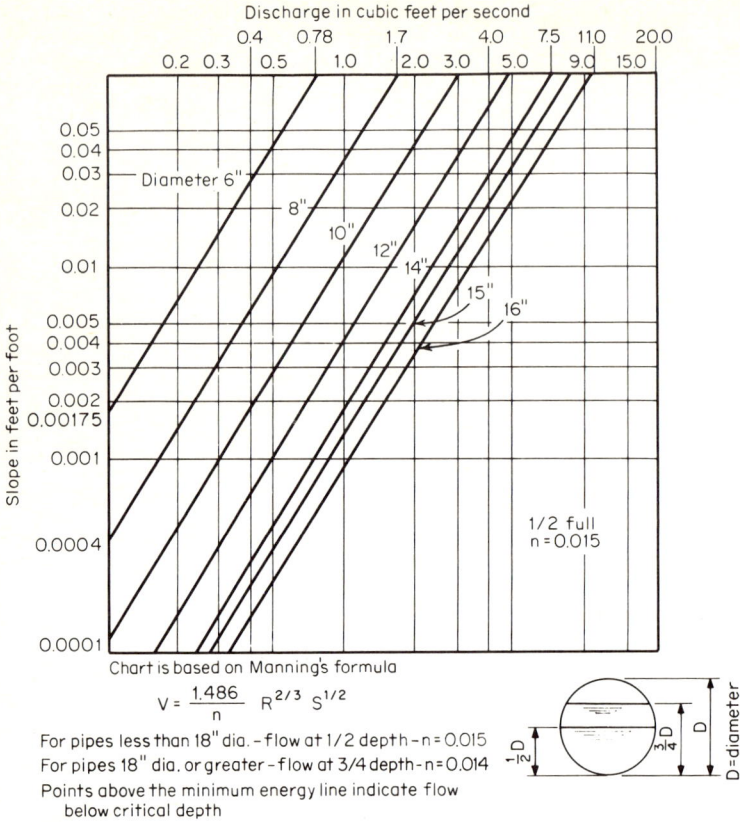

Fig. 6-10 Chart of design capacity for circular sewers. Sewer pipes under 18-in diameter are normally designed to be half full. This is to allow for future expansion of the system, to allow for gas accumulation, and to provide for sudden increases of flow for short periods of time, and it is good engineering practice.

V = velocity, ft/s

n = coefficient of roughness

R = hydraulic radius or $\dfrac{\text{cross-sectional area in ft}^2}{\text{wetted perimeter in ft}}$

S = slope of pipe, ratio of fall in feet per 100 ft of horizontal distance

always be used. Each horizontal drainage pipe should be provided with a cleanout at its upper terminal. Cleanouts are also required on straight runs 50 ft in length. For single 45° change of direction or single 45° offsets, no cleanouts are necessary. Lines requiring more than one change of direction, with angles exceeding $22\frac{1}{2}°$, do require additional cleanouts. See Chapter 7 for additional data on cleanout design for gravity sewer lines.

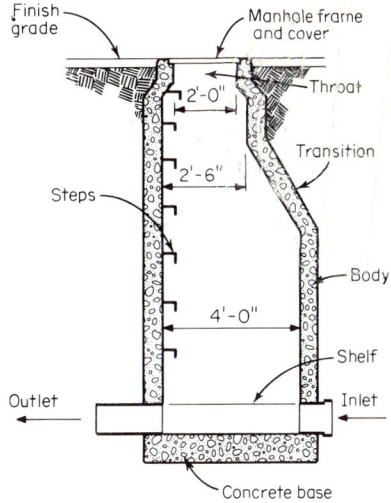

Fig. 6-11 Typical eccentric concrete manhole.

Manholes

In sewers up to 60 in in diameter, recommended practice is to locate a manhole at every change of direction, grade, elevation, or size of pipe. In sewers too small for a man to enter, it is the practice to space manholes not more than 300 to 500 ft apart. However, with closed-circuit television inspection service now available and with modern rodding and bucket cleaning machines, the necessity for such frequent intervals is open to question.

Manholes usually are made of brick or concrete. The latter may be cast in place in forms placed in the trench, but more commonly they are of precast circular segments. A typical eccentric concrete manhole is shown in Figure 6-11.

Precast Concrete Manhole. Precast reinforced manholes combine the convenience and economy of installation with the strength and durability of concrete and steel. These manholes are manufactured in a variety of sizes and shapes. The shapes include both concentric and eccentric cones. Some are made with a flat top for special locations. The shafts are manufactured in 36-, 48-, 60-, and 72-in diameters.

The installation is made in four simple steps:

1. A bed of grout is placed on the poured-in-place concrete footing.

2. The shaft units are placed upon this grout bed. Each riser is seated in grout upon the top of the previously installed riser.

3. The transition section is installed.

4. The manhole cover frame is set into a grout bed.

The entire installation time for a 7-ft manhole can be as short as 15 min. Backfill immediately follows the section placement to minimize "open trench" time.

Precast reinforced concrete manholes must meet the requirements of the ASTM C-478 specifications. The minimum compressive strength of the concrete must be 4000 lb/in² at 28 d. Manholes are manufactured with galvanized steel steps or without steps.

Special Manholes.   Manholes are made for many special purposes. These include the trap manhole, Figure 6-12a and b; the weir manhole, Figure 6-13; and the drop manhole, Figure 6-14.

The gas trap manhole provides a water seal to the upstream side of the manhole. This prevents sewer gases or vapors from passing up into the manhole or downstream pipe. It is accomplished by use of a removable metal elbow and channel so constructed that the water level seals the outlet. This type of manhole is particularly useful where there is danger of combustible liquids in

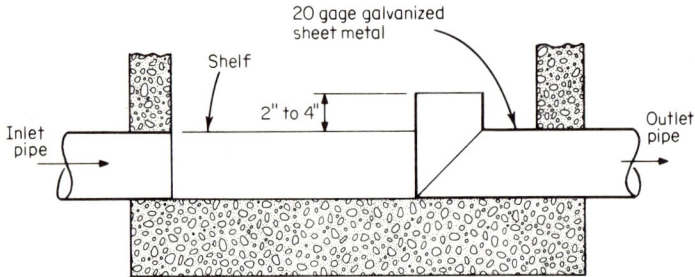

(a) Sand trap

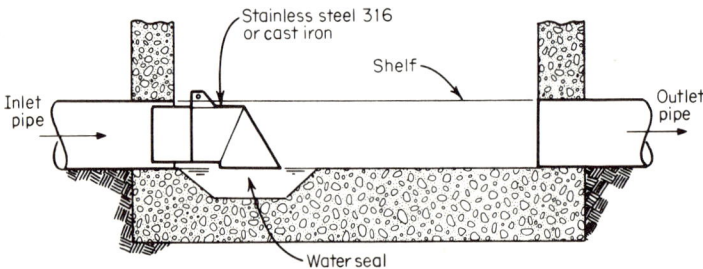

(b) Gas trap

Fig. 6-12   Special manholes. (a) Sand trap. (b) Gas trap. The sand trap provides for suspended solids to settle out before being carried downstream. The gas trap is provided with a movable elbow at the inlet pipe which forms a water seal for the upstream line. This reduces the escape of sewer gases into the manhole and also tends to isolate fires.

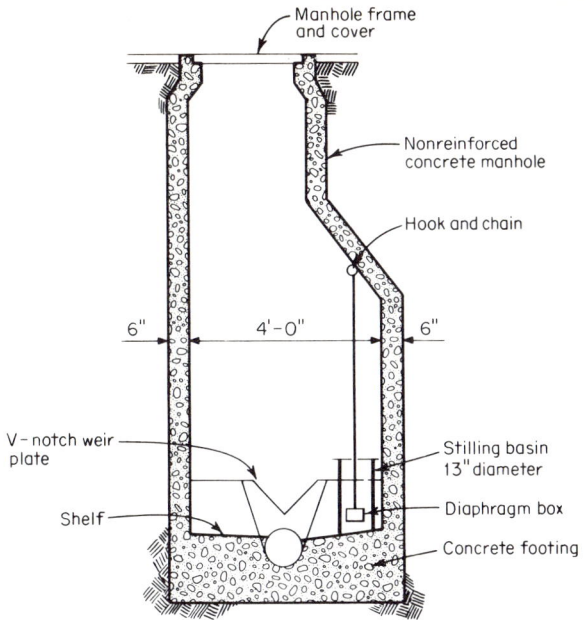

Fig. 6-13   Typical V-notch weir manhole.

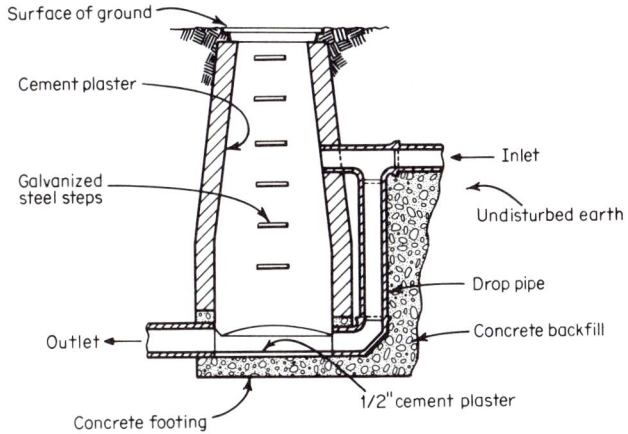

Fig. 6-14   Typical drop manhole.

the wastewater which may become ignited. The gas trap manhole tends to limit the spread of fire.

As its name implies, the shallow manhole is used when the depth of the sewer is too shallow for a conventional manhole shape. This manhole is similar to Figure 6-11 but of course shallower.

The weir manhole has a special bottom configuration to provide a removable V-wedge weir. The flow of water can be determined by measuring the difference in levels of flow upstream of the weir and at the inlet of the manhole. This measurement can be made manually by a rod or automatically by various types of instruments.

The drop manhole is another special manhole, with the inlet much higher than the outlet. A drop pipe is installed outside the wall which discharges the waste liquids at level of the bottom of the manhole to maintain a laminar flow. This prevents creating a turbulence which would release unwanted gases from the sewage.

PRESSURE SEWER DESIGN

Although gravity flow is ideal as far as economics is concerned, it is often found impractical. This may be because of excessive depth of a trench excavation, changes in topography, or the necessity of lifting sewage from a basement to the street level. In these cases a pressure system is required. The lines of such a system are designed as force mains, discharging at some free outfall such as a manhole. The pressure is usually not over 15 to 20 ft of head.

Whenever possible pressure pipes or force mains should be designed with a single continuous rise. In cases where high and low points are unavoidable, precautions should be taken. At high points an air relief valve protects the flow characteristics by relieving trapped air and gases. And where complete drainage of the line is required, a drain valve at the low point is recommended.

The pressure of the system may be calculated from the work done by the pump. If we assume that the temperature of the fluid remains essentially constant, and that the fluid does not pass through another pump, all energy terms at one point in the pipeline must be equal to the sum of the energy terms at another point in the pipeline. See Figure 6-15. Bernoulli's formula states:

$$\frac{P_1}{w_1} + \frac{V_1^2}{2g} + z_1 = \frac{P_2}{w_2} + \frac{V_2^2}{2g} + z_2 = \frac{P_3}{w_3} + \frac{V_3^2}{2g} + z_3$$

where P = pressure, psf
$\quad V$ = velocity, ft/s
$\quad w$ = weight of water, lb/ft^3
$\quad g$ = acceleration due to gravity, ft/s^2
$\quad z$ = height of pipe above the datum, ft

P/w can be called the "pressure head." z is defined as the potential head, and $V^2/2g$ is the velocity head. All three terms can have linear dimensions. The Bernoulli terms may be visualized and the Bernoulli equation pictured geometrically. The line drawn through the free surfaces in the columns connected to the pitot tubes is called the "energy line" (EL). The energy line is seen to be horizontal for frictionless flow, and its distance from the datum plane is the direct measure of the total energy in the flow. The "hydraulic grade

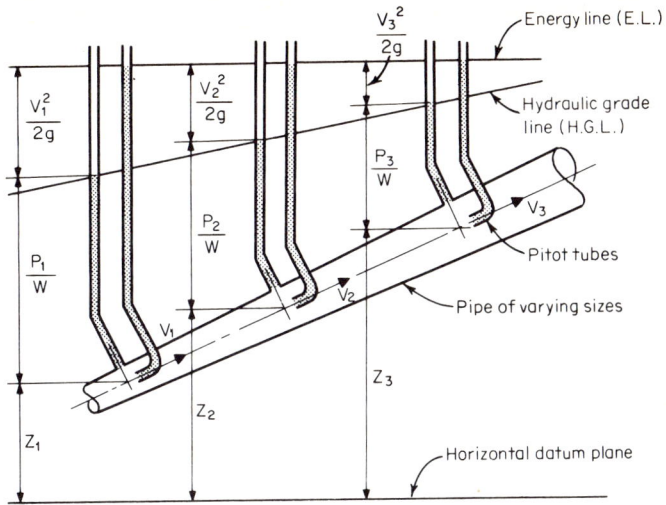

Fig. 6-15 Graphic representation of Bernoulli's equation.

line" (HGL) drawn through the tops of the piezometer columns gives a picture of the pressure variation in the flow; evidently, (1) its distance from the centerline of the stream tube is a direct measure of the pressure in the flow, and (2) its distance below the energy line is a direct measure of the square of the velocity.

Whenever a force main terminates into a gravity line, a special manhole is required. Figure 6-16 illustrates such a manhole. Two elbows are used to discharge flow in a downward direction so as to dissipate its energy. The bottom of the manhole has a channel into which the flow is guided toward the inlet of the gravity main. A manhole cover is required to provide access for cleaning the gravity line and any accumulation of solid material in the bottom of the manhole.

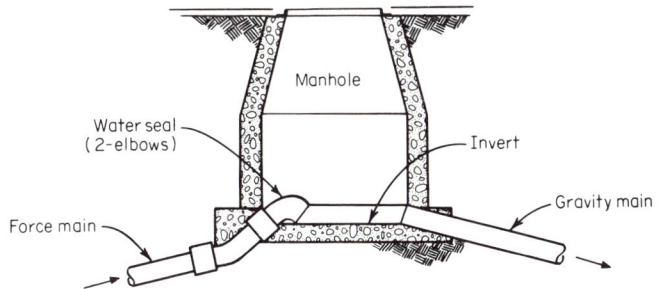

Fig. 6-16 Recommended terminal manhole of force main. This discharge of the force main must be controlled into the channel at the bottom of the manhole. There should be no free fall or splashing at this point. The entire outlet of the force main should be submerged below the waterline of the channel. There should be adequate access for cleaning of the gravity line.

The flow itself in a force sewer line is based on the conventional Chézy formula:

$$H = f \frac{l}{d} \frac{V^2}{2g}$$

where H = the loss of head due to friction in the pipe
$\quad l$ = length, ft
$\quad f$ = coefficient of friction of pipe interior
$\quad d$ = diameter of pipe, ft
$\quad V$ = mean velocity of water, ft/s
$\quad g$ = acceleration due to gravity, 32 ft/s/s

By rearranging this formula and adding the area A of the pipe, the flow Q in cubic feet per second can be determined. Therefore with a known head, whether by difference in grade or pump capacity, the velocity V can be determined. Then it is an easy step to multiply by A for Q.

But use of these formulas is not necessary because charts and tables are readily available which easily provide velocity and flow in any size of pipe with a known head.

Pumps

Of course, the most crucial element in a pressure system is the pump. The most common pump used for handling industrial waste is the centrifugal pump which may be oriented with its axis in the horizontal position or vertical position. These pumps are known as horizontal centrifugal pumps or vertical centrifugal pumps.

The horizontal centrifugal pump is used when water is taken from a tank or dry well installation. A horizontal centrifugal pump can be used also as a booster pump in a pipeline. A vertical centrifugal pump is normally used in a sump, pit, or basin. The motor may be located above the waterline at the floor level or at the top of the basin cover. A submersible vertical pump contains the pump and motor together in a waterproof casing. This type of pump is either suspended by a cable or rested upon the floor of the sump.

All pumps are classified by their capacity in gallons per minute, corresponding pressure head in feet, and the size of the object that the impellers can pass along. Other characteristics of the pump include the material of construction, the type of motor, the rpm, and the type of liquid the pump was designed for.

Pumps in raw sewage service require special design to avoid clogging. Generally they have impellers designed to pass 3-in solids, but unless they are preceded by a screen or a comminutor, they need periodic attention.

Selection of a Sump Pump Installation. The pump discharge head required in a given operation can be calculated by determining the distance from the lowest level of water in the basin to the liquid discharge level. To this value should be added a reasonable factor of safety, accounting for the possibility of back pressure and the friction loss in the discharge pipelines and fittings. Example:

Distance from lowest level of water to discharge level	26 ft
Friction of discharge pipe (150 gal/min, 4-in pipe, 150 ft)	2 ft
Friction of pipe fittings (elbows, check valves, gate valves, Y branch, etc.)	2 ft
Total discharge head	30 ft

Determining Tank or Basin Size

The size of the tank or basin should be carefully selected to prevent both short cycling of the pumping operation and excessive settling out of sediment. A draw-down ratio of 3:1 is recommended. The basin volume between the low liquid level and the high level should accommodate three times the capacity of the pump before the pumping cycle begins. A minimum distance of 1 ft between the lowest liquid level and the basin floor should be allowed. In addition, the distance between the maximum liquid level and the basin cover should be one-third of the storage volume between the high and low liquid levels, but not less than 2 ft. Where conditions limit the basin depth, the required capacity may be obtained by using a larger basin diameter. For fast calculations, use the following example:

Total pump capacity	150 gal/m
Basin to hold three times pump capacity	450 gal/m
From calculations a 48-in-diameter basin will hold	95 (gal/ft depth)
Therefore, basin depth is (450/95)	4.74 ft
Add approximately one-third to depth for storage of liquid and basin cover or min. of 2 ft	2 ft
Add approximately 1 ft for distance between lowest level and basin floor	1 ft
Total pit depth required	7.74 ft
Recommended pit depth (next standard 6-in increment)	8 ft

Calculating Square Pit Capacities.   To determine the size of a square or rectangular pit commonly used in concrete construction, the above example may be followed.

Submergence.   Air may be entrained in the pumped liquid if the pump suction is located too close to the free liquid surface in the suction source. Pumping liquid with entrained air can cause a reduction of capacity, vibration, loss of efficiency, and wasted power. Excessive wear of close running parts, bearing stresses, and shaft damage are also subsequent effects. If the capacity in gallons per minute and the suction inlet size or area are known, the minimum height of the liquid above the suction inlet (submergence) can be determined.

Pump in Wet Basin.   Figure 6-17 shows the important components of a typical wet basin pump station (also called sump pump or lift station). The sump basin, Figure 6-18, can be made of cast iron, carbon steel, fiber glass, or concrete.

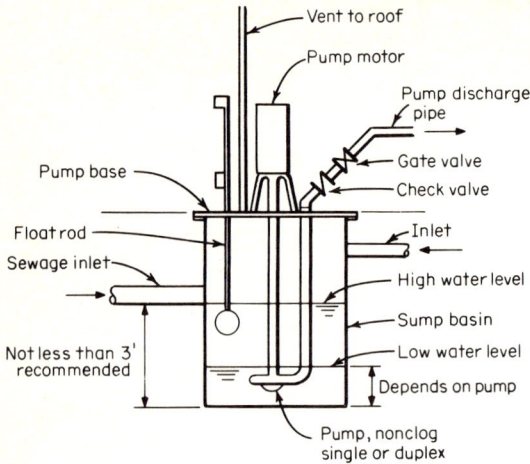

Fig. 6-17   Diagram of a wet basin.

For corrosive liquids, cast iron is generally used. Carbon steel is used for alkaline or noncorrosive liquids. The basin comes equipped with one or more inlet couplings and a basin cover. A gasket is used to keep a tight seal. The cover is prepared with openings for the pump, vent, discharge outlet, and float control. Some of the covers are manufactured for a single-pump installation and others for a duplex installation.

Many prefabricated wet basin pump stations are manufactured. These are easily installed into a properly prepared excavation. They require only electrical hookup to complete the installation.

Pump in Dry Basin.   The dry basin is a superior type of pumping station compared to the wet pit or wet basin. See Figure 6-19. The main advantage of the wet pit is its low cost. The dry basin has many other advantages. The pump inlet is always submerged and therefore priming is not required. A horizontal centrifugal pump can be used. Maintenance can be made on the pump and motor in place. All piping connected to the pump is easily accessible. Access manways are provided over both the pump area and the wet chamber. Another popular type of dry basin is the sewage lift station, illustrated in Figure 6-20.

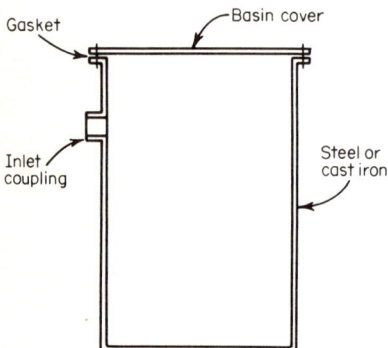

Fig. 6-18   Detail of a sump pump basin.

Manhole

Grade

2'-0"

Discharge

Pump chamber

Gate valve

Pump

Inlet

Check valve

Gate valve

Concrete base

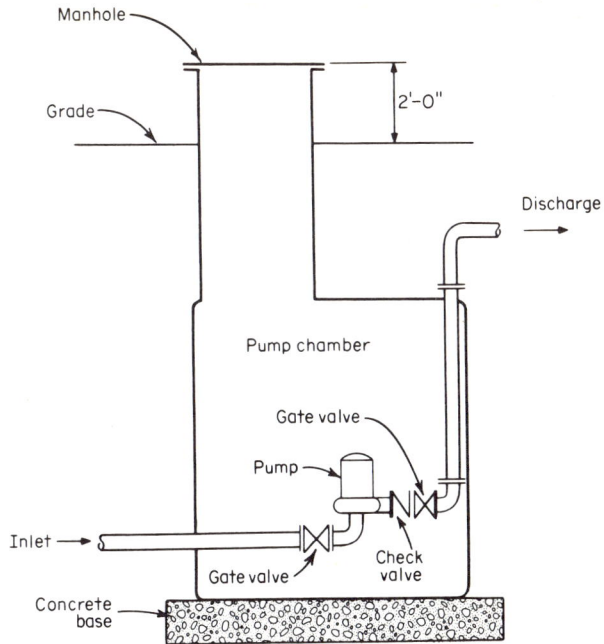

Fig. 6-19 Diagram of a horizontal-type nonclog pump in a dry basin.

Duplex Pumps. Many engineers design a system for two pumps. The advantage of installing two complete pumps, with automatic alternation, in one basin, is that it permits the removal of one for servicing with continual complete protection. In case of an emergency the second pump cuts in, increasing the capacity during peak flows. By alternating the use of the pumps every cycle, both pumps are kept in operating condition and the wear on the moving parts is equalized.

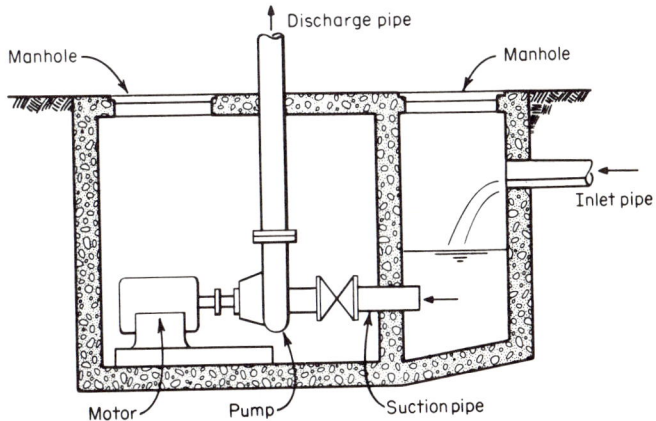

Discharge pipe

Manhole

Manhole

Inlet pipe

Motor

Pump

Suction pipe

Fig. 6-20 Diagram of a sewage lift station in a dry basin.

Governmental regulations usually require that all sanitary sewage lift stations have duplex pumps.

Pneumatic Sewage Ejectors

Figure 6-21 shows a typical pneumatic sewage ejector. This type of sewage lifting device operates in the following manner. The sewage flows by gravity from the

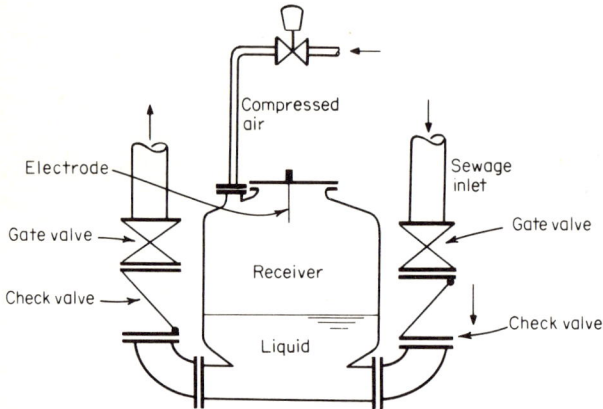

Fig. 6-21 Diagram of a pneumatic sewage ejector. (*Weil Pump Company.*)

collection system into the receiver. When the liquid makes contact with an electrode in the receiver, a three-way valve is energized, allowing air to flow from the storage tank, ejecting the sewage. The period of discharge is controlled by a field-adjustable time delay holding circuit. The three-way valve is deenergized at the end of the discharge cycle and returns to its normal position, allowing the receiver to vent to the atmosphere during the filling cycle. In duplex stations electrical interlocks prevent both receivers from ejecting simultaneously. The air compressors are automatically controlled by a lead pressure switch, a two-circuit alternator, and a standby pressure switch. A test circuit can be provided to check operating cycle of receivers.

VACUUM SEWER SYSTEM

The third type of sewer system is the vacuum sewage transport and collection system. As its name implies, it is a mechanized system of wastewater transport. It utilizes differential air pressure to create flow, as opposed to gravity-induced flow of conventional wastewater collection systems. In this respect a vacuum system is essentially a pressure system.

The vacuum system requires a main collection or pumping station. It requires vacuum pumps in this station to maintain a vacuum on the collection lines feeding the station. The collection lines are small-diameter plastic pipe laid independent of hydraulic grade lines. The system requires a normally closed valve at each sewage input point to seal the vacuum ·collection lines so that vacuum can be maintained on them. This valve opens automatically when a given quantity of sewage has accumulated on the upstream side. The sewage is admitted and the valve then closes. This valve is entirely pneumatic in its control and operation. The differential pressure between local atmospheric pressure and vacuum pressure on the immediate downstream side of the valve, controls and operates the valve automatically. A vacuum sewer sewage collection system can be made to approach a water distribution system in dependability of operation.

PIPE MATERIAL

A sewer system is only as good as the pipes carrying the wastes. In any system the economic life of the pipe depends upon the selection of the correct pipe material. The choice is governed by several factors, including the location of the pipe (underground in yards or under building floors) and the nature of the liquid being transported (corrosive, hot, or abrasive).

Vitrified Clay Pipe

Vitrified clay pipe is a very common type of piping material. Its use has been found to go back thousands of years before Christ. It was used in drains by the ancient Babylonians, Cretans, Greeks, and Romans. These pipes were then used mainly for storm drainage.

Clay pipe was used for sewage as early as the 1500s, when the making of the pipe became mechanized and handmade pipe became obsolete. Today the pipe is used to carry underground sanitary sewage provided the lines

Straight pipe

Perforated pipe

Wye – tee combination

Double wye

1/4 or 90° bend

1/16 or 22-1/2° bend

1/8 or 45° bend

Fig. 6-22 Vitrified clay pipe fittings.

do not run too close to building foundations or other structures. Typical vitrified clay pipe fittings are shown in Figure 6-22.

The clay pipe is manufactured from clay, shale, or combinations of these materials. It is finely pulverized and tempered with water to render a plastic mixture. The mixture is then forced through dies under high pressure into its final shape. The pipe sections are fired to vitrification in beehive or tunnel kilns.

After a gradual cooling period the pipe is removed from the kiln. The pipe is now ready for use, or ready for the application of resilient joint material

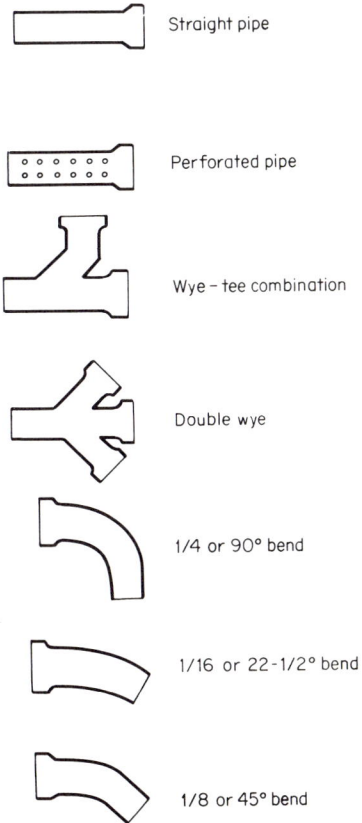

bonded in the factory. The preformed ends form a positive compression joint. One type of joint material is polyurethane.

Concrete Pipe

Concrete drainpipe is manufactured as nonreinforced and reinforced pipe. It is available in various strength classifications. Several types of tests including 3-edge bearing test, sand bearing test, absorption test, freezing and thawing test, permeability test, and compressive test of concrete are used to determine the classification.

Concrete pipe is usually used for storm drainage. Joints and jointing materials include bell and spigot, bell end type, tongue and groove, rubber gasket, cement mortar, oakum cement mortar, and several types of jointing compounds.

Steel Pipe

For any type of installation there are generally five types of steel pipe materials from which to choose. For ordinary use, the most common pipe is black or hot-dipped zinc-coated (galvanized) welded or seamless pipe conforming to ASTM A120, ASA B36.20. This pipe is not intended for close coiling or high temperatures. The black or galvanized pipe is used to carry noncorrosive wastes of ambient temperature.

If the pipe is to be heated, the designer must specify welded or seamless, black or galvanized pipe conforming to ASTM A53, ASA B36.1. A third type of pipe is the welded wrought-iron pipe conforming to ASTM A72, ASA B36.2. This pipe is available as "standard weight," "extra strong," and "double extra strong." For high-temperature use, seamless carbon steel pipe, conforming to ASTM A106, ASA B36.3, is recommended. Pipe conforming to ASTM A135, ASA 57T is used where the pipe must be electric-resistant welded steel.

Cast-Iron Pipe

It is well known that the oldest cast-iron water main in the world is still in service in Versailles, France. It was installed in 1664. Cast-iron pipe has long been the standard conduit to convey sewage in municipal, utility, and industrial piping systems. Cast iron is essentially an alloy of iron, carbon, and silicon, containing appropriate amounts of manganese, sulfur, and phosphorus. Although cast iron has excellent corrosion resistance to water and domestic sewage, care should be taken when it is used to handle some industrial wastes.

Cast iron is normally made as gray cast iron and ductile cast iron. There is also "sand spun" pipe, which is cast centrifugally in sand-lined molds. This method eliminates all casting strains, gas pockets, blowholes, and inclusions. For many years it was considered standard practice to cast in vertical sand-lined molds.

Joints are available in push-on type, mechanical, bell and spigot, ball flanged and threaded IPS (iron pipe size). The most popular connections used for wastewater are the push-on type, which is similar to bell and spigot but more economical as far as labor is concerned.

All pipes and fittings are manufactured in various classes based upon the maximum working pressures. The 150 lb/in² class is the one usually used for drain lines.

Perry's "Chemical Engineers' Handbook," 5th ed., is a valuable source of recommendations of piping material or selected fluids. In regard to cast iron, the book states:

> Gray iron castings are not usually considered corrosion-resistant although they do resist atmospheric corrosion as well as attack by natural or neutral waters and neutral soils. However, dilute acids and acid-salt solutions will attack this material. Cast iron, plain and low alloy, is considered poor to fair for acid solutions such as phosphoric acid.

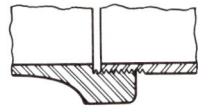

Shoulder type drainage fitting thread

Cast-iron Drainage Fittings. The main characteristic of this type of fitting is that it is designed for unobstructed flow. When a joint is made, the end of the pipe practically meets the shoulder at the back of the thread chamber, forming a smooth passageway to prevent catching or clogging of foreign matter. This can be seen in Figure 6-23. All fittings having openings at right angles have threads pitched to maintain the pipe slope of ¼ in/ft.

Duriron Pipe

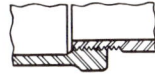

American Standard and API pipe thread

Fig. 6-23 Detail of threaded joints.

Duriron is a high silicon iron alloy that is resistant to practically all corrosive wastes. The silicon content is approximately 14½ percent and the acid resistance is in the entire thickness of the metal. Acidproof joints are secured by the use of specially impregnated pure asbestos rope packing rather than hemp or oakum. Duriron is very hard, resistant to abrasives, and corrosion- and erosion-resistant, and it can be machined by grinding. Two types of joints are shown in Figure 6-24.

Plastic Pipe

Pipe made by the extrusion of thermoplastic compounds is used for gravity, force sewer mains, building drains, waste, vent, and sewer piping. The kinds of plastic pipe particularly suited for gravity sewer and drain piping applications are ABS (acrylonitrile-butadiene-styrene), PE (polyethylene), PVC (polyvinyl chloride), and SR (styrene rubber). Those commonly used for force main use are ABS, PE, and PVC. The thermoplastic materials are suitable for use with most domestic wastes and for many chemical waste system applications. The pipe is light in weight, it will not corrode, and its smooth interior surface exhibits minimal resistance to flow.

ABS pipe and fittings are often described as tough, hard, and rigid. The materials exhibit good impact resistance, tensile strength, flexural rigidity, and creep resistance. They are not attacked by weak acids or strong alkalies.

PE pipe is a tough, resilient, flexible material with excellent resistance to impact stresses and chemical environments. It has low heat resistance but remains ductile at temperatures well below 0°F. The mechanical strength of PE pipe is somewhat less than other thermoplastics but quite adequate for sewer and drain system service.

Another type of corrosion-resistant pipe is known as "Saran lined" drainage pipe. Dow Saran process and drainage piping products are manufactured from Q2 polyvinylidene chloride plastic. This is a standard iron pipe with an inner surface lined with Saran, manufactured by the Dow Chemical Company. Complete "Saran pipe" is also available for use where structural strength of the steel is not required. The Orangeburg pipe is also popular for drainage use and is of nonmetallic material. This pipe is widely used for underground installations. Koroseal pipe is a high-temperature, rigid, lightweight, chemical-resistant pipe which is capable of handling hot corrosives up to 210°F.

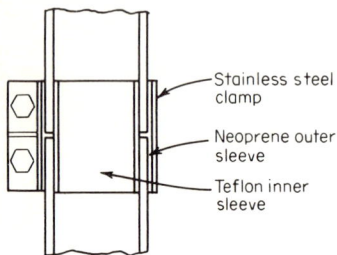

(a) Mechanical joint

- Stainless steel clamp
- Neoprene outer sleeve
- Teflon inner sleeve

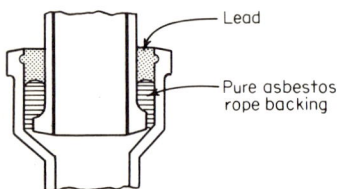

(b) Bell and spigot

- Lead
- Pure asbestos rope backing

Fig. 6-24 Detail of joining of cast-iron pipe. (*a*) Mechanical joint. (*b*) Bell and spigot.

Plastic Pipe Fittings

Industrial drainage fittings are available in PVC and polypropylene. Joining of this pipe is by thermal bonding for polypropylene, solvent-cemented joints for PVC and PVDC, and threaded joints for all materials provided the pipe is Schedule 80.

Johns-Manville Ring-Tite joint for PVC pipe is rated for 125 lb/in² for nonpotable water systems. It is recommended for oil field and other industrial applications and is made from unplasticized polyvinyl chloride. This pipe can be easily cut with a carpenter's handsaw. Double bell couplings are also available to join two spigot ends. See Figure 6-25.

Cement Asbestos Pipe

Johns-Manville Transite Pipe or asbestos cement pipe is resistant to corrosion, immune to electrolysis, and free from tuberculation. It is rated as 100, 150, and 200 lb/in² and is available in sizes from 3- to 36-in diameter. Joints are

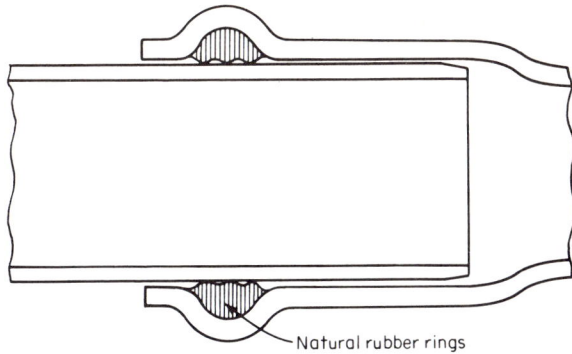

Fig. 6-25 Detail of PVC Pipe Ring-Tite joint. (*Johns-Manville Company.*)

Natural rubber rings

Ring-Tite with an asbestos cement coupling. There are cast-iron fittings with Ring-Tite joints so that a multitude of other types of pipe can be connected to the asbestos cement pipe.

Johns-Manville Epoxy-lined Transite Pipe is used where aggressive industrial liquids are handled. The epoxy lining provides extra internal protection. The lining is available in 20- and 40-mil thickness. Both types are made as nonpressure and pressure types. When pipe is field-cut, a coating of polyurethane is applied. One type of joint of asbestos cement pipe is shown in Figure 6-26.

Pipe Couplings

The jointing of dissimilar pipes is a continual problem to the designer. One type of coupling can connect clay sewer pipe to cast-iron pipe. It can also be used as a bushing to connect different types and sizes. Sealants are not required. This coupling is a rubber gasket and stainless steel straps which are extremely corrosion-resistant. This type of coupling is usually limited to 3- to 8-in pipe sizes. For pipes 4 to 42 in, a "sealed in" mechanical compression joint can be used. It comes equipped with an outside cover for receiving the sealant.

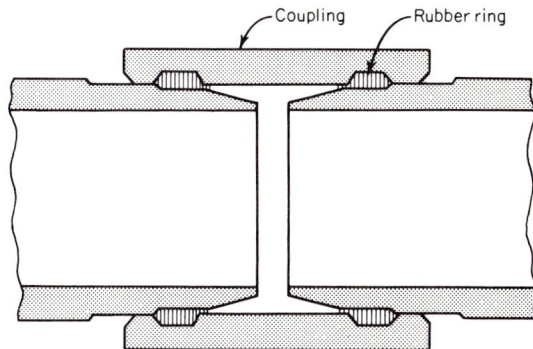

Coupling Rubber ring

Fig. 6-26 Detail of cement asbestos pipe joint. (*Johns-Manville Company.*)

Symbol	Abbr.	Description
	W	Waste above grade
	W	Waste below grade
	V	Vent above grade
	V	Vent below grade
	D	Indirect drain
	SD	Storm drain
	PD	Pump discharge line
	GV	Gate valve
	GLV	Globe valve
	ANV	Angle valve
	BFV	Butterfly valve
	GBV	Gate valve with by-pass valve
	BV	Ball valve
	CV	Swing check valve
	NCV	Nonslam check valve
	BWV	Back water valve
	E	Any line with E means existing
	MH	Manhole
	POC	Point of connection
	AD	Area drain
	PCFI	Plumbing contractor to furnish and install
	SVA	Solenoid-operated valve (electric)
	MVA	Motor operated valve (electric)
	PRV	Pressure reducing valve with needle valve
	RV	Pressure relief valve
		Pitch down in direction of arrow

Symbol	Abbr.	Description
		Flow in direction of arrow
		Riser down
		Rise or drop
		Branch connection out of top
		Branch connection out of bottom
		Branch connection out of side
	COP	Cap on end of pipe
	PT	Plugged tee
	CR	Concentric reducer
	ER	Eccentric reducer
	UN	Union
	CO	Cleanout plug
	FCO	Floor cleanout
	WCO	Wall cleanout
	COTG	Cleanout to grade
	VIR	Valve in riser
	AN	Pipe anchor
	GU	Pipe guide
	FC	Flexible connector
	STR	Strainer-wye type
	T	Any line with T means temporary
	FS	Floor sink
	FAG	Fixed air gap
	FD	Floor drain
	RI&C	Rough in and connect
	E.P.	Elevation of pipe/bop and/or ℄

Fig. 6-27   Symbols and abbreviations for piping and valves.

Piping Drawings

As in all disciplines of engineering, the language used is a system of recognized symbols and terminology. This nomenclature is generally accepted by the designer, draftsman, purchasing agent, vendor, and piping contractor. Without such a common language any piping job would be chaotic. Piping symbols most commonly used are shown in Figure 6-27.

Pipe Description

On complex projects a complete description should be made of all pipelines. This should contain the line number, the pipe size, and the type of flow, whether continuous or intermittent. Also the schedule should indicate the maximum flow in gallons per minute, type of fluid, and temperature. And finally, the description of the pipe material and the type of service should be noted. The drawings should indicate by abbreviation the type of service.

Selection of Valves and Valve Designs

Proper selection of valves will make the operation of the wastewater treatment facility more efficient with less maintenance cost. For manual operation only a few types of valves are available. Considering metal valves only, there are gate valves, globe valves, and check valves. The gate valve is for full shutoff or full turn-on and is not recommended for throttling. This valve also offers the least resistance to the flow of liquid. The globe valve is mainly for throttling purposes. It has a high resistance to flow. For manual operation of flow it has the advantage in that volume flow is proportional to the number of turns of the hand wheel. The check valve is used as a companion to the gate and globe valves. A swing check valve should be used when the gate valve is used, and a lift check valve should be used with the globe valve. Figure 6-28 illustrates a suggested hookup for a pump installation, valves included.

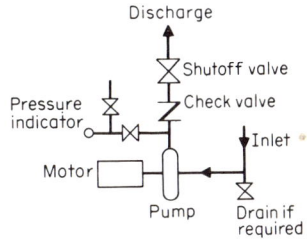

Fig. 6-28 Typical pump piping hookup.

Floor Drains

In most process areas there is a need for floor drains of different types. Each drain serves a specific need. Figures 6-29, 6-30, and 6-31 illustrate various kinds of floor drains.

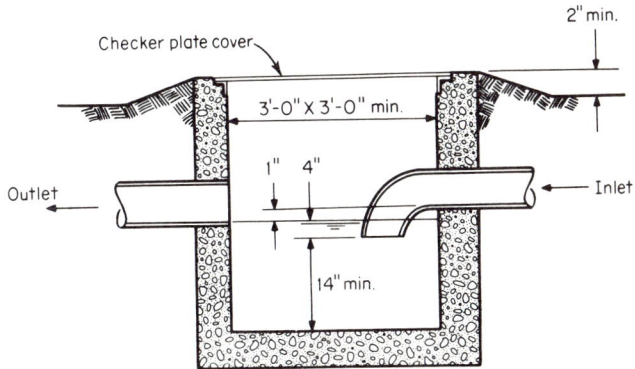

Fig. 6-29 Typical oil drain.

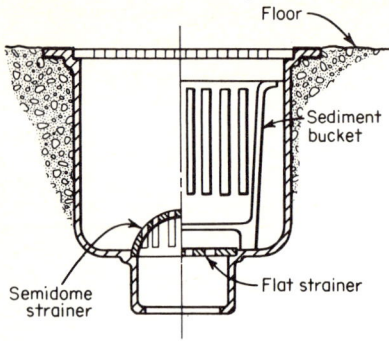

Fig. 6-30　Floor sink.

CONTROL EQUIPMENT

If the system is to be automated, more sophisticated devices are called for. The type and extent of control equipment will be governed by the treatment method, equipment units, complexity of operation, and governmental requirements.

Flowmeters

These may be any of the normal units for this class of work, magnetic type or those using nuclear isotopes for measurements. For general flow measurements which may be required by a municipality to determine flows as a basis for sewer service charges, the Parshall flume and venturi types are commonly used.

Level Indicators

To prevent accidental tank overflow and to indicate to plant personnel the need to haul away wastewater, most wastewater storage tanks require some type

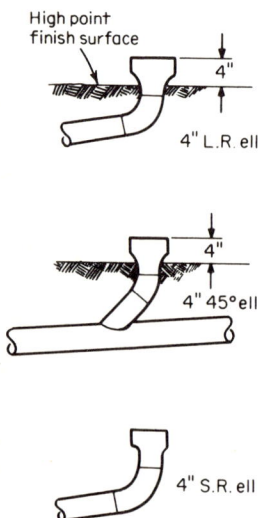

Fig. 6-31　Typical process drains.

of outside level indicator. For added protection an alarm can be installed. The indicator may be locally or remotely mounted. Components include the transmitter or controller, indicator, alarm, and recorder. These can be pneumatically or electronically operated.

Electronic buoyancy transmitters can measure liquid level, liquid density, or interface level by sensing the buoyant force exerted by a displacer element and then transmit a proportional microamp dc signal to the receiver.

The bubbler-type level instrument provides continuous indication of the depth of liquid in a tank of any size and shape when freely vented to the atmosphere. This type of instrument can be installed up to 500 ft from the measurement point. It is ideal to use when centralized instrument readings are desired. No part of the instrument is in contact with the liquid. The standard indicator or recorder can be used for the measurement of acids or other corrosive liquids.

The bubbler-type level is connected by a $\frac{1}{4}$-in bore impulse pipe to a dip tube immersed vertically in the liquid. A small supply of air is fed into this pipe and continuously bubbles out of the end of the dip tube. The pressure developed in the dip tube and in the connection to the instrument is equivalent to the head of liquid on the dip tube. As the head varies, so does the pressure to the instrument. This information is indicated on the instrument in terms of liquid depth, or tank capacity in gallons.

The approximate cost of such a level indicator in 1973 was

Level recorder	$500
Level indicator	350
Miniature air compressor	100
Alarm contact	50

Another type of device is the level switch. It is manufactured in stainless steel or buna N and can be installed in pressurized vessels. The switch can be actuated at high or low level. It operates on an electrical "make or break" system and can activate a warning light or an alarm, open a valve, or start a pump.

Continuous liquid depth indication can be provided by a system with a magnet-equipped float. Located on a vertical shaft, the float moves with the liquid level. The float operates a series of hermetically sealed, magnetic top switches on a voltage divider within the shaft. The remotely located receiving meter is connected across the shaft. The reading is directly proportional to the voltage drop. The system is simple to maintain since the float is the only moving part.

Aboveground tanks can be equipped with a float and cable type of liquid level device. The gauge is mounted on the side of the tank. A cable is connected to a reel within the gauge and extends over the top of the tank and vertically into the tank to a float. The float guide rises and falls with the liquid level. This movement causes the cable to reel in and out, operating the indicating portion of the gauge.

pH Control

In the treatment of chemical wastes by other chemicals, it is often critical to maintain a definite pH in the reaction tanks. Wastes above or below the optimum pH will require additives. Since time is of the essence in determining the proper pH, it is recommended that an electrode unit be installed at the holding or homogenizing tank for each waste that may need adjustment, with instant transmission of the reading to the register-recorders in the operating building. In one large plant this system required thirteen electrode stations scattered around the area. On the main control panel in the operating building (Figure 6-32) space was provided for three register-recorder units, but the chief operator had them installed on the wall of his office instead. Two of these registers had six buttons each, and the third had one button. In order to ascertain the pH of the waste at any point, the operator had only to push the proper button on the register and the reading at that point immediately showed on the gauge. This system eliminated the need for an operator to travel the several hundred feet between the operating building and some of the field electrode stations. The system required an attendant to tour the area and clean the thirteen electrodes about once a week. The total cost of the equipment for this system, not including installation, was $7830 in 1956.

Effluent discharged by the industry must also meet specific pH limits. pH

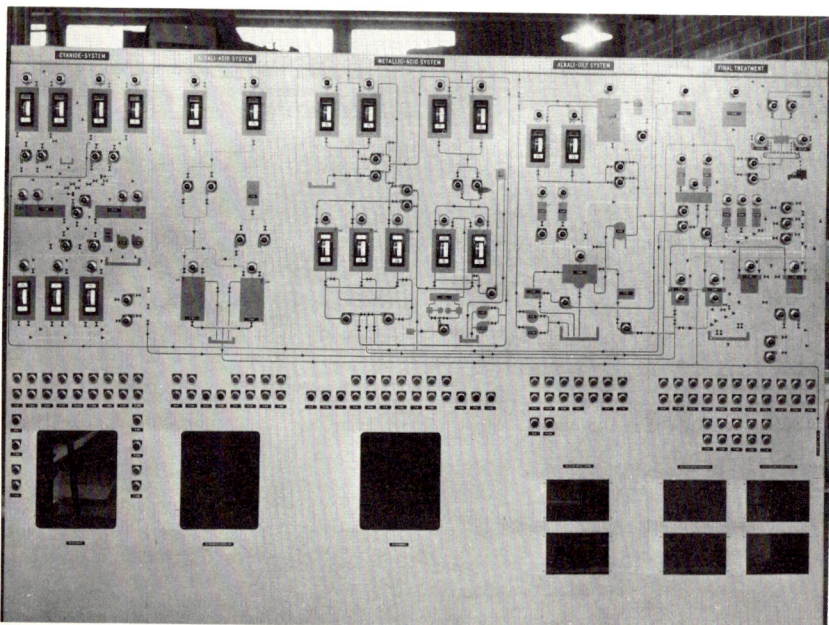

Fig. 6-32 Main control panel, waste treatment plant, Pontiac Motor Division, General Motors Corporation, Pontiac, Mich. (*K & H Engineering Company.*)

meters are used to maintain a permanent record. They also serve to alert the operator to a malfunction in the neutralizing system.

Automatic Analyzers

The success of the treatment depends upon an accurate knowledge of the concentration of toxic elements in the waste. In both raw and treated wastes tests must be made continuously. Tests are best done by using one of the several makes of automatic analyzers now on the market. These units have automatic sampling and analytical units which take samples of the wastes constantly, make the analyses in a matter of minutes, and record the results on charts in the operating building. Before the advent of these units the process of taking a sample of wastes, transporting it to the laboratory, and making the test by accepted methods required about 2 h. In this interval improperly treated wastes might be discharged into the stream or sewer, resulting in severe contamination downstream of the plant. The units cost from $2000 to $10,000 each, depending on the size and number of sampling points and the number of analyses to be made. It is important to remember that if installation saves the salary of one laboratory technician, it will pay for itself in a year or so. Manufacturers of such units include Technicon, Inc. and Beckman.

Again it is important to remember that many municipalities require periodic testing of effluent. Such analyzers meet their requirements if placed in a control manhole after the last treatment phase. Analyzers can therefore control treatment and meet governmental regulations.

Control Panel

This is a very important feature of an industrial waste treatment plant. It should be installed in a prominent place in the operating building, preferably in its own room or cubicle. The chief operator or superintendent can observe the operation of the various plant units on the control panel and be warned of any malfunction immediately. The panel should be of the graphic type, as illustrated in Figure 6-32. In those plants that treat a variety of wastes separately, the panel should be divided into sections, with the controls for each type of waste concentrated in one area, and the final section for the phase in which all flows and/or sludges may be combined. The tank level gauges should be mounted on the main panel, with push-button controls for every equipment unit using an electric motor or any electrically actuated item. A green light indicating proper functioning and a red light indicating trouble are useful. The attendant should be able to push one button to start a unit and another to shut it down. The pH register-recorder and other indicating devices should be installed on this panel or mounted in the office of the superintendent. The alarm devices for each mechanical unit, such as bells or horns, should also be placed in or adjacent to the office. The name of each unit on the panel should be clearly indicated, and the start and stop lights and controlling buttons should be immediately below the graphic indication of the unit they control. If the panel is divided into sections, the function of each section should be clearly identified at the top of the panel. The panel should be constructed with all instruments so mounted on it that they are within reach of a normal person. It should be mounted on supports above the floor and with free passageways entirely around it.

Chlorine Storage and Dispensing.   In several states the state health department requires that all chlorine storage and dispensing equipment be installed in a room accessible only by a door from the outside of the building and with no connection between the chlorination area and the operating building. Adequate ventilation by fans or by quick-opening windows controlled from outside must be provided, as well as equipment for detecting leaks of chlorine. Where a chlorination room is an integral part of the operating building, it is convenient to have the upper part of the wall between the two areas consist of a heavy glass window, tightly sealed, to enable the operators to observe the dials and gauges on the chlorinators from the outside.

Although the main control of mechanical units should be from the main control panel, switches should be provided at every unit in the field to enable an operator to shut them down for observation, lubrication, maintenance, or repair, without reference to the main panel. These field switches may be placed in lockboxes to protect against unauthorized use.

Marking of Units and Pipelines

Although not directly related to automation and automatic controls, marking of the individual units and pipelines will provide better control of the operation. For safety of personnel and proper operation and rapid control, every pipeline and valve should be clearly marked as to the normal contents of the line. This may be done by several means, such as painted color bands or stripes and named designations applied as stickers at intervals along the pipeline, with arrows indicating the direction of flow in the line. Every valve on a line should be painted with the color designated for that line. Each valve should also be clearly numbered. These indications should be in print large enough so that anyone with normal eyesight can read them from 10 to 15 ft away. Stickers, striping material, etc., are obtainable from a number of manufacturers, or the superintendent can use his own system. Large charts showing the designations of the lines and the valves that control them should be placed at several locations in the area so that the operator need not depend upon his memory of the contents of the several lines.

SPECIFICATIONS FOR CONSTRUCTION

The writing of specifications is as important as the preparation of the working drawings. The specifications should clearly describe the quality, manufacturer, and other characteristics of all material and equipment going into the project. In addition to the technical specifications, "general conditions" cover the legal and contractual aspects and the quality of work. A suggested outline of the general conditions of specifications is as follows:

General Conditions

1. Name of company

2. Scope of work

3. Definitions
   Owner
   Engineer
   Contractor

4. Work included

5. Contracts

6. Proposal

7. Examination of premises

8. Drawings

9. Record drawings

10. Variations of work

11. Lines and grade

12. Permits

13. Utilities

14. Materials and workmanship

15. Inspector

16. Compliance

17. Protection of workmen, public, work, and property

18. Workmen's compensation insurance

19. Indemnity to owner

20. Public liability and property damage insurance

21. Risks

22. Taxes

23. Payments

24. Changes in work

25. Waivers of lien

26. Arbitration

27. Bond

28. Tests

29. Public utilities

30. Fences and barricades

31. Cleanup

32. Guarantee

Types of Specifications

Two types of specifications may be used, open and closed. Engineers are restricted when preparing plans and specifications for a project for a municipality. Usually, the ordinances of cities require that specifications for a project to be paid for from public funds must be so worded as to enable contractors to include in their bids equipment and materials different from the basic intentions of the designing engineer. This frequently imposes a heavy responsibility on the engineer, who has specified certain types of equipment and apparatus with an "or equal" or "of approved equal quality" clause.

It is frequently difficult to prove to the municipality that a unit which is manufactured by a reputable company is not equal in all respects to the one which the engineer believes to be best suited for the particular project. But in the case of an industrial waste treatment facility paid for by an industry, there is no such obligation. The industry has the right to use whatever equipment the engineer deems best, so that the engineer can order units that he wants to use and write a completely closed specification.

Since a great deal depends upon the performance of the equipment in an industrial waste treatment plant, the author believes that closed specifications in such projects are entirely logical and justified. In fact, the Water Pollution Control Board of New York State announced that it is not their concern to dictate for or against any particular type of equipment that is specified or desired by the engineer for an industrial waste treatment project. Their stated primary interest and responsibility is to determine that the method of treatment selected for a given case will produce the degree of treatment required. The engineer may use and specify any demonstrated equipment that will produce this result.

LOCAL CONDITIONS

Before the final general plan and design have been firmly decided, there are three additional considerations to be taken into account. A survey of the availability of local materials is specified. If the plant buildings are of a type common to the vicinity, this should be taken into consideration to maintain uniformity.

If concrete materials must be brought from long distances and good brick is made locally, this should also be considered. In some instances concrete blocks have been used economically for tank structures. In other cases redwood tanks offer economies in construction.

The effect of the environment on the system and the effect of the system on the environment are two sides of the same ecological problem. Normal climatic conditions in the area must be taken into consideration. Severe winter or summer climates, humidity, rainstorms, etc., will greatly affect the economy of some designs.

In one case in New Zealand, the authorities were concerned about the digestion of sludge from a municipal waste treatment plant. The sludge was to be in contact with heated soil near underground hot springs. Actually, the high temperature was beneficial to the process, as it eliminated the need for heat-producing equipment normally required to maintain the optimum environment for good digestion of organics.

The effect of discharging warm effluent into local streams must be carefully determined. In some cases it may be necessary to cool the effluent before discharge. Care must be taken so as not to upset any of the delicate ecological balance.

Frequently, an industry faced with a waste treatment problem will place an order for equipment with a manufacturer who has advised them on what units they will need in the new facility. When the engineer called in to design the plant is confronted with this fact, he has two alternatives. He can try to adapt the purchased equipment to his proposed design. At the same time he might probably advise the client that because of this forced use he cannot be responsible for the functioning of the system or for its maintenance and repair cost. The second alternative is to refuse to use the equipment and demand that the industry cancel the order.

In one case in the author's experience a large oil company engaged his firm to design a waste treatment plant. After the contract was signed, the management advised that they had already purchased the equipment from a certain manufacturer for a large sum. Although this equipment was not what the engineer would have specified, it could be used. The plant units were designed to accommodate the purchased units.

In another case, the author was asked to give an opinion as to the best means of destroying a small amount of cyanide in the wastes from the plating department of an electronics equipment manufacturer. Inquiry showed that the total amount of cyanide in the wastes was not over 5 lb/d. After the engineer made his recommendation, he was informed that the company had already purchased equipment for this project. When the order was inspected, it was found that the equipment was sized for over 400 lb of cyanide per day. Thus the plant was being saddled with an installation many times too large and costly.

The wisest plan for an industry is to leave the selection of the type of equipment for any industrial waste project to the engineer. He has the experience with many different types, and he knows the faults and the best points of all of them. Economy in the initial and operating costs will be best served by this method. The engineer has been engaged to advise; he should be permitted to do so.

IN-PLANT POLLUTION CONTROL

In summary, conventional treatment methods are generally capable of removing up to 90 percent of the BOD from organic industrial wastes. Wastes can also be readily treated for the removal of sediments, particulate matter, COD, heavy metals, acidity, alkalinity, etc. The technology is available for the removal of most types of harmful ingredients. The limitations on existing capabilities are more economical than technical. In other words, an industry can provide all treatment for which it is willing and able to pay. Most of the existing pollution of streams can be eliminated if the fullest possible use of present knowledge and technique is made, with the proper study of the conditions of the case in hand and the actual needs. In some cases, such as refractory organics, present processes may be inadequate and costly, but research should be directed toward the modifications or improvement of these methods to render them suitable for existing conditions.

In this phase the engineer should carefully investigate some of the newer methods and techniques included in this text. Not all available technology that could be applied to the problem of pollution abatement is being fully used today. This is partly due to the lack of proper dissemination of the knowledge among those who would use it. But the primary reason is economic, the desire to reduce the initial cost of industrial waste treatment plants. Too little attention has been given to in-plant changes and use of different process materials and methods. Whereas in earlier years changes in the use of materials in a factory would involve expensive changes in equipment or routine, it is now realized that although the cost may still be there, the reduction in waste volume and concentration achieved may be a factor of major importance for the industry. Economy can be accomplished by reducing the size and complexity of the waste treatment facilities, or perhaps by entirely eliminating treatment. Instances of this are the use of hydrochloric acid instead of sulfuric acid in the pickling of steel and the adoption of the dry method of descaling.

Another instance of the benefits of in-plant changes and "good housekeeping" is related in the book "The ORSANCO Story," which discusses a report of the metal-finishing industry action committee. This committee concluded that as much as 80 percent of the polluting potential of some types of waste metal-plating liquids could be eliminated by observance of certain precautions and operating procedures.

But despite this, industry is reluctant to permit the loss of time and production which might be involved when internal or process changes are recommended. The long-range economy possible by eliminating waste treatment entirely, or at least reducing its cost by in-plant changes, has not been given the mature consideration it deserves.

It is recorded that a steel company in Wisconsin has completely shut off its connection with the local sewerage system because internal arrangements now enable the mill to operate entirely on recirculated water. By process modifications the company is said to have eliminated 94 percent of the suspended solids, 75 percent of the waste pickle liquor, and 45 percent of the water formerly returned to the river. The methods adopted treat the wastes to the extent necessary for the reuse of the water in plant processes. This procedure has been found to be much less costly than the treatment required for disposal of the waste effluent into the stream.

The potential economies in following the above-mentioned approach instead of blindly accepting the order to "treat the wastes" are so evident that any engineer who does not investigate this phase of a project is failing in his responsibility to his client. In one instance in the author's experience, the water supply for a steel mill was being drawn from the company-owned wells. Fear was expressed that the capacity of these wells was being exhausted. Over 6000 gal/min was being used in the mill and discharged into the stream. Treatment to eliminate pollution was indicated. It was found that by moderate treatment of the wastewaters, much of it from cooling operations, a large volume could be returned again and again to the mills. By this means the draw from the wells was reduced to 2000 gal/min, and much of the former polluted discharge to the stream was eliminated without waste treatment facilities of any kind.

Chapter Seven

The construction, operation, and maintenance

The engineer has completed the plans and specifications and received approval from the regulatory authority. It is time to submit the package to the client for final approval. If the client is satisfied, the next step is construction. Several options are available to the client. He may send out for bids from a number of contracting firms; he may negotiate a contract with a contracting firm known to provide satisfactory services at a reasonable price; or he may choose to do the construction in-house.

Regardless of the option selected, the project may require at least one full-time supervisor. It goes without saying that the engineer, responsible for the design of the plant, should inspect the progress of the work periodically. He should also be available for clarification of plans and specifications for emergency changes as they occur. In one case in the author's experience, the plans called for two large final sedimentation units to be built of concrete and set in the ground. When the excavation was begun, it was found that the legs of two high-tension power line towers extended below ground about 8 ft and rested on 6-ft-square steel plates. The utility company immediately objected to more excavation in that area and to the placement of the tanks in the ground so close to the towers. Since the area did not permit locating these tanks at any other site, an entire change in design was made. The new design had steel tanks set on concrete foundations above the ground. The engineer's assistance during the construction stage is important, and it must be negotiated in the initial contract.

Generally, though, the engineer cannot be available during all job working hours. Often a responsible engineer in the engineering firm is contracted to do the supervision of the installation. The resident engineer would then be on

126

hand or available at short notice during the entire day. Arguments between the supervisor and the contractor can be settled by a bit of diplomacy.

Management may want an outside contractor to install the facility but will provide a company supervisor, perhaps the plant manager, to oversee all activities. Such a supervisor has more authority, being a representative of the firm that pays the bills. It is assumed that such a plant manager will be as familiar with the design as the engineer who designed it.

A third possibility is the hiring of an independent construction supervisor. Such a man would have no allegiance and no distractive side duties, and he could devote his full energies to supervision.

No matter who is selected as supervisor, the individual will inevitably have to smooth out differences between the contractor and management as they arise. In one case where the available site for the waste treatment plant had formerly been used for storage of parts of the company product, the work was delayed because the management was dilatory in finding other storage grounds for these parts. This prevented the contractor from following an orderly sequence of operations. Workmen for that particular phase were laid off and attention was diverted to another section of the project. It is important that the contractor's workmen and the labor force in the factory be kept out of each other's way. Each of these parties feels that his work is paramount. It will often be the duty of the supervisor to resolve these difficulties and keep all the wheels in motion.

If the engineering firm provides the resident engineer, it is not in his province to tell the contractor how to do his job. It is the responsibility of the engineer to maintain cordial relations with the superintendent of the contractor. He should provide the contractor with an ample supply of plans and specifications so that they may be discussed and settled at the site without having to refer every argument to the head office of the engineer. In other words, the resident engineer should be a responsible member of the staff of the designing engineer, and his judgment should be trusted to resolve difficulties amicably and with the minimum of extra cost to the client. Pleasant relations between the engineer and contractor will get a better job done in the desired time.

CHANGES AND DELAYS

If changes are necessary as a result of management decisions, slow delivery dates, etc., they can cause delays. The added cost for stoppage of work, additional material, extra labor, and the loss of time should be recorded. A signed order for such work should be obtained from the supervisor to forestall later arguments over legitimate compensation for the changes. Too often a representative of the management will make a casual remark about the desirability of doing something otherwise than as the engineer had planned, but when the bill for the extra work is presented, no one knows anything about it. See Chapter 2 for coordination with the engineer.

Equipment must be purchased as soon as possible so as not to delay installation. Again in purchasing, the client may choose to have the contractor take care of all the necessary purchases, paying him progressively during the job. The engineer in coordination with the client's purchasing department may write the requisitions, or the plant manager may handle this aspect of the project, working closely with his purchasing department and seeking aid when needed from the engineer.

PAYMENT FOR EXTRA WORK

Practically every job involves extra work beyond the contract cost. The payment for extra work usually requires that materials, equipment rental, and other costs be substantiated. This can be done by vendor's invoices. If vendor's invoices are not submitted, the owner might establish the cost of such items at the lowest current prices (in quantities required, delivered to the work site).

To the totals computed for labor, materials, equipment rental, and other services and expenditures, authorized by the engineer or owner, may be added the following suggested percentages:

Labor	15%
Materials	15%
Other services and expenditures	15%
Equipment rental	None

Labor, materials, and equipment may be furnished by the contractor, subcontractor, or others on behalf of the contractor. The client should only deal directly with, and make all payments to, the prime contractor.

PURCHASE ORDERS

Regardless of who handles requisitions and purchasing, a general outline or a recommended checklist is helpful. Each purchase order should contain the following:

1. Specifications, descriptions, performance, definition of item purchases, and reference list of drawings should be included.

2. Price of item as given in vendor's proposal.

3. Freight FOB and location of delivery should be clearly noted.

4. Use tax, if required.

5. Delivery date and penalty if any.

6. Spare parts and special tools.

7. Painting, shop or field. If special paint or color is required, this should be specified.

8. Installation supervision, engineering service, and amount furnished; cost should be stipulated.

9. Methods of payment.

10. Discount terms.

11. Method of shipping.

12. Warranty.

13. Vendor's drawings, shop drawings, and type of schedule.

14. Affiliation, small vendors.

15. Job and purchase order number.

16. Account distribution.

17. Vendor's correct legal name, address, and phone number.

18. Package marking: item or make number.

19. Check for omissions (what other material and labor are necessary to complete installation?).

20. Check vendor's financial and performance reputation.

21. Inspection test necessary.

22. Samples to be submitted.

23. Method of computing extra work.

24. Method of computing escalations.

25. Acceptance and approval by regulating agencies.

26. Control and appurtenances (who furnishes?).

27. Has bid abstract been made for comparison between vendors?

28. Approval by owner and engineer.

SPECIFICATIONS

The engineer spent many months preparing plans and specifications. They along with other contract documents govern all work. It is therefore imperative that the contractor keep a copy of the plans and specifications at the work site. This allows the supervisor and any inspectors easy access.

Anything in the specifications and not on the plans, or on the plans and not in the specifications, should be considered as though shown or mentioned on both. If reference specifications and standard plans are referred to, they should be considered part of the contract documents. While it is usually the intent to include on the plans and specifications much of the information pertaining to the conditions which may affect the cost of the proposed work, the owner and his engineer are not expected to warrant the completeness or accuracy of such information. It is therefore considered the contractor's responsibility to ascertain the existence of any conditions affecting the cost of the work which would have been disclosed by reasonable examination of the site. Upon discovery of any error or omission in the plans or specifications, the contractor should immediately call it to the attention of the owner or his engineer.

Occasionally there is a conflict between the plans and specifications. The order of precedence should be as follows:

1. Requirements of local ordinances of all governing agencies.

2. Plans; detailed plans have jurisdiction over general plans.

3. Standard plans.

4. Specifications.

It should be noted that change orders, supplemental agreements, and approved revisions take precedence over plans and specifications.

PLANS

A typical plan in an industrial wastewater installation is the plan for sewer pipes. On long sewer lines it is usually required to prepare fully detailed plans and profiles for the construction crew. Profiles should have a horizontal scale of not more than 100 ft to the inch and a vertical scale of not more than 10 ft to the inch. It is customary that the plan and profile be drawn on the same sheet so that features on the profile are shown directly under the plan view. A typical plan and profile drawing should include:

1. Location of all streets, sewers, buildings, or other important features that may affect construction

2. Size and type of sewer pipe

3. Horizontal distance measured in stations between manholes

4. Vertical distance measured in feet and hundredths of a foot between invert of manholes and top of manhole covers

5. Grade of pipe between manholes given as a ratio of vertical difference per 100 ft horizontal distance

6. Location of all special features such as invert siphons, concrete encasements, and elevated sewers

7. Details for construction of all appurtenances such as manholes, cleanouts, and inspection chambers

CONSTRUCTION PURCHASE ORDERS

With the plans in hand, construction of the sewer line can begin. The purchase order for the construction project should include the following items:

1. All labor necessary to complete the work.

2. All material required, such as vitrified clay pipe, branches, jointing material, concrete, manholes, and manhole rings and covers.

3. The contractor should supply all necessary tools, equipment, etc., required to do his work.

4. The contractor should pay for and obtain a permit from the governing agency.

5. All work should conform to governing codes and be inspected and approved.

6. System should be tested before pipes are backfilled.

7. All trenching shall be braced in accordance with OSHA and local regulations.

8. All backfill of trenches shall be in accordance with soil engineering practice for maximum compaction.

SURVEYS

To begin the actual construction, a surveyor is normally called upon to stake out the new sewer line on plant property. The first requirement in staking out the line is to run a transit line parallel to the centerline of the sewer where it will not be disturbed by construction or excavated soil. Hubs are driven at intervals along this line flush with the ground. Profile levels are then run along the transit line, and the elevations are noted to the nearest hundredth of a foot. A grade sheet is prepared listing the distances from the top of each hub to the bottom of the pipe trench, or the top of the pipe at a point adjacent to each hub. Excavation and pipe placement are then controlled throughout the job. See Figure 7-1 for illustration of pipeline survey stakes.

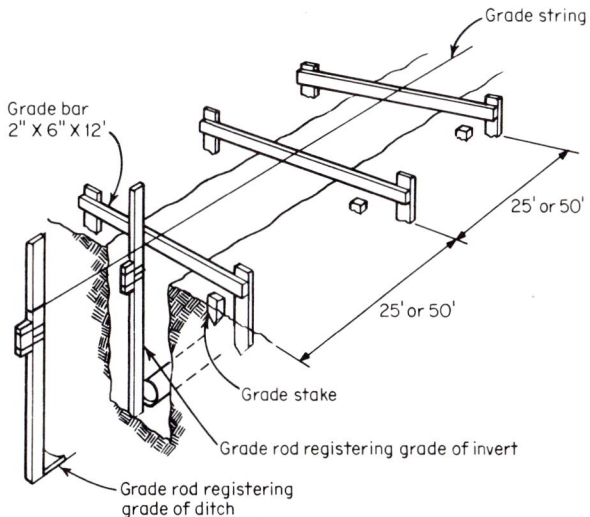

Fig. 7-1 Grade staking for the construction of a sewer line.

(a)

(b)

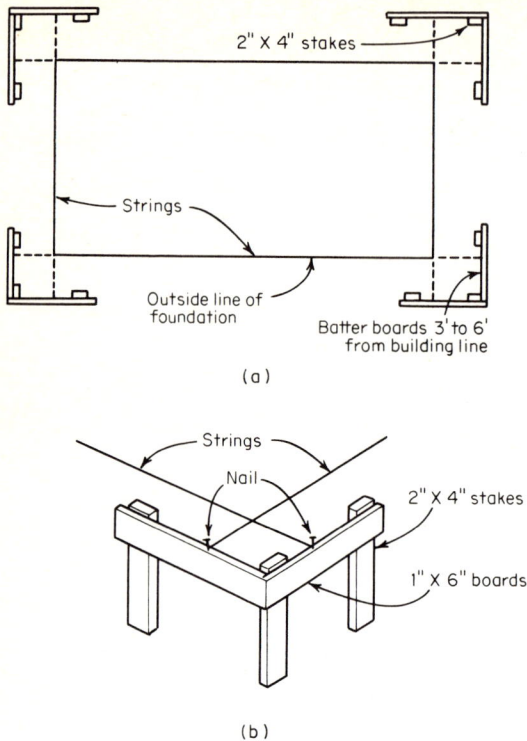

Fig. 7-2 Batter boards for a structure. Construction batter boards, used as a guide for excavation, forming foundation walls, and establishing grade to top of wall. (*a*) Plan showing arrangement of batter boards and string lines. (*b*) Detail of corner of batter boards.

Concrete structures to be constructed must also be staked out. Usually three stakes are placed at all corners of the structure far enough back so that they will not be disturbed during construction. Batter boards are nailed to these stakes with their tops horizontal and set at a definite elevation. Nails are driven into the tops of the boards on the projected line of the building. Strings stretched between nails at opposite ends of the lines provide both the line and the grade of the structure. With these established reference points, the building can be constructed exactly as designed. See Figure 7-2 for typical batter board installation.

TRENCHES

Before the trench is started, it is important to remember that all trenches and pits over 5 ft deep usually require a permit. This permit is obtained from an agency of the state, county, or city which will inspect the work.

The walls and faces of all deep excavations which expose workmen to danger from moving earth must be effectively guarded by a shoring system, sloping ground, or an equivalent method. Bedding is also important. The structural strength of clay pipe is very much dependent upon how the pipe is bedded. The bedding varies with the type of excavation. There are four types of bedding, as shown in Figure 7-3.

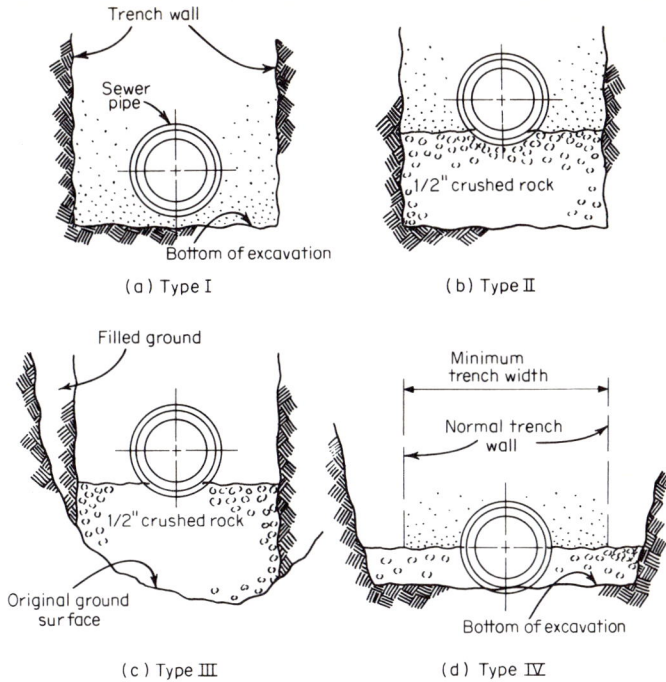

Fig. 7-3 Recommended bedding of vitrified clay pipe.

Bedding Types

Type I bedding is for the normal trench, excavated in good soil, whose width is within allowable limits. The bedding material supporting the pipe or conduit should be gravel, crushed aggregate, or native granular material. There should be at least 1 in of bedding below the bell and 6 in at each side of the pipe. The bedding should extend 1 ft over the top of the pipe.

Type II bedding is for excavations into wet or spongy ground which provides poor support. The bedding requires a minimum of 2 in of ½-in crushed rock under the pipe.

Type III bedding is required for excavations into inadequately compacted, previously filled ground. This also requires 2 in of crushed rock below the pipe, to extend to the original ground surface. The bedding needs to be extended sidewise only 6 in on either side of the pipe.

Type IV bedding is used when the excavations have exceeded the allowable limits. Such a bedding requires a concrete cradle for the full width of the trench. Its depth should be equal to half of the pipe diameter. The concrete cradle is placed on undisturbed soil free of clay or silt. Conventional bedding can then cover the pipe to normal trench width. The rest of the backfill can be native soil, adequately compacted. In another case, if the cover over a clay pipe sewer is less than 42 in, measured to existing street or ground surface, or to a lower surface resulting from grading operations, the pipe should be installed with type IV bedding. See Figure 7-3.

Trench Dimensions

The trench width minimums and maximums are dependent upon pipe diameters. Figure 7-4 shows recommended minimum widths for various pipes. The minimum

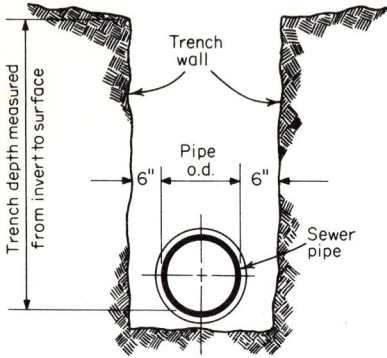

Fig. 7-4 Minimum trench width for the installation of sewer pipe.

width is determined by adding 12 in to the outside diameter of the pipe. Since the pipe should be centered in the trench, there will be 6 in on each side of the pipe.

Table 7–1

Pipe size	Depth of trench					
	18–20'	*16–18'*	*14–16'*	*12–14'*	*10–12'*	*Less than 10'*
	Width of trench					
4″ and 6″	2′ 2″	2′ 2″	2′ 2″	2′ 2″	2′ 2″	None
8″	2′ 3″	2′ 3″	2′ 4″	2′ 4″	2′ 6″	None
10″	2′ 5″	2′ 6″	2′ 7″	2′ 8″	2′ 9″	None
12″	2′ 8″	2′ 9″	2′ 9″	2′ 11″	3′ 1″	None
15″	2′ 11″	3′ 0″	3′ 2″	3′ 3″	3′ 6″	None
18″	3′ 3″	3′ 4″	3′ 6″	3′ 8″	4′ 0″	None
21″	3′ 7″	3′ 8″	3′ 10″	4′ 1″	4′ 6″	None
24″	3′ 9″	3′ 11″	4′ 2″	4′ 5″	4′ 10″	None
27″	4′ 1″	4′ 3″	4′ 6″	4′ 10″	5′ 4″	None
30″	4′ 4″	4′ 7″	4′ 10″	5′ 3″	5′ 10″	None
33″	4′ 8″	4′ 11″	5′ 2″	5′ 8″	6′ 5″	None
36″	4′ 11″	5′ 2″	5′ 6″	6′ 1″	6′ 11″	None

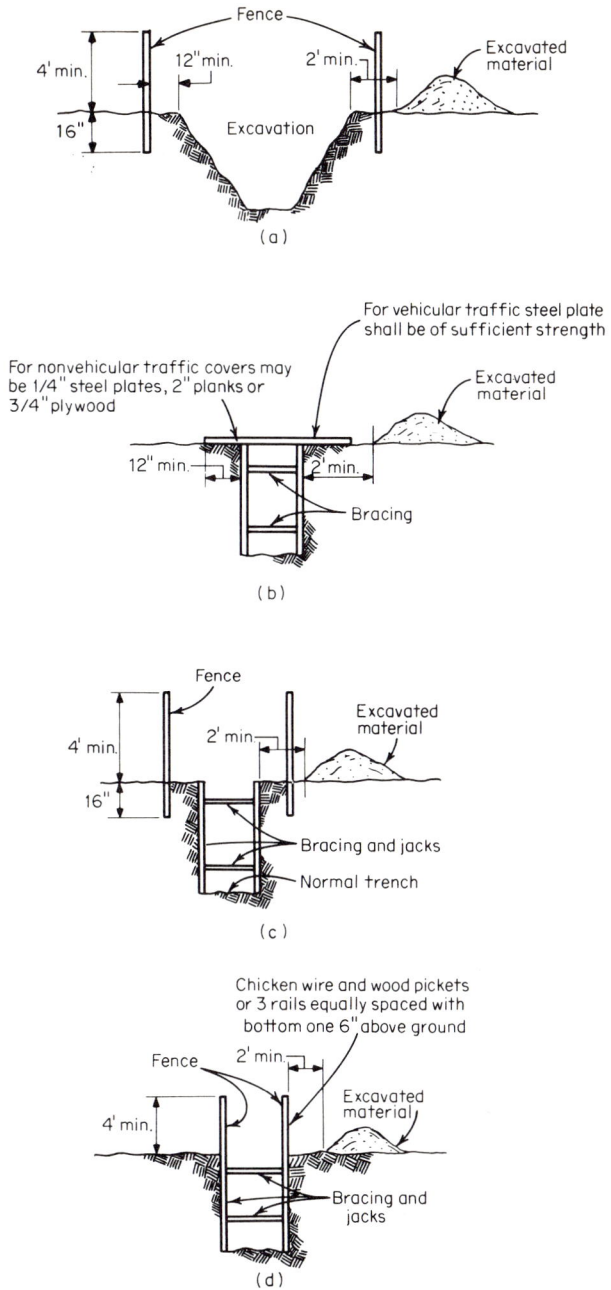

Fig. 7-5 Recommended protection for trench excavation.

The maximum width of the trench is dependent upon its depth. The depth of a trench is the vertical distance from the surface of the ground directly over the pipe and the invert of the pipe. When the maximum allowable width is exceeded, a special bedding or cradling procedure should be followed. Table 7-1 indicates the maximum recommended trench widths for pipe 4 to 36 in in diameter and trench depths up to 20 ft.

Safety considerations for the workers is an important factor in all pipeline trench excavations. These are governed by Occupational Safety and Health Administration (OSHA) and most local city and county agencies. Figure 7-5 illustrates a recommended procedure for shoring of trenches.

WORK ON EXISTING FACILITIES

In most industrial wastewater installations, work must be done also on existing sewer facilities. The sewer contractor should be required to control odors during the work. To ensure this, sewers should be open to the atmosphere for the least possible time. Enclosures, seals, and forced ventilation may be necessary to prevent odor nuisance and should be provided. If the existing line is still in use, the contractor must provide bypasses. His proposed method of bypassing should be approved by the engineer.

Where a manhole on an existing sewer line is to be remodeled, a part of the manhole channel should be removed. The depth should be sufficient to permit construction of new channels and shelves at least 3 in in thickness. When a manhole is to be constructed at an existing sewer, the pipe should be entirely removed within the area of the new manhole and a new footing should be constructed according to the detail plans. If the manhole is within public property, it must be constructed according to the applicable standard plans of the governing agency. Sewerage should be conveyed across the manhole by pumping or by bypassing through a tight-fitting metal sleeve having substantially the same dimensions as the corresponding inlet and outlet pipes. Sewage should not be permitted to flow over new concrete channels until they are 5 d old.

When new work is to be constructed inside a sewage structure against an existing brickwork or concrete surface that has been exposed to sewage atmosphere, the existing surface should be prepared as follows:

1. All soft and loose material should be removed and the surface cleaned by sandblasting or wire brushing.

2. Joints of existing brickwork should be raked. Loose brick should be replaced with new brick if the surface is to be plastered.

3. The surface should then be washed with a 3 percent solution of hydrochloric acid (HCl) followed by a final rinsing with clear water.

All other piping should be promptly installed after any required excavating or cutting has been done. This ensures that openings for piping will be closed as soon as possible after pipe has been tested and approved.

Occasionally a new pipeline is required to pass under an existing clay pipe. The excavation for the new line necessarily eliminates the earth support of the old pipe and may cause damaging settlement. Therefore, it is recommended to support the old pipe at the crossing with a concrete support, as shown in Figure 7-6. The concrete support should be poured in undisturbed soil on the bottom and should extend the full width of the excavation. The concrete mix should be at least of 2000 lb/in² compressive strength. This type of support should be used for all existing concrete or clay pipes which are not tunneled. The bottom side of the oil pipe should receive a coating of asphaltic emulsion to prevent bonding to the concrete.

The junction of one gravity pipeline with another is a critical point in a sewerage system. This is especially difficult when a new line is to be connected to an existing pipe. Figure 7-7 illustrates the procedure to be followed in connecting to an existing pipe. When both lines of a junction are to be installed at the same time, the connection can be made as shown in Figure 7-8.

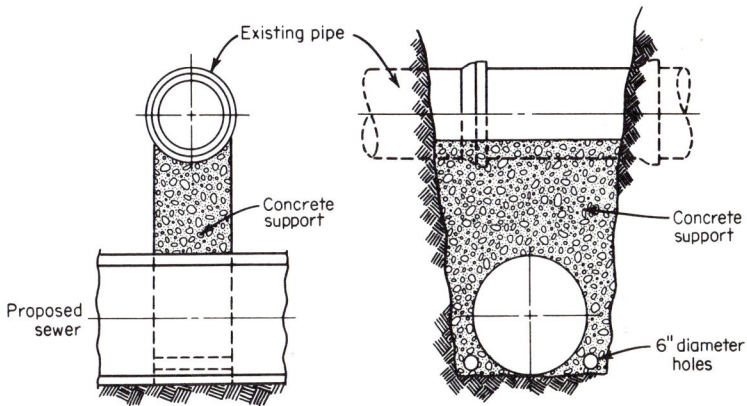

Fig. 7-6   Recommended pipe support where new pipe passes under existing pipe.

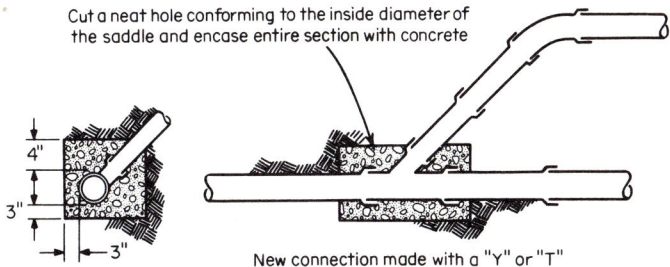

Fig. 7-7   Field joining of clay pipe.

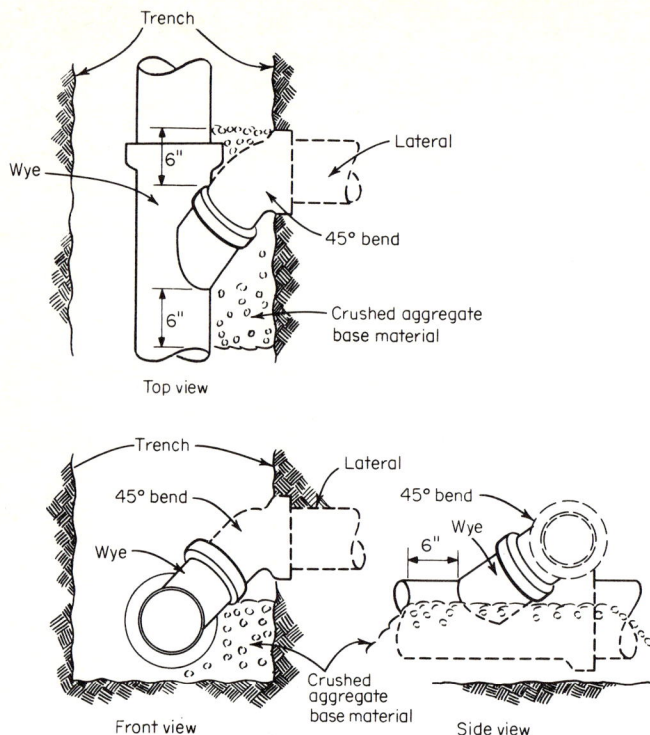

Fig. 7-8 Recommended wye or tee support of lateral to main sewer pipe.

INSTALLATION TOOLS

An invaluable tool for installing new pipelines in an existing facility is the "pipe detector." This is an offspring of the military mine detector. The use of a detector might prevent accidental severing of a buried conduit and a consequent sudden plant shutdown. Pipe detectors have a 6-ft depth penetration range. Some are powered by a single 9-V transistor radio type of battery.

PIPING INSTALLATION

All piping, except pipes requiring a necessary pitch, should be run as straight and direct as possible and should be installed parallel to walls, beams, and girders. Vertical stacks, vents, and risers should be plumb and true. The various systems of piping should be installed in orderly, parallel groups, neatly spaced with clearances allowed for coverings and maintenance.

Nipples are usually of the shoulder type and of the same weight and material as the pipe connected. Close nipples are used only when they are of extra strong weight. No bushings of any type should be used in any piping system. When easy maintenance or removal of piping or specialties is necessary, flanges or ground face unions can be installed. Eccentric reducing fittings should be installed on horizontal piping when they are necessary to afford complete drainage of the piping system.

Swing joints, expansion loops, and anchors are often provided to control expansion of piping when these lines are exposed to thermal changes. Piping should be so graded and valved as to allow for complete drainage and control of the system. Joints between ferrous and nonferrous metallic piping require insulating couplings. Where exterior walls are penetrated by piping, they should be properly sealed and flashed against the weather.

SAFETY REGULATIONS

The design of industrial wastewater treatment facilities must be done with the safety of personnel in mind. In fact, violations of the Federal Occupational Safety and Health Standards can result in severe fines and even cessation of operations.

Open surface treatment tanks containing corrosive or toxic liquids can be a hazard. The hazard potential is classified on a scale of A to D. This indicates the severity of the hazard associated with the substance contained in the tank. Fumes, vapors, or mists from certain liquids are highly dangerous. To protect the personnel, either the tank must be covered or a properly designed ventilation system using hoods must be employed.

All employees working around open surface tanks should be instructed as to the hazards of their particular jobs and the personal protection and first aid procedures applicable to these hazards. Workers in areas which have wet floors are required to wear rubber or other impervious boots or shoes. And when required to handle work wet with a liquid other than water, employees must wear impervious gloves, aprons, or other protective clothing.

Whenever there is a danger of splashing, for example, when additions are made manually to the tanks, or when acids and chemicals are removed from the tanks, the employee should wear tight-fitting chemical protection goggles or an effective face shield.

Near each tank containing a liquid which may burn, irritate, or otherwise be harmful to the skin there should be an emergency shower and eyewash.

In the case of cyanide-bearing wastewater, dikes or other storage structure arrangements must be provided to prevent the possibility of intermixing of cyanide and acid in the event of tank rupture.

Floors and platforms around the tanks should be designed in such a manner as to prevent them from becoming slippery, and they should be cleaned by frequent flushing. All tanks should be provided with drain connections and access doors for cleaning and inspection purposes.

Maximum concentration of various corrosive materials in vapor or gas form in working areas above open tanks is as follows:

Sulfuric acid	1.00 mg/m^3
Sodium hydroxide	2.00 mg/m^3
Phosphoric acid	1.00 mg/m^3

Open tanks containing hot or corrosive liquids should have a handrail 3 ft 6 in high to prevent personnel from accidentally falling in. Open tanks containing hot or corrosive liquids, so located that the operator cannot see the contents from the floor or working area, should have emergency shutoff valves controlled from a point not subject to the danger of splash.

Carboys containing acid should be provided with inclinators, or the acid should be withdrawn by means of pumping without pressure in the carboy, or by means of hand-operated siphons.

There should be means provided for handling caustic soda or caustic potash containers to prevent operators from coming in contact with the caustic. These workers should be provided with special safety equipment such as protective clothing, respiratory devices, protective shields, or goggles. Wherever acids or caustics are used, provision should be made for an adequate supply of flowing fresh, clean water.

The design of a new treatment facility must provide for normal leakage and accidental spillage as well as catastrophic breakage. Any area in which solutions are stored, pumped, or treated should be diked in some manner. This is to prevent concentrated contaminants from spilling into floor drains. Leakage can come from deteriorated packings at agitator shafts and pumps. Particularly hazardous to maintenance personnel is the leakage of corrosive liquids. Workers will be reluctant to work on a floor upon which concentrated caustics or acids can injure or destroy shoes and clothing.

Each item of equipment, be it pump, tank, valve, or pipe fitting, should be recognized as requiring regular inspection and maintenance. Pipelines should be clearly indicated to alert workmen of dangerous material. Floor drains in contaminated areas should not be directly connected to the sewer systems but should discharge into a holding tank. Contents of the holding tank should then be pumped back into the treatment facility, hauled away by commercial tank truck, or, if found uncontaminated, pumped to the sewer system.

Aboveground storage of corrosive, toxic, or highly contaminated liquors is particularly dangerous since they are subject to catastrophic spillage. A break in a pipeline, failure of a check valve, or accidental disconnection of a fitting can cause sudden spillage of thousands of gallons of dangerous water. This sudden flood can spill into the local storm drain system, the sewer system, or the storage area, damaging vast amounts of finished products. The problems of maintenance and catastrophic spillage control should be seriously considered by management when the treatment facility is being designed.

LIFE EXPECTANCY OF EQUIPMENT

Management usually considers industrial waste treatment plants an expensive addition to their normal operations. They are highly interested in how long the equipment installed can be expected to last. Longevity, with the least maintenance, is the test of the integrity of the manufacturer, whether equipment is built for lasting life without regard to cost or whether it is a cheap unit built to serve an immediate purpose at low cost. Just as human mechanisms are subject to accident and breakdown, so mechanical items suffer from the same difficulty, usually because of improper care. There are unforeseen contingencies, but if recommendations of responsible manufacturers are followed by the engineer when designing the plant, the contractor in installing the units, and the operating staff in running and caring for it, the life expectancy can be considerable. It has been general practice in the equipment field to figure, for tax purposes, on a normal life of 20 years for a unit manufactured by any of the well-known manufacturers. However, according to the late Charles E. Keefer whose long career was connected with the sewage treatment problems of the city of Baltimore, in most cases the 20-year expectancy was too short. The list below gives the useful life of a number of the mechanical units used in many industrial waste treatment plants. "Useful life" means the time in which a mechanical unit may be expected to operate, with care, without major breakdown or major replacement of parts due to normal wear and tear.

There are numerous instances in the author's experience of units that have

Equipment	Estimated useful life, years
Grit collectors:	
Rotary	30
Chain flight	10
Sedimentation units:	
Rotary sludge collectors	35
Chain and flight type	15
Distributors for trickling filters:	
Rotary	25
Fixed nozzles	35
Blowers, for activated sludge:	
Aeration units	25
Chlorinators	15
Digesters, sludge:	
Mixing units	
Sludge collection equipment	25
Filters, vacuum:	
Mechanism	25
Filter media (cloths)	2–4
Incinerators, sludge:	
Rotary-drum type	20
Flash driers	20
Motors and generators	25
Pumps, centrifugal	25
Ejectors, pneumatic	45
Pumps, vacuum	25
Compressors, air	25
Conveyors, belt	15
Screens, fine, rotary	30

outlasted their expectancy because of unusual care on the part of attendants, but the periods listed above will serve for general purposes.

The periods given in the list are for the mechanical units only. The structures in which they are installed, usually of concrete, would be expected to have a longer life, except that when a plant is put out of service because of expansion, or construction of a new facility, the tanks and structures of the abandoned plant may be completely destroyed.

MANUAL ATTENTION
VERSUS AUTOMATION

If the operation of an industrial waste facility will require several full-time attendants, it is wise to leave certain details of the operation to them, to encourage them to remain alert and active.

Automation

The basic need for automatic control is in those operations directly connected with the procurement of the results planned from the actual treatment processes, the safety of the personnel, the conservation and proper use of chemicals and additives, the rate of flow of the individual wastes, and the volume of the final effluent from the plant.

The instrumentation on any plant should be in proper ratio to the cost of the plant. If the management is concerned about the cost of operation, it will pay to use automatic chemical feeders which will produce charted records of the amount of chemical used per period of time. This information, plus the cost of the chemical, will provide the management with a true record of the full cost and will serve for the use of tax authorities when they are checking the deductions made by the cost department of the industry. Automatic control of chemical and other additives will also serve to reduce the cost of the treatment since the supply will be adjusted to the demand, thus eliminating waste. Comparison of costs between manual and automated operation is shown in Figure 7-9.

The engineer, then, must consider the overall costs of his total design, as well as the initial cost. If, with mature study, it can be shown that by an initial expenditure of a few more thousands of dollars there may be a saving in the daily operating cost, it is well worth consideration. For example, if an engineer can show that the daily operating cost of an industrial waste facility can be reduced by using some items of equipment or apparatus of a higher grade, he should explore and explain that possibility. If the saving in operating cost is $1 per day, the annual saving is $365. This sum represents the interest at 5 percent on $7300. Thus, to obtain this saving, the engineer would be justified in recommending initial expenditures up to several thousands of dollars. Also, if the better unit will operate with less breakdown and maintenance, further economies will be obtained.

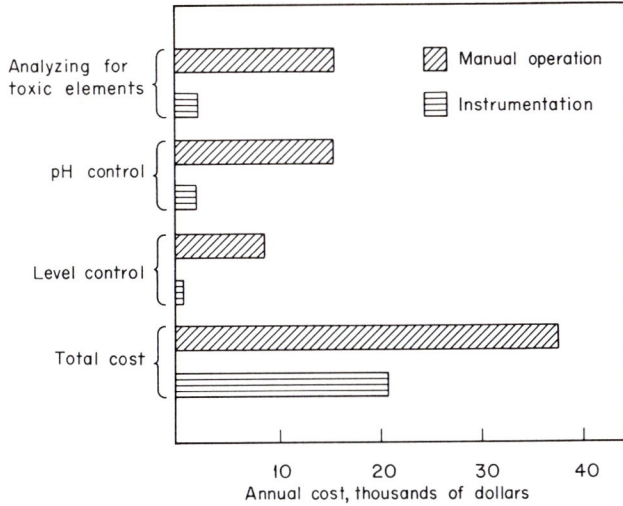

Fig. 7-9 Comparison of operating costs (prepared by E. B. Besselievre, 1962).

Manual

Maintenance is a crucial detail of any industrial wastewater treatment operation. Proper maintenance will extend the life of equipment and eliminate many of the problems associated with an industrial facility.

Cleanouts.   Probably the most important feature in a sewerage system relating to proper maintenance is the number and location of cleanouts. Because of the nature of most industrial wastewater there is a continuous accumulation of solid material in the pipelines which must be cleaned out. The only access to the interior of these pipes is through the cleanouts. Figures 7-10, 7-11, and 7-12 illustrate the various types of cleanouts.

Sewer Cleaning.   There are many devices designed and used for cleaning plugged or fouled sewer lines. The simplest are various forms of the "snake"

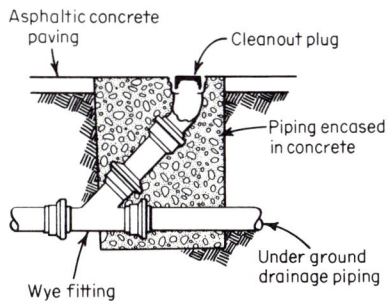

Fig. 7-10 Detail of a cleanout to grade in a heavy traffic location.

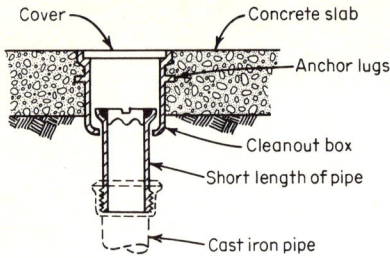

Fig. 7-11 Detail of a cleanout in an area surfaced in concrete.

which are hand-operated. These have different types of heads, as shown in Figure 7-13 which include (*a*) the square stock corkscrew, (*b*) the auger, (*c*) the root saw, and (*d*) the sectional rod.

The power units are available equipped with electric motors for small lines. But for major pipe cleaning there are truck- or trailer-mounted high-velocity sewer cleaners, equipped with cleaner hose, reel, high-pressure cleaning gun, and pump. See Figure 7-14. These units include water tanks of 30- to 1000-gal capacity for storage of water needed for the high-velocity spray from the nozzle of the hose. There are two spraying actions, one forward and one backward, each at 1200 lb/in² at 30 gal/min. The back spray pulls the hose and the nozzle through the line. The forward jet in the nozzle opens a path for easy hose passage. No internal pressure is required—only high-velocity water.

The operation of the high-velocity sewer cleaner is as follows. The vehicle is positioned at the manhole with hose reel above the opening. A high-velocity hose, fitted with a steel jet nozzle and a nozzle extension, is introduced directly into the sewer. Water pressure, up to 1200 lb/in² pressure at the pump, pulls the hose up the sewer against the flow. The forward jet in the nozzle breaks up accumulation in front of the hose to allow easy passage. When the hose is withdrawn under pressure, the rear water jets scour the sewer clean. Light materials are flushed down the line, and the heavier materials are deposited in the manhole for easy removal. A final jetting of the interior of the manhole with a hand gun completes the cleaning operation.

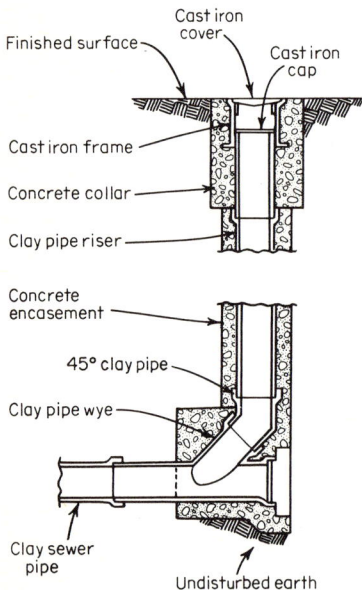

Fig. 7-12 Detail of a terminal cleanout structure. This type of structure provides for future extension of the line if required. This can be done by removing the clay cap. The concrete encasement and collar are to protect the clay pipe from damage. The cover and frame are made of gray cast iron.

Miscellaneous Tools. Every plant maintenance department should have on hand some of the more basic maintenance tools. The basic

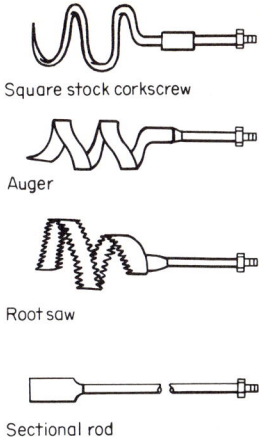

Square stock corkscrew

Auger

Root saw

Sectional rod

Fig. 7-13 Sewer maintenance tools.

Fig. 7-14 Portable sewer cleaner. (*Myers High Velocity Sewer Cleaner.*)

pipe-cleaning tool is the electric eel. This is a continuous self-feeding cable driven by an electric or gasoline engine. Some models are made to clean 50 ft of 1¼- to 3-in pipe; 200 ft of 3- to 10-in pipe, or 500 ft of 4- to 14-in pipe. These eels are equipped with quick couplings and tempered-steel cleaning tools. See Figure 7-13 for common maintenance tools.

Pipeline tools also include portable pipe saws and pipe cutters. These are available for manual, gasoline, electric, or hydraulic power. Also required for bell-and-spigot pipe are repair clamps, lead wool, lead ingots, pouring pots, ladles, lead joint runners, jute packing, and caulking tools. Threaded pipework will require pipe wrenches, chain pipe tongs, taper pipe taps, and threaders.

OPERATION AND MANAGEMENT

An industrial waste treatment plant designed by the best engineer in the world, using the latest and best methods of treatment for the wastes in question, and constructed by the best contractor available, can be an utter failure if not operated by competent, reliable, and dedicated personnel. On the other hand, a poorly designed, indifferently constructed plant may frequently be made to produce satisfactory results by the proper kind of operating staff.

Good management and control are so vital to the success of a waste treatment project that the author believes when a contract to design an industrial waste treatment plant is signed, the first demand of the engineer should be that the management select the person whom they will put in charge of the completed facility. This person must be available to work with the engineers. He will assist in the collection of data relative to the character of the wastes. Factory operations which may have an important effect on the character or volume of the wastes will also be investigated. The designing engineer should explain the theories and practices of the methods of treatment and handling of materials that he proposes to use. The materials of construction of tanks and buildings as related to factory policy and practices and possible obstructions to the placement of units should also be discussed. The operator, on his part, should work with the engineer all through the period of design, so that he knows the object of each unit of the plant, the purpose of the various items of equipment, and the final location of each one.

Such cooperation will redound to the credit of both the engineer and the industry, resulting in approval of the operation by the regulatory authority. This procedure has worked out so well in the case of one large project with which the author was connected as the plant designer that after more than 11 years of plant operation, the original superintendent has reported that the plant is functioning as originally designed.

Instruction of Operators

It is to the advantage of the client that the designing engineer of an industrial waste treatment plant educate the operators of the completed plant on the general principles of the design, the schedule of operations, the control of methods, and the solution of problems of malfunction of units, etc. In the author's opinion it should be a contractual item that the engineer would be engaged to initiate the operation and to supervise it for a period of at least 1 year from the start of operation. In the best-designed and best-constructed plants emergencies will arise which will require intelligent solution. If the designing engineer is obligated to attend to these, the result will be satisfactory. If they are left to chance, no one will be responsible.

Instruction of operators does not mean that the engineer should personally instruct everyone on the operating staff of a large facility, but it does mean that he should instruct in detail the person who will actually be responsible for the successful operation of the plant. The chief operator or superintendent should then instruct the individual members of his staff on their duties and responsibilities.

The designing engineer should be responsible for the preparation of a manual of operation. This should be complete in detail, describing the purpose and function of every unit and item of equipment and apparatus in the design—the size, motor power, type of control, etc., for every mechanical unit. Starting and stopping operations must be detailed, especially in cases of malfunction of a unit. The individual to call upon in case of trouble with any unit or its breakdown should be designated. This person should be someone in a local office of the manufacturer, not in the main office, perhaps hundreds of miles away. When service is needed, it is needed fast. Precautions to be observed in the handling of chemicals should be outlined.

In one large project the operating manual prepared by the author consisted of 233 pages, divided into five sections, each section being devoted to one of the five different classes of wastes, and each of these sections referring to the instruments for that section on the main control panel. The manual was loose-leaf, so that each section could be given to the operator in charge of that particular waste, and he would not have to look through the entire book for information about the item for which he was responsible. Time being precious when an emergency arises, it should be the responsibility of the engineer to expedite operations. In the manual referred to, sketches of the flow plan of each type of waste were included, and also a reference by number to the individual drawings and operating manuals of the manufacturers of the individual units of equipment. Copies of these manufacturers' manuals were provided. In order to assure that the individual operators receive the designing engineer's manual, at least twelve copies must be prepared. One of these should be delivered directly to the chief operator of the waste treatment plant, and the others placed in the hands of a responsible party in the plant engineering section. In practice it is desirable, after the construction is completed, for the engineer to go over the operation of the plant in detail with the chief operator.

A further duty and responsibility of the engineer is to formulate a record sheet for the pertinent operating data. Much of the data on the record sheet is needed for the information of the cost department of the factory since the costs of operation are deductible items. The amount of treatment materials used enables the purchasing department to assure that sufficient quantities are on hand at all times to prevent interruption of plant operation. These records will bring to light examples of bad and wasteful operation, and result in their correction.

An interesting example is a water treatment plant for an industry in Ohio. The water from the river was treated by accepted conventional means. Unusual costs of operation having been reported, the author was called in. The plant had been properly designed to treat a certain volume of water, and the usual chemical-feeding equipment was provided. The records of operation for over 1 year indicated that during the first 6 months the operation had been properly controlled, the amounts of chemical used each day varying in ratio as the characteristics of the raw water changed.

After 6 months, it was noticed that the amount of treatment chemicals indicated on the record sheets never varied by 1 lb for the next year. Since this was an improbable situation for such a long period, it was evident that something was wrong. An investigation was ordered, and revealed that the superintendent of the plant, who had wanted equipment different from that installed, was engaged in a deliberate program of sabotage to compel the management to remove the unwanted equipment and substitute his choice. The solution to this problem was obvious—a new superintendent.

The basic information needed for the regulatory authorities is the rate of flow, or total volume of the wastes treated per day, and the effluent content of the polluting elements that are to be removed or reduced. On the other hand, the operating staff and factory management are primarily interested in the cost of operation. These several items may all be incorporated on one record sheet, but in the author's opinion they are best kept separate.

In many cases the regulatory authority will indicate the facts that are of major interest to them, and may even provide sample forms of a log. See Chapter 10 for examples.

**Table 7–2 Log Sheet for a Complex
Industrial Waste Treatment Plant**

Item	*Daily records, 31 d*

Sludge
  Wet filter cake, 1000 lb
  Dry solids, 1000 lb
  Wet sludge, 1000 lb
  Wet sludge, gal/d
  Wet sludge,
    Percent flow of total plant effluent
Chemical usage
  $FeSO_4$ ppm feed to final clarifier
  $Al_2(SO_4)_3$ feed to final clarifier
  $FeSO_4$ ppm feed to Sediflotor
  $AL_2(SO_4)_3$ ppm feed to Sediflotor
Cyanide
  CN from strong cyanide, lb/d
  CN, total lb/d, all sources
  Cl_2 lb/lb of CN
Heavy-metal constituents of wastes
  Ni, lb/d
  Cu, lb/d
  Cr, lb/d
  Hg, lb/d
  Au, lb/d
  Ag, lb/d
Oily wastes
  Oil, lb/d from alkali-oily holding tanks
  Oil, lb/d in total plant effluent
Chemicals fed
  Lime, lb/d
  $FeSO_4$, lb/d
  $Al_2(SO_4)_3$, lb/d
  Cl_2 lb/day
Waste flows, volume/d
  Alkali, acid, 1000 gal
  Alkali, oily, 1000 gal
  Metallic, 1000 gal
  Cyanide, 1000 gal
    Total of all as treated

CERTIFICATION OF OPERATORS OF INDUSTRIAL WASTE TREATMENT PLANTS

It has been the custom in practically every state in the Union and in a number of provinces of Canada to require the certification or licensing of the operators of water and sewage treatment plants. Surprisingly, there is practically no mandatory certification of operators of industrial waste treatment plants. In the author's opinion this is completely wrong and should be corrected as soon as possible. Industrial wastes contribute a very large part of the pollution of our streams and other bodies of water. It is known that many industrial wastes contain toxic, and even lethal, ingredients, which should not be discharged into the streams. It is also recognized that the total extent of pollution by industrial wastes is far in excess of that caused by the discharge of municipal sewage. In many cases, where the criterion of the extent of pollution is measured by the "population equivalent" (PE) of the sewage, the PE of the industrial waste load will far exceed that of the sewage. In one instance, a small city with a population of a little over 2000 was found to be receiving a load of pollution from the wastes of the principal industry in that town equal to 25,000 persons. Since the wastes of this industry were discharged into the municipal sewerage system, the operator of the treatment works had a major problem to deal with.

If the function of the operators of water and sewage treatment plants is considered so important that they must be certified by law, and since it is recognized that industrial wastes are more detrimental to health, it is entirely reasonable to expect that the operators of industrial waste treatment plants be put under the same control.

In a study of this matter by the author, information was furnished directly by a number of state health departments. It was found that at the time of the study (1966) only one state actually had compulsory certification of the operators of industrial waste treatment plants. Some others reported that studies were being made on the subject; others that the legislatures of their states were reluctant to act; and still others that laws on this subject had been passed but were so circumscribed with conditions that they were ineffective. There seems to be an attitude among the legislators and health authorities that industry will resent control of its privately owned industrial waste facilities, and of course it is not good politics to earn the resentment of industry. The author sincerely doubts that any responsible industry will object to an action that in the end will ensure that its expensive industrial waste treatment facilities will be worth all that was invested in them.

The operation of the industrial wastewater facility is in the final analysis the responsibility of the industrial wastewater control engineer. His duties are all-encompassing; they can be summarized as follows:

Develop a total water management philosophy

Interpret and anticipate legislation

Advise management on policy matters related to new legislation

Ensure company compliance with new and existing legal controls

Supervise permit taking and mandatory reporting to numerous control agencies

Deploy measuring instruments to provide information

Ensure effectiveness of control measures

Analyze and interpret data

Design and install effective control equipment

Maintain a competent staff

Maintain preventive maintenance program

Conduct education programs

Participate in related professional societies

Implement, manage, and enforce the total water management program

Chapter Eight

Methods of treatment of industrial wastes

In today's complex society, all wastes can no longer be treated alike. Some scheme must be applied to the various types of industrial wastes so that one can talk reasonably about the necessary methods of treatment. Table 8-1 shows one schematic breakdown of possible industrial contaminants. Wastes from living or previously living sources are called biological wastes. Such wastes are generally treated quite differently from wastes from nonliving, nonbiological sources.

Biological wastes include discharge from municipalities, tanneries, paper mills, food manufacturing plants, slaughterhouses, and textile plants. Table 8-2 lists pollutants and typical industrial sources.

Nonbiological wastes include wastes from metal-plating industries, aircraft industries, sheet metal industries, and printed circuit board manufacturers. Most of these industries discharge what is commonly referred to as "spent chemicals" into the plant effluent. It is toward "spent chemicals" that new laws are directed. A cursory look at a plant's monthly chemical consumption will give an idea of the types of chemicals being discharged.

Most of the chemicals used are substances that are soluble in water. They form aqueous solutions and consequently cannot be separated by ordinary physical means of separation (aside from evaporation). Any material dissolved in water can be viewed as a contaminant. Some substances are lethal at any concentration, others a nuisance. Any substance in considerable excess will upset the balance of nature. So moderation is the first guideline.

Typical industrial contaminants include acids such as hydrochloric, sulfuric, phosphoric, and hydrofluoric; bases such as sodium hydroxide, calcium hydroxide, and ammonium hydroxide; cyanides; chlorine; phosphates; metals such as copper, iron, magnesium, mercury, and chromium; salts such as sodium bicarbonate,

Table 8–1   Types of Wastewater Contaminants

Biological
  Suspended solids
  Dissolved solids and liquids
Nonbiological
  Suspended solids
    Colloidal
      Emulsified oils and greases
    Noncolloidal
      Floating oils
      Metallic sludge (viscous or granular)
  Dissolved solids and liquids
    Ionic
      Acids
      Bases
      Metals
      Cyanides
      Carbonates
      Chlorides
      Ammonia
      Sulfates
      Phosphates
      Sulfides
      Nitrates
      Thiocyanides
    Nonionic
      Chlorine
      Volatile
        Organics
        Gasoline
        Benzene
        Phenols
        Naphtha
        Solvents
      Nonvolatile (hydrocarbons)
        Chlorinated
        Phosphorated
        Metal-complexed organics

potassium chloride, copper sulfate, sodium carbonate; hydrocarbon solvents such as benzene and ether; and chlorinated hydrocarbons such as carbon tetrachloride.

Each industry has its own typical discharge. Metal-finishing plants have great use for strongly basic solutions, strongly acidic solutions, and solutions containing a high concentration of chromium. Metal-plating plants use acids, bases, and copper salts, and cyanide compounds in their processes. The list of typical industries and their discharges goes on. Each chemical discharged by an industry has its own characteristics, concentration, nomenclature, and harmful limits. Some typical contaminants are included in the following discussion. Various states

Table 8–2   Substances Present in Industrial Effluents

Substance	Present in wastewaters from:
Free chlorine	Laundries, paper mills, textile bleaching
Ammonia	Gas and coke manufacture, chemical manufacture
Fluorides	Scrubbing of flue gases, glass etching, atomic energy plants
Cyanides	Gas manufacture, metal plating, case hardening, metal cleaning
Sulfides	Sulfite dyeing of textiles, tanneries, gas manufacture, viscose rayon manufacture
Sulfites	Wood pulp processing, viscose film manufacture
Acids	Chemical manufacture, mines, iron and copper pickling, DDT manufacture, brewing, textiles, battery manufacture
Alkalies	Cotton and straw kiering, wool scouring, cotton mercerizing, laundries
Chromium	Metal plating, aluminum anodizing, chrome tanning
Lead	Battery manufacture, lead mines, paint manufacture
Nickel	Metal plating
Cadmium	Metal plating
Zinc	Galvanizing zinc plating, viscose rayon manufacture, rubber processing
Copper	Copper plating, copper pickling, cuprammonium rayon manufacture
Arsenic	Sheep dipping
Sugars	Dairies, breweries, preserve manufacture, glucose and beet-sugar factories, chocolate and sweets industries
Starch	Food processing, textile industries, wallpaper manufacture
Fats, oils, and grease	Wool scouring, laundries, textile industries, petroleum refineries, engineering works
Phenols	Gas and coke manufacture, synthetic resin manufacture, petroleum refineries, textile industries, tanneries, tar distilleries, chemical plants, dye manufacture
Formaldehyde	Synthetic resin manufacture, penicillin manufacture

SOURCE: By permission from A. D. Aldrich, Attracting New Industries to Florida: Water Pollution Considerations, *J. WPCF*, vol. 38, no. 10, October 1966.

publish works listing contaminants, documented experiments on the harmful effects to organisms and industry, and recommended maximum concentrations by various world health agencies.

DISSOLVED SOLIDS

The broadest category of pollutants is dissolved solids. Dissolved solids in natural waters consist mainly of carbonates, chlorides, sulfates, phosphates, and some nitrates of calcium, magnesium, sodium, and potassium. The mineral content is raised by the addition of chemical wastes, dissolved salts, acids, alkalies, gas, oil well brines, and drainage waters from irrigated lands. Industrial sources of these pollutants include metal platers, metal finishers, breweries, cooling towers, boiler bleed-off, and overflow of any tank containing mineral salts.

All substances in solution change the physical and chemical nature of the water. They also exert osmotic pressure that affects aquatic life. The USPHS sets 500 mg/l as the maximum total dissolved salt concentration since amounts in excess of this are discernible by taste. But dissolved solids in industrial waters can cause foaming in boilers and interfere with clearness, color, and taste of many finished products. High concentrations of total dissolved solids also tend to accelerate corrosion. For this reason, many industries deionize their water before using it.

Since dissolved solids are generally ionic, their presence is determined by the conductivity reading of the water. Conductivity will indicate the concentration of the dissolved solids which can then be converted into a parts per million or mg/l value.

SUSPENDED SOLIDS

Natural waters contain suspended solids as well as dissolved solids. Natural suspended solids may consist of erosion silt, organic detritus, and plankton. Industry and communities increase the natural amount of suspended solids by increased erosion, gravel washings, mine tailings, steel mill wastes, and metal plating and finishing wastes. Food- and paper-related industries also discharge waters containing suspended solids.

There is a distinction between suspended and settleable solids. But strictly speaking, until they have settled to the bottom of the water course (or laboratory sample bottle), all settleable solids are suspended solids. Only a fraction of suspended solids is settleable. This fraction depends upon quiescence, temperature, density, flocculation, and other factors. These factors are controlled in order to transform suspended solids into settleable solids before industrial wastes are discharged.

Suspended solids are measured by laboratory tests performed on the percentage of settleable solids in a particular sample of water. Limits are also placed on the particle size and turbidity caused by the presence of the suspended solids.

ACIDITY

Acidity can be described as the power of a sample of water to neutralize hydroxyl ions. Acidity is therefore determined by titrating with sodium hydroxide (0.02 N NaOH). Depending upon the indicator used, "minimal acidity" or "total acidity" will be determined.

pH is frequently thought of as the measure of all acidity; but pH and acidity are not necessarily the same thing. pH measures the hydrogen-ion concentration (pH = the negative log [base 10] of the hydrogen-ion concentration). If a solution is buffered, it may contain much acid without having that reflected in the pH value. Therefore, acidity depends upon the state of the water and substances therein.

The hydrogen ion in acidic solutions is highly reactive. The zinc outer coating of galvanized-steel pipes is dissolved in acidic solutions. Almost all metals are eventually affected by the presence of acids. Concrete decomposes in the presence of acids of very low pH.

Because it is the hydrogen ion that is reactive, most industrial discharge of acids is limited by the pH values of the discharge rather than by the total acidity of the discharge.

ALKALINITY

Alkalinity is caused by the presence of hydroxides (sodium hydroxide, potassuim hydroxide), bicarbonates, carbonates, and some borates. Total alkalinity is measured by titration with 0.02 N sulfuric acid (using phenolphthalein as an indicator).

When not buffered, the hydroxide ion will raise the pH substantially. Limits on alkalinity are set in terms of pH rather than total alkalinity.

In itself, alkaline water is not harmful. In fact a moderate amount inhibits corrosion of pipes. Excessive hardness (caused by the presence of dissolved carbonates and bicarbonates) may cause scaling, which is detrimental to industrial processes.

CYANIDES

Cyanides generally occur in metallic salts such as copper cyanide and sodium cyanide. These salts readily dissolve in water. Electroplating baths make great use of cyanide salts. Other uses include extracting gold and silver from ores and generating HCN gas for fumigating.

When dissolved in water, the salt ionizes to the metal ion and cyanide ion. The cyanide ion itself is not dangerous unless the pH is lowered to 6 or less. The ion then complexes with hydrogen to form HCN, the toxic principle of the substances. From 50 to 60 mg of HCN ingested by a human is considered lethal.

Limits on cyanide concentrations, expressed in mg/l, are based on the possibility that the wastewater containing the cyanide may become acidic.

Cyanides in water decompose, but only very slowly. Cyanides can also be oxidized to a relatively inert ion, the cyanate ion. Cyanates later undergo hydrolysis to ammonia carbonate (in aerobic water) or ammonia and formic acid (in anaerobic water).

The cyanide ion is generally found in effluents from gasworks and coke ovens, from the scrubbing of gases at steel plants, from metal-cleaning plants, and from electroplating plants.

CHLORINE

Chlorine gas is the elemental form of the element chlorine. The greenish-yellow gas dissolves in water to a maximum concentration of 7000 ppm. In water, chlorine gas hydrolyzes to HOCl and Cl^-. It is affected by pH since HOCl dissociates to OCl^- and H^+.

Combinations of chlorine and organics, or chlorine and cyanide, may be detrimental since chlorine is very reactive. Free chlorine is generally added in tertiary treatment of sewage to destroy harmful bacteria. Chlorine is also used in the treatment of cyanide wastes. Industries try to avoid overtreatment of the cyanide which results in the discharge of water containing free chlorine. Such free chlorine could destroy bacteria necessary to the proper functioning of the treatment plant.

Limits placed on the amount of chlorine in a wastewater stream are in no way related to the concentration of *chlorides* allowed in the same water stream. Chloride is the ionic form of the element. It generally forms ionic compounds, many of which are soluble in water.

COPPER

Since copper metal is insoluble, most copper enters the water system by copper salts such as copper chloride, copper sulfate, and copper nitrate. Such salts are used in the electroplating industry, engraving, and photography. Wastes from rinse tanks and filter cleanouts also find their way into the sanitary system.

Copper ions, as a pollutant, do not accumulate in the human body. In massive amounts, though, copper can cause illness or even death. The most serious effect of copper, in the sanitary sewer system and subsequent discharge into bodies of waters, is its destruction of bacteria in treatment plants. Without optimum numbers of bacteria, a treatment facility cannot properly treat sewage for which it was designed. Used by industries, contaminated water can harm production. Aluminum utensils are harmed by trace amounts of copper. Also, trace amounts of copper in metal-plating baths affect brightness and smoothness of metallic deposits.

Because copper does not decompose when discharged, and because of its detrimental effect on bacteria, it is usually removed by precipitation of an insoluble copper compound before the industrial waste is discharged.

MERCURY

Mercury occurs in elemental form and as mercuric salts. Elemental mercury is insoluble in water. It is used in scientific and electrical instruments, dentistry, power generation, solders, and the manufacture of lamps. But the most frequent source of mercury is mercuric salts.

Mercuric chloride, a very soluble salt, is used in embalming, disinfecting, preserving, printing of fabrics, tanning, electroplating, manufacturing ink, and numerous other processes. Wastes from any of these industries and also wastes from lead mining may contain mercury.

Mercuric cyanide is also very soluble in water. It has been used as a diuretic, as a topical antiseptic, and as a disinfectant. It is likely to be part of the wastes of chemical plants.

Other mercuric salts are used in the manufacture of explosive caps, as herbicides and fungicides, and to control slime in paper mills. Mercury is likely to appear in the waste discharge of these industries also.

Mercury has been determined to be more toxic than copper, hexavalent chromium, zinc, nickel, or lead to giant kelp. Mercury poisoning in humans has also been documented. Fish who survive in mercury-containing waters absorb mercury and concentrate it. Humans eating contaminated fish also concentrate the mercury, with resultant birth defects, insanity, severe intestinal problems, and possible death. Because of the serious health hazard posed by mercury, and the fact that its effects are accumulative and not reversible, stringent limitations have been placed on the concentration of mercury allowed in industrial discharges.

CHROMIUM

Chromium enters the waste stream generally in the form of hexavalent salts such as sodium chromate, potassium chromate, and sodium dichromate. The salts are used extensively in metal pickling and plating operations, in anodizing of aluminum, in leather tanning, and in several other industries in sparser amounts. Trivalent chromium salts such as sodium chromite or potassium chromite are used much less extensively. Generally they are found in the effluent of the ceramic and glass industries. Because chromium compounds are corrosion inhibitors, they are used to treat cooling-tower water. Chromic acid in very dilute quantities is used to protect metals from corrosion. Discharge of such treated water is also a major source of chromium.

Hexavalent chromium is very reactive in solution, being a strong oxidizing agent. In dilute amounts, chromium itself does not appear to harm the human body. But extensive tests on long-term effects have not been carried out. Both trivalent and hexavalent chromium are toxic to various species of fish. Therefore chromium-bearing wastes are harmful to aquatic life. Chromium is also harmful to lower forms of life, including bacteria. Most industrial waste must first pass through a sewage treatment plant. Since the bacteria in such plants are destroyed by trace amounts of chromium, the treatment plant becomes no longer effective. To prevent injury to treatment plants, chromium limits are set around 0.05 mg/l.

DISSOLVED OXYGEN

Dissolved oxygen in natural waters cannot be designated as a primary pollutant, but it does come under the category of a corollary pollutant because excessive dissolved oxygen may adversely affect the waters. Generally, it is with a deficiency or complete absence of dissolved oxygen that the governing agencies are concerned.

Inadequate dissolved oxygen in surface waters may contribute to an unfavorable environment for fish and other aquatic life. Oxygen may be removed from waters by chemical combination with pollutants or by aerobic bacterial action. Pollutants that lower the dissolved oxygen in a waste stream are limited by law. If all oxygen is removed, the growth of odoriferous products of anaerobic decomposition is promoted.

"Too much" dissolved oxygen is harmful to industrial pipelines since additional oxygen accelerates the corrosion in pipes. Both extremes, excess and deficiency of dissolved oxygen, are regulated by law. Regulation of minimum dissolved oxygen is determined by allowable BOD and COD levels.

DYES

Dyes are used predominantly in the textile industry. They are classed as direct, acid, basic, sulfur, vat, and miscellaneous. The direct, acid, and basic dye wastes are all highly colored and commonly have higher BOD than domestic sewage.

Direct dyes are readily soluble and can be used without mordant. Where they are used mordants and adjuvants may prove to be more toxic than the dye itself. Used mostly on cotton, rayon, and vegetable fibers, direct dyes are azo compounds and are frequently derivatives of benzidine and tolidine. Because of their great solubility, direct dyes are difficult to remove from industrial wastes.

Acid dyes, used on animal fibers such as wool, are salts of color acids and are also azo compounds of benzidine and tolidine. Basic dyes are salts of color bases capable of dyeing animal fibers directly but requiring a tanning mordant for vegetable fibers. Derived principally from aniline and its homologues basic dyes are mainly triphenylmethane dyes.

Sulfur dyes, so called because they contain sulfur compounds and are used in conjunction with sodium sulfide in the dye bath, are used exclusively on vegetable fibers. They are difficult to treat and their effect upon streams is frequently quite detrimental. Vat dyes are those that acquire their dyeing properties as a result of the reducing action in the dye vat just prior to dyeing. Vat dyes are reduced to a soluble form in the vat by means of a strong reducing agent such as hydrosulfite in conjunction with an alkali. They are generally used with cotton, and they may be anthracene or indigo compounds.

The amount, strength, and character of wastes from a textile or paper mill depend on the way in which the dye is applied and how often dyes are changed in the operation of each process. When dyes are changed, vats must be dumped and cleaned. At other times, the wastes consist only of the uniform flow of rinses and wash waters. Because dyes affect the BOD, are toxic to various species of aquatic life, and generally affect the appearance of water streams, the discharge of dyes is severely limited.

PHENOLS

Phenolic wastes are produced in the plastic, coke, and petroleum refining industries. They are highly soluble in water, alcohol, benzene, and other organic solvents. Phenolic wastes arise from the distillation of wood, from gasworks, coke ovens, oil refineries, chemial plants, and sheep dips, and from human and animal refuse.

Despite the fact that they are used as a bactericide in strong concentrations, weak phenol solutions are decomposed by bacterial and biological action in streams. The ingestion, by humans, of concentrated solutions of phenol will result in severe pain, renal irritation, shock, and possibly death. A total dose of 1.5 g may be fatal. It is not likely, however, that harmful concentrations of phenol will be consumed in drinking water because such concentrations of phenol are much higher than taste considerations allow.

Since phenol in contact with chlorine accentuates a disagreeable taste in water, the requirements of the various states are very rigid as to the phenol content of the final waste discharges. The limits may be 15 ppb or less.

The effects of phenols and phenolic compounds are varied. Generally phenolic compounds are toxic to fish, but they also impart a disagreeable taste to the flesh, far below toxic limits. Therefore even nondrinking water must be essentially free of phenolic compounds.

TYPES OF EQUIPMENT

Today the mails are full of advertisements for pollution abatement equipment. Equipment that was once considered for "in process" use only is now advertised for "pollution control." This should not be surprising since many industrial processes require water of drinking-water purity. The technology is available for pollution abatement. It remains to select the right equipment or combination of equipment to produce the desired effects.

There are two very general categories of pollutants. The first category can be described as dissolved solids and liquids. Such pollutants are found in the molecular or ionic state in industrial discharge waters. Removal or elimination of these contaminants can be accomplished by any one or a combination of the following treatment processes. The particular process depends upon the nature and concentration of the dissolved solid or liquid. Each unit is discussed fully in the body of the chapter.

1. Neutralization

2. Oxidation

3. Reduction

4. Reverse osmosis

5. Ion-exchange

6. Carbon adsorption

7. Recovery of acids from pickling liquor

8. Phenol destruction

9. Evaporative stripper

The second major category of pollutants may be classified as suspended solids and liquids. Such pollutants exist as particles, larger than a single molecule. Suspended solids and liquids include settleable solids, which, if allowed to quiesce, would settle to the bottom of the water bed or holding tank. The category includes also oils that would rise to form a layer on the surface of a body of water. The particle size of suspended solids and liquids can range from a few micrometers to large fractions of an inch.

The means to control the emission of suspended solids and liquids include the following treatment packages.

1. Filtration
   a. Activated carbon filters
   b. Activated carbon plus polymeric chemical filters
   c. Moving-bed sand filters
   d. Diatomaceous earth filters
   e. Fiber glass screens
   f. Microstraining
   g. Reverse flow
   h. In-depth filters

2. Electrodialysis or dialysis

3. Micro floc

4. Centrifuges

5. Concentrator

6. Vibrating screens

7. Incineration

8. Atomized suspension technique

9. Oil-water separator

NEUTRALIZATION

Since most municipal sewage treatment regulations restrict the pH of industrial wastewater, it is necessary to pretreat acid wastes. The municipalities require that pretreatment be completed prior to any discharge to the municipal sewer system.

In-plant Mixing

Frequently, plants such as metal-finishing plants that use and discharge highly acidic rinse waters also use and discharge caustic rinse waters. A metal-finishing plant may use iron phosphate to pretreat the metal before painting, but to remove excess paint from the part racks, a hot alkaline solution may be used. The overflows from both rinse tanks can be combined in a neutralizing chamber. The chamber may be the first two compartments of a four-compartment clarifier. The pH of the plant's discharge can be raised to acceptable limits without any expenditure for additional chemicals. Simply combining acid and base chemical wastes will meet the restrictions on pH limits.

Sump Neutralization

Other plants or industries may produce only acid wastes. If no spent alkaline wastes are available, the neutralization must be accomplished by the addition of purchased chemicals. Industries with low acid discharge rates can make use of a collection sump containing solid calcium carbonate (marble chips or limestone). The acid reacts chemically with marble chips to produce a neutral salt. The process is inexpensive since 1 lb of pure sulfuric acid requires approximately 2 lb of calcium carbonate. Recharging the unit requires only a new load of marble chips. But in order to be effective, the acid must remain in contact with the solid for some length of time. The length of time depends upon the concentration of the acid and the size of the chips. The chips should be no smaller than 1 in and no larger than 3 in. Small chips tend to lump together or flow out of the sump; large chips do not have sufficient surface area to facilitate rapid reactions.

A neutralizing sump such as the one described consists of a pump (or two if a pump will be running continuously), a sump sized for peak flows, and a source of calcium carbonate.

Combination Neutralizing Sump

The system shown in Figure 8-1 is a two-stage acid waste neutralization system. Its purpose is to neutralize acid-bearing wastewater by the use of a limestone sump and by automatic caustic addition and mixing.

The three sumps are made of steel shell with a ceramic liner. The first sump is a combination manhole and interceptor. The second sump contains the neutralizing medium, which can be either lump limestone or marble in 1- to 3-in size. For sulfuric acid wastes, a dolomite is preferred.

The third sump contains a pH probe, an agitator, and feed pipe from the caustic storage tank. The control valve on the caustic feed is controlled

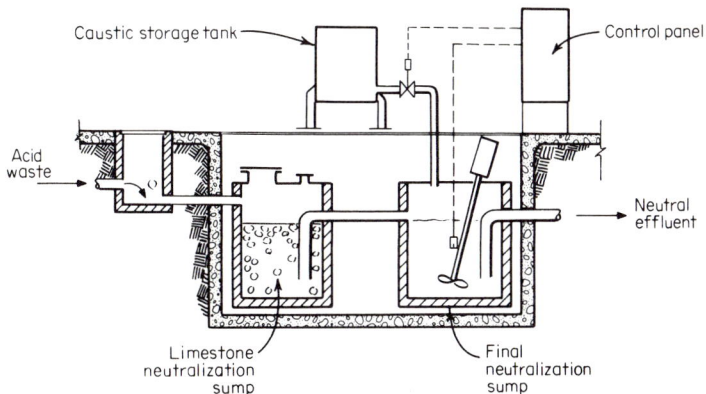

Fig. 8-1 Acid waste neutralization system. (*Maurice Knight Co.*)

by the pH probe in the solution. The controls can include a pH meter, pH alarm, start and stop button for the agitator, and indicating lights. Maintenance includes inspection, removal of foreign material collected at the top of the bed, and replenishment of caustic. In order to avoid the use of pumps, the entire installation is usually in a concrete pit.

Packaged Neutralizers

For large volumes of discharged acid or alkaline wastes there are available automatic neutralizing units. These units are preassembled and designed to continuously neutralize acid or alkaline wastewater up to 12,000 gal/h. The unit's effluent is a neutral salt solution.

In large plants the flow of acid wastes, while continuous, is not constant with respect to concentration. Addition of too little alkali could result in damaged pipes or sewage treatment facilities. Addition of too much alkali is a waste of reagents and money. For these reasons, the preassembled unit is generally fully automatic.

The unit consists of a reaction tank and a reagent tank. These can be seen in Figure 8-2. The level controls are available with the necessary pumps to provide automatic filling of reagent tanks or for automatic sump systems. The control system includes a pH meter, pH controller, and motorized logarith-

Fig. 8-2 Package water treatment plant.

mic valve. The meter reads the pH of the acid and adds exactly enough alkali to bring the waste to the required pH value. A pH recording device may be added for companies needing or wanting a record of the pH of the final effluent. Other options include alarms to indicate insufficient neutralization or additional valving for plant shutdown in case of treatment failure.

The system can be installed so as to be completely manual; partly manual, partly automatic; or completely automatic. The choice depends upon plant volume and plant economics.

ACID MINE WATERS

Ever since coal mines have been open to air, water, and iron, in the form of pyrites, acid wastes have been forming. The pollution from this combination has been present in the streams in coal mining areas for generations. It is a natural process. Attempts have been made for many years to prevent the junction of the three elements, by sealing mines to prevent the exit of the acid waters and damming up the outlets to abandoned mines, but the problem has been so widespread and the cost so enormous, that little permanent benefit has resulted. In the case of abandoned mines the question has always been, who is to pay for the attempt to stop the pollution? The owners of the mine are gone. There is left only the municipality or the state, and funds have not been available to do the job.

Now with the demands by the public and the federal authorities for the ending of pollution of all sorts, expedited efforts have been made to eliminate this form. In Pennsylvania, the state in which there has been the most pollution from this source, the problem has been intensified by the enactment in that jurisdiction of its own Clean Streams law, in 1965, which prohibits active mine operators from discharging acid in any stream in the state, whether the stream is clean or already polluted. This, the first legislation of its kind in the United States, will undoubtedly be followed in other states which have the same problem. Passing a law, however, does not provide the remedy for the problem. Someone still has to pay the heavy bill. Unquestionably, the greatest part of the burden will fall upon the owners of active mines. But the problem of the abandoned mines is still unsolved for the present.

The state of Pennsylvania, again acting on its own precept, has authorized the construction of neutralizing plants on a permanent basis at its own expense. Several manufacturers of equipment who have had experience in the neutralization of acid wastes have developed the means to meet this challenge. Dorr-Oliver, Inc., built the first installation under this plan and put it into operation in 1966. This is a unique system of neutralization with lime, but instead of the lime being mixed with the acid wastes, the stream itself is neutralized by the addition of a lime slurry. Two other similar installations will be built to do the same thing on other streams. This is the first constructive step in a program which is estimated to cost Pennsylvania about $150 million in the course of the next 20 years.

In the average case, particularly of an abandoned mine, the iron pyrites, which are common minerals found in pockets in the layers of rock adjacent to the coal seams, are exposed only as the coal is removed. When air and water then come into contact with these deposits, sulfuric acid and ferrous sulfate are formed. This acidic combination then drains into the nearest stream. A survey of this problem has estimated that about 14,000 mi of the streams in the coal mining areas of this country are polluted in this way and that a grand total of 3.5 million tons of sulfuric acid finds its way into the streams in 1 year.

In enforcing the Pennsylvania law the state has ruled that all existing mine-drainage permits issued before January 1, 1966, would expire in 1966, and a mine owner, in order to obtain a new permit for the discharge of mine water, must prove to the satisfaction of the state that his proposed mining operations will be conducted in a manner that will prevent pollution. That, of course, is a large comprehensive order. In addition, mine owners must submit plans to the state for the methods they propose for closure of played-out mines or for the restoration of the ground surface when applying for new permits.

An interesting solution of an acid mine water problem under unusual local conditions is reported in "The ORSANCO Story." In Indiana, coal deposits in a flat terrain are overlain by deposits of limestone. In operation this layer is stripped off. The pits resulting from the removal of the coal become filled with water, which normally becomes acid. It was suggested that if small dams were to be built at the outlets to these excavations, after the coal was removed, the sulfurous materials would be submerged, and thus oxidation would be prevented, and any acid that might be formed would be neutralized by the alkalinity in the water in contact with the limestone. The results have been rewarding, since not only has an acid-drainage problem been alleviated, but the small lakes created are providing highly prized recreation centers. Fish abound in the neutralized water, and there is a strong demand for sites for summer cottages on the shores.

In another use of a natural neutralizer the U.S. Bureau of Mines has demonstrated that coarse limestone, which is inexpensive, successfully removes iron and neutralizes sulfuric acid in mine waters in a sustained reaction. The Bureau admits, "it has been known for years that limestone is an effective neutralizer," but says that by continuously agitating a mixture of small pieces of limestone in mine water, the limestone is abraded into smaller particles. In the past, the Bureau states, the limestone would lose its reactivity because it became coated with water. This process allows limestone to be used for the first time in a sustained neutralization of mine waters.

In Ohio, engineers at the state university conducted studies to determine exactly what factors are responsible for the formation of acid in the drainage from mines. In other places, because of the trend to compulsory treatment, the operators and owners of mines are building treatment plants to enable them to comply with imposed regulations for discharges from active mines. This problem is unique and unlike that of other industries.

In general, when an industrial plant is shut down, the discharge of waste ceases, and pollution ends. But in an abandoned mine this is not so. As long as the air, water, and pyrites can get together, acid waste is produced, and instead of improving with time, the pollution may get worse, for there is no control of the discharge. Because of the enormous extent of this problem, an incentive has been offered to inventors and manufacturers to develop reasonable and practical means for handling these wastes. Neutralization of acid wastes is not a new process, but it has been said that, even if a practical plan were to be developed, there would not be sufficient lime available in the country with which to implement it. Treatment of the acid at its source or the prevention of its formation is a more realistic approach. This has resulted in such suggestions as backfilling abandoned mines with whole rock; diverting drainage water; chemical grouting to make the rock impermeable and to bind the sulfur by using plastic bubbles to fill abandoned mines; sealing them against air; and spraying bacteriophage viruses into mines to kill acid-forming bacteria. Whether any of these experimental ideas is practical or not, they are all very expensive.

Sealing of abandoned mines was first attempted by the federal government in 1933. In many cases sealing resulted in a 70 to 80 percent reduction in acid, but the sealing materials were allowed to disintegrate, thus nullifying the closure. In the Pennsylvania case, the Dorr-Oliver Company, operating under a grant from the state, has used a mobile demonstration plant at various sites to obtain design and economic data. The process employs lime neutralization with aeration of the mixed liquor and mechanical dewatering of the slurry. From their test work they estimate that effluents conforming to the Pennsylvania Sanitary Water Board standards can be produced at costs of from 1 cent to $1.09 per 1000 gal, depending upon local conditions. The effluents from the plants would be neutral and contain less than 2 ppm iron.

Other Methods

Another method for acid mine wastes is to feed these wastes to a bed of pulverized coal. A layer of solids mixed with the coal would be continuously removed and burned. The pulverized coal is passed countercurrently through the liquid. It is stated that the effluent is better than that from an activated sludge system of the conventional type. Coal consumed would be about 5 tons per million gallons of waste. This process has been developed by the Rand Development Corporation of Cleveland, Ohio.

The Westinghouse Corporation has developed a flash-distillation system for acid mine wastes. It is estimated that to treat 5 million gal/d, a plant would cost approximately $10 million. The estimated cost of treatment would be 33 cents per 1000 gal. The process would eliminate stream pollution by these wastes.

Another process chemically treats the wastes, aerates them, thickens the sludge, and finally centrifuges them.

The Bethlehem Mines Corporation at Marianna, Pennsylvania, is using acid mine water as a cleaning medium for raw coal. The system, put on stream in September 1965, is neutralization. The raw coal contains about 30 percent calcium and, with the magnesium carbonates present, provides more than enough alkalinity to neutralize the acidic mine water. The final waste wash water has a pH of 6.7 to 7.1 and an iron content of 0 to 1 mg/l, which is well within the limits set by the regulatory authority.

An addition to the suggested methods of treatment of acid mine waters is the Chem-Seps continuous countercurrent mixing-bed ion-exchange system. In treating these wastes a strong-acid resin is used. This system is said to be capable of removing 95 to 99 percent of iron; the resin is regenerated with calcium chloride, and the strip solution then contains iron chlorides plus excess calcium chloride. See Figures 8-3 and 8-4. This is treated with lime to get ferrous and ferric hydroxide precipitates and calcium chloride solution. Thus the chlorides are used again and again, and lime is cheap. One disadvantage of this system is the raising of the hardness of the final effluent, which may be detrimental to the stream in which the final discharge is made. The principal advantage of this scheme is that, the operation being a vertical one, the entire plant occupies a very small ground area. Figure 8-5 is an illustration of this system as used in the recovery of phosphoric acid.

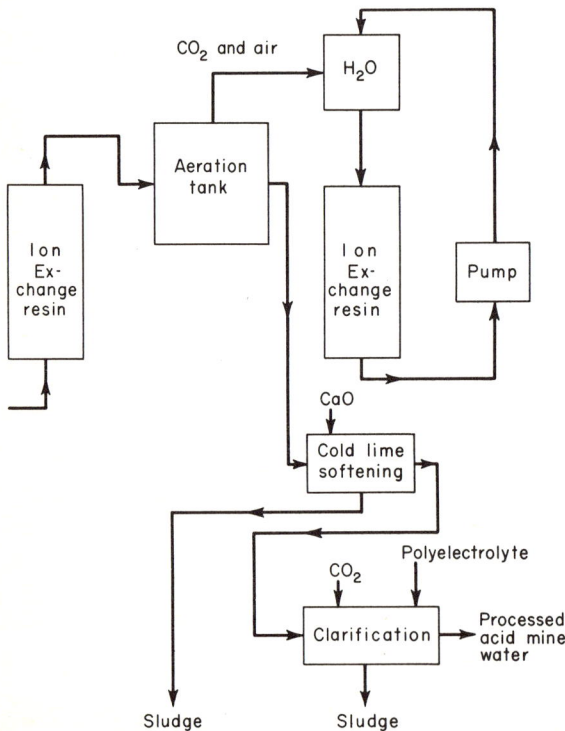

Fig. 8-3   Ion-exchange resin process for treatment of acid mine drainage waters.

Fig. 8-4 Chem-Seps continuous countercurrent exchange contactor, Production Plating Company, Adel, Ga. (*Chemical Separations Corporation.*)

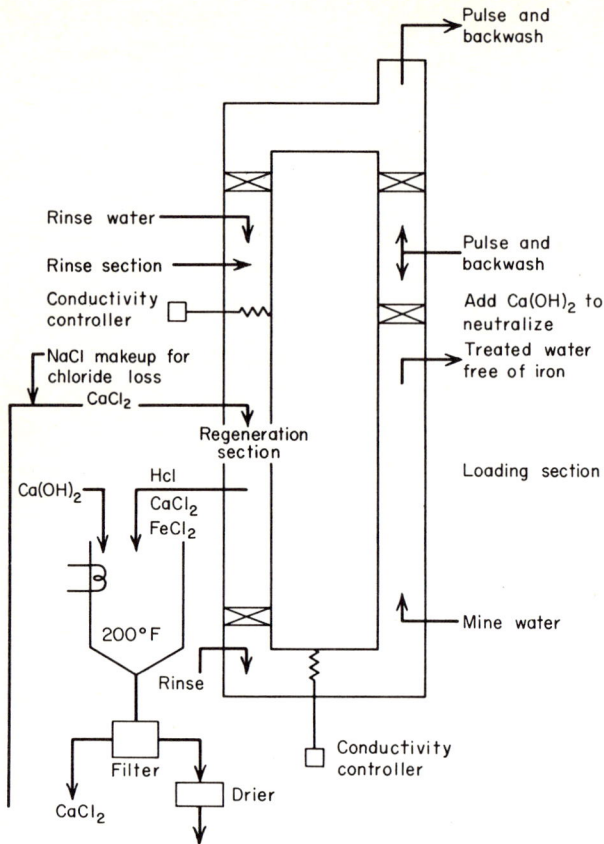

Fig. 8-5 Continuous strong-acid-type ion-exchange treatment of mine water (patents applied for). (*Chemical Separations Corporation.*)

OXIDATION UNITS FOR CYANIDE DESTRUCTION

Cyanide is one of the most strictly regulated contaminants. It is exceptionally lethal when combined with acidic wastes. For this reason metal-plating and metal-finishing companies keep cyanide-bearing wastes including copper cyanide, zinc cyanide, and cadmium cyanide separated from other plant wastes. Oftentimes these wastes are hauled away to dump sites at considerable expense to the plant.

Preassembled units are now available to industrial plants. These units will oxidize the cyanide to its relatively inactive cyanate specie. The metal ion present precipitates as an insoluble hydroxide.

The basic unit consists of four tanks, an agitator, solenoid valves, and pH and oxidation-reduction potential probes. Two of the tanks are used as reaction chambers for the oxidation and subsequent precipitation and neutralization. The other two tanks contain the reagents.

The units can handle flows up to 3100 gal/h, depending upon concentration of the contaminants and absence of complex iron and nickel salts. The cyanide concentration usually ranges from 50 to 200 ppm.

A measured amount of cyanide-bearing wastewater is pumped into tank 1. A pH meter controls the valve and allows chlorine gas and sodium hydroxide (or more commonly, sodium hypochlorite–bleach) to enter the tank and raise the pH to 10 or 11. The oxidation-reduction potentiometer (ORP) maintains the potential at +600 mV. The following reaction takes place in the first tank:

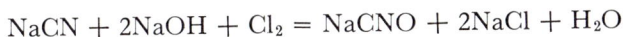

$$NaCN + 2NaOH + Cl_2 = NaCNO + 2NaCl + H_2O$$

For this first-stage reaction, the ORP controller maintains the appropriate amount of chlorine or hypochlorite necessary to complete the reaction of cyanide to cyanate. The second-stage reaction is controlled by addition of chlorine or hypochlorite in proportion to that which was required by the first-stage reaction. The equation for this reaction is:

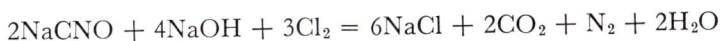

$$2NaCNO + 4NaOH + 3Cl_2 = 6NaCl + 2CO_2 + N_2 + 2H_2O$$

The second-stage pH is controlled between 7.5 and 8.5 by addition of acid, if hypochlorite is used. This reaction further oxidizes the cyanate to harmless carbon dioxide, nitrogen gas, and salt water. The metal precipitates as a hydroxide. The precipitate settles and must then be dewatered.

Oxidation of cyanide-bearing wastes is essentially a batch operation. For continuous flow, two parallel units are necessary. Essentially, the mechanics of the unit are quite simple. The pH and oxidation-reduction potential are set in advance, to optimum settings. Automatic controls regulate the time and amount of reagents necessary to maintain these values.

Various reagents may be used in the oxidation process. The most popular is sodium hypochlorite (bleach), with acid needed for neutralization. If acid is available from plant discharge, well-planned plant piping may result in great saving in acid reagents.

Other reagents used in the oxidation of cyanide include chlorine gas and sodium hydroxide. If chlorine gas is used, the concentrations are 6.82 parts available chlorine per part of cyanide. Along with the chlorine, sodium hydroxide is needed to raise the pH. 2.25 parts of NaOH are required per 1 part of cyanide. But when using free chlorine, special care must be taken because chlorine gas itself is lethal.

Hydrogen Peroxide. Hydrogen peroxide is a new reagent being used for oxidation of cyanide. Peroxide has advantages in that it is more compact than bleach. A great volume of bleach is needed to accomplish the oxidation of any sizable batch of cyanide. Hydrogen peroxide is very unstable and fairly expensive but requires less space. Reaction of concentrated peroxide and cyanide wastes results in a foaming action that requires a holding tank four-thirds the volume of the wastes to be treated.

Ozone.   The use of ozone in the treatment of water is an old process, but the application of this chemical for treatment of wastes from the metal-plating industry is fairly new. The first commercial-size plant using ozone for the destruction of cyanide was installed by the Boeing Company, manufacturers of aircraft, at their plant in Wichita, Kansas. The plant is capable of producing 120 lb of ozone per day. The dosage of ozone used in the treatment is about 20 mg/l. In this process the cyanide is first oxidized to cyanate, which is about one-thousandth as toxic as cyanide, and the cyanate is then hydrolyzed to CO_2 and NH_3.

It is generally believed that recovery of cyanide wastes is impractical, but through regulated reactions with various reagents, the toxic cyanide can be oxidized and the metal precipitated. The more efficient the collection method, the more concentrated the cyanide-bearing wastes and the more compact the treatment equipment.

CHROMIUM REDUCTION

Hexavalent chromium is found in chromic acid and chromate salt rinse waters of metal-finishing plants. The concentration, in the range of 100 to 200 ppm, gives the water an orange hue.

Up to 10 lb of chromium per hour can be treated by preassembled treatment units. These units consist of reaction tanks, reagent tanks, an agitator, pH probes, ORP probes, and automatic valves.

The chemistry of chromium treatment is based on the reduction of hexavalent chromium to trivalent chromium. The trivalent chromium will precipitate as chromium hydroxide. There are at least three reducing agents from which to choose. Sulfur dioxide, a gas, can be added to the chromium-bearing wastes. If the gas is used, 1.9 parts are required per 1 part of Cr^{6+}. Sodium bisulfite, generally used as a liquid, can also be used as a reducing agent. In a 44 percent solution, 3 parts of the reagent are used for each 1 part of Cr^{6+}: Sodium metabisulfite is added to the reaction tank in powdered form. One pound is added for each gallon of solution.

After selection of the reducing agent, the following chemical reactions take place in most chromium reduction units. The hexavalent chromium waste effluent is pumped into a treatment tank where the pH is controlled at 2.5 and the ORP is controlled at approximately -300 mV. Then the chromium waste effluent is treated to reduce hexavalent chromium to its trivalent state. The chemical equation for this reduction reaction is

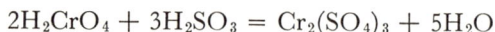

$$2H_2CrO_4 + 3H_2SO_3 = Cr_2(SO_4)_3 + 5H_2O$$

The first step is accomplished by the addition of sulfuric acid and either sulfur dioxide gas, dry sodium metabisulfite, or liquid sodium bisulfite, all of which form sulfurous acid in the presence of sulfuric acid.

In a low pH, the reduced chromium waste next goes into the second treatment stage, where sodium hydroxide is added to raise the pH to 8.0 to 8.5. The chemical equation for this stage is

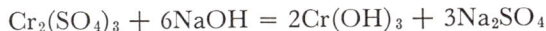

$$Cr_2(SO_4)_3 + 6NaOH = 2Cr(OH)_3 + 3Na_2SO_4$$

In this pH range, the trivalent chromium precipitates as its insoluble hydroxide.

In some small facilities, the chromium reduction process can be accomplished manually. In that case, the chromium waste effluents are pumped directly from the process tank into the treatment tank, where the pH is adjusted to 2.5 and the ORP is set approximately at −300 mV. This is accomplished by the manual addition of sulfuric acid and sodium bisulfite, all of which forms sulfurous acid in the presence of sulfuric acid. The agitator in the tank maintains a homogeneous mixture.

Upon completion of the reaction, which is indicated on the ORP/redox meter, liquid sodium hydroxide is added manually until the solution has a pH of 8.0 to 8.5. In this range, trivalent chromium precipitates as its insoluble hydroxide. The agitator is stopped to allow the precipitate to settle to the bottom of the hopper.

A minimum of 1 h should be allowed to permit the particles to settle. The drain valve at the bottom of the hopper is then opened momentarily until only clear solution flows out. The sludge removed is drained to a holding tank for disposal. The upper drain is opened to allow clarified effluent to be discharged to the sewer.

Fig. 8-6 Prefabricated chrome destruction plant, complete with mixing tank and control panel. (*Sethco.*)

The automatic units accomplish the same succession of operations without the close supervision. In large plants, the volume of chromium-bearing wastes justifies the sophisticated treatment. See Figure 8-6. In smaller facilities, manual operation is sufficient.

Reclaiming Chromium

In industries involved with chrome plating operations, the recycling of chromic acid from the rinse bath is economically advantageous. There are numerous instances of profitable recovery of chromium, even though the initial cost of the plant is high. Water can also be recycled for reuse in rinsing operations. This eliminates the problem of disposal of chromic acid wastewater, since there is no acid to discard. One prefabricated unit is claimed to have a recovery efficiency of better than 99.5 percent. The unit removes undesirable metallic impurities from the waste stream while it provides recycled chromium.

The chromium recovery is done by means of an ion-exchange system. Three short beds are used, one containing anions, and the other two cation-exchange resins. Wastewater is first pumped into a cation pretreater, where cation impurities are removed by the resin. The water is then pumped onto an anion bed. In the anion bed, chromate ions are removed by the resin and essentially pure water is recycled. The anion resin is then regenerated with sodium hydroxide, and this effluent is pumped through a cation bed to be collected as a concentrated chromic acid. The cation beds are regenerated with sulfuric acid.

It is reported that such a system returns useful chrome to the plating system at a cost of $17 per day, which includes regeneration chemicals for the resins, amortization of the plant, and resin replacement, whereas it is estimated that the cost of treatment of wastes containing chromium by normal methods would have been about $63 per day. In another case it is reported that treatment by ion exchange of chromium wastes returned 86,500 gal of water per day to the system. This coupled with the recovery of chromium resulted in a net profit of more than $10 per day.

INTERCHANGEABLE PACKAGE
TREATMENT SYSTEMS

The prefabricated cyanide oxidation, chromium reduction, and neutralization systems are designed to be interconverted easily. These units are interchangeable by merely changing piping and adding or removing valves. Semiautomatic, packaged batch wastewater treatment systems effectively treat acid, alkaline, cyanide, or chromium wastewater before disposal. A unit consists basically of an electronically equipped control panel and a combination of chemical storage tanks and treatment tank(s), all in a compact unit. Controls eliminate the overuse of chemicals and assure treated effluents. PVC-lined steel treatment tanks are corrosion-resistant. None of the units require special floors or pits.

The control panel contains a combination of pH and ORP meter which automatically follows the pH and oxidation or reduction reactions in the treatment tank. Also included in the unit are switches for the mixer, for automatic drain valve control, for the level alarm, and for pumps. The control panel may be mounted in any convenient location.

The treatment tank comes with a conical bottom to allow for a sludge collection area of 250 gal, or 32 ft³. The treatment tank is fitted with a mixer, a high-level alarm, an automatically interlocked drain valve for draining the liquid to the sewer, and a manually locked sludge removal valve. Where the flow rate of the installation dictates, two such tanks may be hooked up to one control panel together with appropriate switching. This is economical and space-saving. For cyanide destruction, vented tank covers are required.

These units may be equipped with accessories. Such equipment includes pumps for the chemical additives required to accomplish neutralization, chromium reduction and neutralization, and cyanide oxidation and neutralization. A high-quality mixer equipped with a totally enclosed chemical plant motor is also optional. High-level solid-state alarm and switching are generally standards with the preassembled packages, as are all necessary interconnecting cables for the probes.

IN-PLANT CHANGES

One of the conditions which may have a decisive influence on the cost and extent of industrial waste treatment is changes in the factory itself. This may reduce the volume and concentration of polluting elements in the final discharge, and the extent and cost of the waste treatment facilities. It may also affect the cost of production of the factory item.

As stated earlier, an attempt to change methods in a factory may meet with resistance and refusal, but it is worth trying.

The possibilities of such changes are listed in this section; they show important economies which have actually been made in certain industries. Undoubtedly, there will be many more possibilities as the value of this kind of study becomes more apparent.

One of the most fully developed in-plant changes which has enabled metal-plating plants to reduce their treatment costs, and in some cases to eliminate them entirely, is the Lancy system of integrated treatment. In this system, illustrated in Figure 8-7, the wastes from the plating tanks containing small amounts of cyanide and chromium are treated as part of the actual productive process, and the resultant waters, freed of their toxic elements, are returned to the process. This system not only eliminates the costly outside treatment plants usually required for these wastes, but also conserves the water. The system operates as follows. The depleted plating solution still containing cyanide or chromium is directed into a tank, located either directly in the plating line or immediately adjacent to it. In this tank, provided with an agitator or mixer, chlorine is added to break down the cyanide or chromium. After this process is completed, the water is returned to the system. In most cases there is no waste discharge, and the industry is freed from surveillance by a regulatory authority.

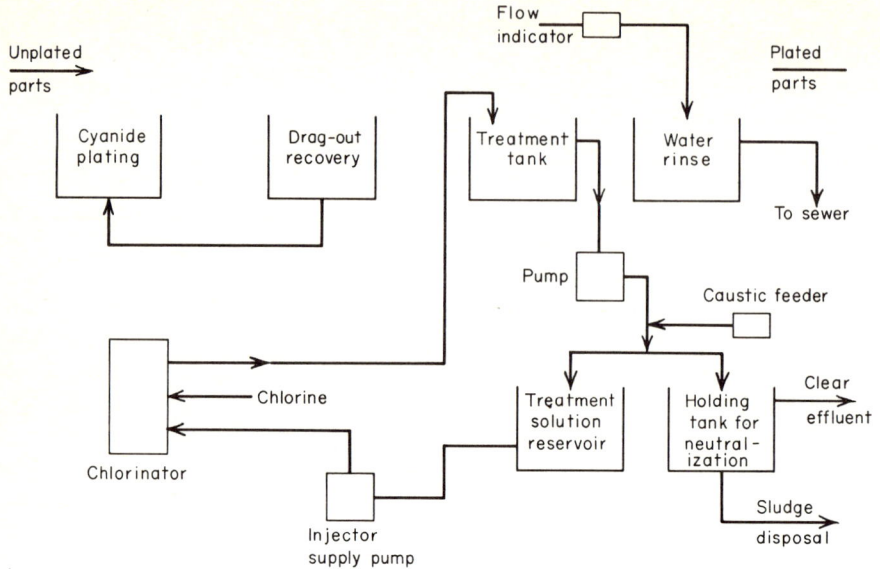

Fig. 8-7 The Lancy integrated system for treatment of cyanide-bearing waste in plating shop.

A somewhat similar arrangement, installed by the C. Howard Hunt Pen Company in their plant at Statesville, North Carolina, is described in an article in the bulletin *Multi-Matter* published by the Multi-Metal Wire Cloth Company. This is also an integrated system, suitable for small plants, and contemplates the installation of the plating wastes treatment tanks in the plating line or area, with only a final discharge of properly treated effluent going to the local sewerage system.

An interesting prediction about the reduction or elimination of pollution by industrial wastes was offered by James K. Rice, in a discussion presented at the American Society of Mechanical Engineers Petroleum Engineering Conference in 1966, in which he stated:

I'm sure that the next five years will see the advent of deposit-control agents that will allow the use of raw water in boilers, if need be. It is obvious that with such agents it will be possible to use very-low-quality waste streams as makeup for boilers and thus use the evaporation process as a means of concentrating refractory wastes in water as well as generating useful steam. A reduction in waste-water by a factor of 10 to 20 would greatly help incineration, calcining, deep-well disposal or impoundment of refractory wastes.

REVERSE OSMOSIS

Reverse osmosis is a fairly new treatment method that separates pure water from its contaminants rather than removing contaminants from the water. This membrane filter operation is sometimes called "superfiltration" or "hyperfiltration." It is distinct from the dialysis type of operation in that high pressures are applied to the wastes to be treated to force the liquid through the semipermeable membranes, leaving the dissolved solids behind. The membranes consist of very thin films of cellulose acetate. The pressures necessary for operation run as high as 1500 lb/in². Rates of flow range from 20 to 40 gal/ft² of membrane. Flows up to 2000 to 100,000 gal/d have been treated. A number of manufacturers are now offering special membranes for this system. The Desalination Systems, Inc., offers a cellulose acetate membrane, in sheet or roll form, up to 26 in in width and in rolls of 200 lineal ft.

Reverse osmosis is a phenomenon which occurs naturally whenever a dilute liquid and a concentrated liquid are separated only by a semipermeable material, such as a diaphragm, which selectively permits only one kind of molecule to pass through it. Under normal conditions pure water would diffuse through the membrane into adjacent higher-concentration liquid, but by the application of pressure to the concentrated solution, the flow goes through in the opposite direction, that is, by reverse osmosis.

This system, originally developed for the desalination of seawater, has been found applicable to certain industrial wastes, such as the removal of sulfuric or hydrochloric acid from pickling liquors, separation of high molecular weight hydrocarbons, separation of heavy metals, salts, and in general separation of dissolved ionic and nonionic materials from pure water.

It is not a cheap process, but it may well prove to be the best practical solution for the final polish of an industrial waste discharge into waters where an especially rigorous restriction exists or where an especially high-grade water is required for recycle for process work.

The influent of the system does have certain restrictions. The total dissolved solids concentration should be within 2 to 5 percent. The pH range is 4 to 8. The concentration of manganese and iron must remain below 0.3 ppm. The allowable free chlorine is less than 0.2 ppm. The temperature should be between 35 and 85°F. Since reverse osmosis units work on the principal of osmotic pressure, the suspended solids must be less than 10 μm in diameter.

If the influent criteria are met, and the total dissolved solids is low, the effluent will represent 75 percent of the influent and consist of essentially pure water. The contaminated waste has been concentrated to 25 percent of its original volume.

The reverse osmosis (RO) unit uses a semipermeable membrane and therefore needs no regeneration. In cases of hard water, scaling may be a problem. In such instances, either acid feed or water softening is used in pretreatment. Periodic sterilization and cleaning are recommended. Package systems are available for dissolving metallic oxides and other contaminants from the membranes.

Posttreatment of the purified effluent is seldom necessary. Only exceptions arise when acid feeding has formed CO_2 gas, which must be allowed to escape. The purified effluent is pure enough to recycle as fresh process water or makeup water.

Aero-Jet General Corporation made tests to demonstrate the effectiveness of reverse osmosis in the treatment of wastes from chemical and petrochemical processes. Industrial wastes that have been treated satisfactorily by reverse osmosis include:

1. Metal plating—removal of toxic metals

2. Chemical—removal or recovery of exotic chemicals

3. Pharmaceutical

4. Separation or concentration of raw or biological solids in wastes with high organic demands

5. Wastes of animal origin

6. Wastes of plant origin

7. Separation or concentration of virus particles

8. Recovery of sulfuric and hydrochloric acid from pickling liquor

To suit specific problems, special membranes may be required. Clogging of membranes is a serious problem, and membrane cost is presently high. Studies indicate that membranes for various industrial wastes can be prepared for from $10 to $20 per square foot, depending on the complexity. Power required for operation varies with use, but is usually from ¼ to ½ hp/1000 gal. Membrane life is not a definitely known factor, but certain laboratory work indicates that 6 to 12 months of useful life may be attained. Reverse osmosis may prove to be suitable for replacing conventional treatment if reliable comparisons of the true cost of each method can be made, but in the light of present knowledge and available cost data, the use of this method is probably questionable for very large volume flows of wastes.

Dorr-Oliver, Inc., has developed a unit employing the reverse-osmosis ultra-filtration membrane concept, which it is believed will replace many steps now used in biological waste treatment methods. This system is said to eliminate all suspended solids, bacteria, and viruses from an effluent. Effluent BOD ranges from 5 to 10 mg/l, which compares favorably with those from conventional secondary treatment phases. The cost of a large membrane-type plant should be less than for a conventional plant designed to produce a high-grade effluent. Tests of this unit indicate usefulness in the treatment of certain types of industrial wastes. The membranes used consist of cellulose acetate strips. Since these do not become heavily caked, high flow rates are possible. This unit is compact enough to be installed inside a building. Each cubic foot of the unit has from 50 to 150 ft^2 of membrane surface.

The Du Pont Company has developed a new concept in the reverse osmosis method of polishing wastes. They have developed a strong, durable hollow fiber of microscopic dimensions which acts as the permeable membrane. Called the Permasep permeator, it is a cylinder into which millions of the fine hollow fibers are packed lengthwise. When the waste, under pressure, is fed into the permeator, it surrounds the entire length of each of the fibers. The pressure forces much of the liquid through the permeable walls of the fibers into the hollow centers, while the dissolved solids remain in concentrated solution on the outside wall of the fibers. The purified waste is then drawn off at one end of the cylinder, while the solution containing the retained solid is washed off at the opposite end. One remarkable aspect of the development is the amount of permeation surface provided, said to be 9000 ft^2/ft^3 of fiber. The experimental cylinder was 12 in in diameter, with 7 lineal ft of active fibers contained in it. The unit resembles a tube heat exchanger and contains approximately 50,000 ft^2 of permeable surface. The normal capacity of a unit of this size is said to be 7500 gal/d when used to desalt water. However, new membranes being studied will, it is hoped, enable a similar unit to handle 75,000 gal/d. Although the unit has been developed with desalination in mind, the author believes that it may have beneficial use in handling certain industrial wastewaters.

Reverse osmosis units are still in the developmental stages but with mass production and an immediate need for less pollutant in the industrial effluent, RO can be a key to solving the problem of industrial pollution.

ION EXCHANGE

Ion-exchange units, or deionizing units as they are frequently called, are generally used to purify raw water or upgrade treated water to an acceptable level. See Figures 8-8 and 8-9. Their major function is to remove dissolved solids. In waste-

Fig. 8-8 Two-step manual demineralizer. (*The Permutit Company.*)

Fig. 8-9 Two-step automatic demineralizer. (*The Permutit Company.*)

water control, this would correspond to a lowering of the total dissolved solids value. Ions such as Na^+, Ca^{++}, and Mg^{++} as well as Cl^-, SO_4^-, and CO_3^- can be removed. In their place are substituted H^+ and OH^- ions.

The influent to the system must be free of substantial suspended solids. Silt has a tendency to foul the resins. Iron also fouls the resin. The deionizing units can take flow rates of 50 to 1800 gal/h.

There are two general types of ion-exchange units. The simplest is a mixed-bed unit. It is rather compact with a single tank that produces water of exceptionally high purity, with less than 1 ppm total ionizable solids. The more complicated unit is a two-bed deionizer. It has separate tanks for cation (positive ion) and anion removal.

A typical two-bed unit consists of four tanks. Two tanks are deionizer tanks. These can be heavy-gauge steel made with a plastic liner to guard against the corrosive effects of the acids and bases used for regeneration. The other two tanks are for storage of regeneration chemicals (one for acid and one for alkali). These tanks can be of steel with a protective coating to retard corrosion. Some units are equipped with lights that glow when the quality exceeds requirements or go out when the quality end point is reached. The units may be equipped with solubridges to give exact readings of the conductivity of the effluent.

The operation of the unit is simple. The wastewater enters the first bed, where half of the ions are removed. It then passes through the second bed, through sensing devices, and back into the plant water system for reuse as recycled water. Effluent purity ranges from 3 to 5 ppm of dissolved solids. The unit has the capacity to treat 12,000 to 60,000 gr of total dissolved solids (based on 10 gr/gal).

The tanks must be regenerated periodically. Sulfuric acid is used to regenerate the cation tank and NaOH to regenerate the anion tank.

The control panel, connected to the sensing device (a conductivity meter), indicates whether the water quality exceeds the standard. When the effluent no longer meets the predetermined standards, a button on the control panel is pushed and regenerative chemicals are automatically injected into the beds. The wastes produced during regeneration cycles are mixed, and a neutral effluent is recycled for reuse or further treatment.

Other ion-exchange units are based on the same principles. Some are not as sophisticated as others. There are manual mixed-bed and two-bed units; semi-automatic mixed-bed and two-bed ion-exchange units; and automatic mixed-bed and two-bed ion-exchange units. Operation costs for each include cation and anion resin replacement, acid for regeneration of cation resins, caustic for regeneration of anion resins, and electric power to operate pumps, valves, control panel, etc.

CARBON ADSORPTION

Activated carbon has been used for many years for the reduction of tastes and odors in the treatment of water and for the elimination of color in the treatment of certain wastes. Carbon adsorption systems are based on the principle of the charcoal filter. Surface tension on the carbon causes molecules to adhere. The greater the surface area provided, the greater the number of adhering molecules. The unit itself consists of a tank or tower of activated carbon through which the plant effluent is passed. Activated carbon particles can be manufactured from a variety of carbonaceous raw materials such as bituminous coal, walnut shells, or hardwood. The process is not new, for adsorption is the principal factor in flocculation.

A carbon adsorption system is best suited to removing dissolved organics such as saturated oils, alkanes, and alkenes, chlorinated hydrocarbons, dyes, phenols, benzene, and many high molecular weight organics. BOD can be reduced from 200 ppm to 1 ppm. COD can be reduced from 200 ppm to 20 ppm. Chlorinated hydrocarbons can be reduced from 5 ppm to 1 ppm. Depending upon the size of the column, virtually all organic impurities can be removed. The column effluent is then ready to be discharged or recycled.

Despite the rather high cost of the carbon, satisfactory results are obtainable at a cost of about 10 cents per 1000 gal of waste in large plants. This rate assumes a waste feed free of suspended solids.

There are several types of carbon adsorption columns. One system is a fixed-bed column that utilizes the down-flow method of operation. As the plant effluent enters the column, it is dispersed over the entire bed. The liquid is removed in a similar manner at the bottom of the column. When the carbon has adsorbed its capacity, the carbon is washed out of the column and either disposed of or regenerated. It may be rejuvenated by incinerating in the absence of oxygen.

Another process considered suitable as a tertiary polishing phase for biological industrial wastes consists of a series of single- or two-stage countercurrent-agitated tank-contact steps followed by polyelectrolyte flocculation and final clarification. For the recovery of the carbon after each step, the settled carbon is dewatered by centrifuging or rotary-drum vacuum filters, dried to about 75 percent solids in an indirect stream drier, and then fed to regeneration furnaces.

A third system is called a moving tower. The plant effluent enters from the bottom and is forced upwards. Periodically, a portion of the media is sluiced from the bottom and a fresh portion is added to the top. The carbon is either rejuvenated or disposed of.

A new process developed by the Worthington Corporation consists of using carbon as the medium for pasteurizing. Two sterilizer stages are also used. The temperature is controlled for given time periods. The new process is called Pasterilizer.

It is stated that intermittent operation of carbon columns reduces soluble organic materials in clarified biooxidized waste effluents to levels below 1 mg/l. Carbon columns do not remove some highly polarized organic molecules. Organic matter adsorbed on a carbon column can go septic, causing a breakthrough of turbidity and organic matter.

Activated carbon is said to be able to adsorb up to 30 percent of its weight in mixed organics in staged countercurrent fixed-bed contactors. Loadings of 4 to 10 gal/ft^2 (min) with a contact time of 20 to 40 min are said to have resulted in 70 to 80 percent removal of gross organic matter from secondary effluents. This would possibly increase the total organic removal in conventional-type treatment to 97 to 98 percent.

Fiber Industries, Inc., of North Carolina, has conducted laboratory investigations using carbon filters as tertiary treatment for polyester fiber wastes. Reported results indicate a BOD reduction of up to 95 percent, with a final effluent of 25 mg/l. Final color is reported as less than 10, odor below threshold, taste negligible, and total solids 500 mg/l.

An interesting idea for reducing pollution has been developed by researchers at the Max Planck Institute in Munich, Germany. This idea is that stems of common reeds, which absorb phenols and decompose them into plant food and harmless gases, will clear coliform bacteria from polluted river water in as short a time as 17 d. It is not clear whether the reeds are to be used as a filter material or the water is to be run through beds of standing reeds. In this connection, it is reported that the city of Krefeld, Germany, is filtering sewage through a screen of vegetable reeds. The leaves of these plants have 23,000 pores/cm^2.

The principle of carbon adsorption can be applied to dissolved pollutants in any waste stream. It is particularly suited for organics and biological pollutants.

RECOVERY OF ACIDS FROM PICKLING LIQUOR

For many years the pickling or scale-removing process employed in steel and wire mills for the final cleaning of the steel has used sulfuric acid for dissolving the iron scale. The resultant problem of reducing or eliminating stream pollution by the use of lime or other alkaline substances to neutralize the acid has been costly to these industries. There have been a number of developments in the past few years, on the theory that there were potential values in these wastes in the sulfuric acid and the ferrous iron and that recovery of these substances would not only reduce or entirely eliminate the stream pollution, but would return a profit to the industry in the sale or reuse of the recovered products. This resulted in patented methods being developed in Europe, involving concentration under vacuum, electrolytic regeneration, and fluid-bed regeneration.

One of the first of these European methods to be given serious consideration in the United States was the Ruthner process, represented in this country by the Blaw-Knox Company of Pittsburgh, Pennsylvania. There being much skepticism in the United States over the economics of this system, since it would apply to American steel mill practice, a group of seven United States steel companies cooperated in the installation of a full-scale pilot-plant operation at Niles, Ohio. The plant was put under construction in 1957, and is said to have cost about $500,000. With respect to plant operation, the process is successful. Sulfuric acid and ferrous sulfate can be recovered by the process.

In one instance in which the author was concerned, the initial and operating costs of such a plant were entirely beyond the financial resources of the company. It was suggested that plants could be built at strategic points in steel mill districts, and the pickling liquors hauled to them for processing. In this case the volume of pickling liquor produced was about 12,000 gal/d. At a quoted cost of 2 cents per gallon for hauling the wastes from the mill to a central processing plant, the daily charge would have been $240, to which would be added the costs of processing and the hauling of the acid and iron back to the mill or some other point. It did not appear to be an attractive solution.

Since the advent of the Austrian Ruthner process, a considerable number of other systems have been proposed for this same purpose. The ultimate economics of any of these systems remains to be conclusively proved, but since they all offer a possible solution of the pollution of streams by pickling liquors, regardless of their economy as recovery methods, a brief description of some of these methods is warranted. Also, because of the doubtful market for the recovered sulfuric acid, there has been a definite move on the part of the steel mills to abandon the use of sulfuric acid and change to hydrochloric acid. Also, because of the pollution problem and the cost of treatment or recovery, there is a trend toward "dry" descaling of steel.

Ruthner Process (Blaw-Knox Company)

The waste pickle liquor is first concentrated in an evaporator. The concentrate is then transported to a reactor, where it is combined with hydrochloric acid gas. In this reactor ferrous chloride and sulfuric acid are formed. The sulfuric acid is then separated from the ferrous chloride by centrifuging. The ferrous chloride then goes to a roaster, when ferric oxide is formed. The gases liberated from the roaster, along with the liquid acid from the centrifuge, go to a degassing chamber. The sulfuric acid is removed as a liquid from the degassing process, and is returned for reuse in the pickling process in the plant from which the pickling liquor came, or may be sold. The remaining gases from the degasser are passed through an absorption and stripping system, after which they are again reused in the reaction chamber.

Zahn Process

This process of reclamation of sulfuric acid is handled in the United States by Koppers-Inland Steel Company. It is said to have been used successfully and commercially in Germany to recover up to 50 percent of all available free sulfuric acid in the pickling liquor. It is reportedly applicable to plants producing 10,000 gal or more of waste pickle liquor per day. It operates as follows: Spent pickle liquor is pumped to a spray head in an evaporating chamber. Here hot air and flue gases from a combustion chamber concentrate the liquor and cause the ferrous sulfate monohydrate to crystallize out of solution. Vapor-laden air is discharged to atmosphere through a mist eliminator and a stack. The slurry is dropped into a crystallizing tank, where fresh sulfuric acid is added from a metering tank. This causes more monohydrate to drop out. The slurry is then separated in a vacuum filter and washed. Salt is conveyed to bins or hopper cars for sale or disposal. Mother liquor, containing about 35 percent acid and 1 to 2 percent iron, is pumped to a holding tank, ready for dilution and return to the pickling tanks. No reheating is required. It is reported that one plant in Germany has operated since 1954, processing 48,000 gal/d. Labor costs are said to be low, only one man being required to operate the plant.

Lurgi Process

This process for the recovery of hydrochloric acid from spent pickling liquors was also developed in Germany. The process regenerates the acid in a fluid bed. In a pickling plant using hydrochloric acid, the acid solution circulates between a pickling tank and a storage tank. As the solution pickles the steel, the acid reacts with iron oxide scale from the steel, producing ferric chloride. The concentration of acid continuously falls off, while the dissolved iron increases.

The purpose of the Lurgi system of regeneration is to restore the balance so that the acid level in the pickling liquor stays constant (usually around 10 percent). The theory of this system is to treat a continuous-bleed stream from the circuit to remove iron from the liquor system at the same rate as it is being pickled in the vats. The bleed stream, or spent pickle, is fed to a preevaporator and heated with gases from the regeneration reactor, which is next in the flow plan. Concentrated liquor from the preevaporator then enters the lower part of the reactor, at an acid concentration of about 13 percent and a ferrous chloride level of 20 percent. This reactor contains a fluidized bed of sand and is fired by oil or gas to maintain an operating temperature around 1475°F. In this bed additional hydrogen chloride is formed. Reaction products leave the top of the vessel, together with unreacted components of the solution plus the combustion products from the fuel. A cyclone removes the ferric oxide from the stream; then the hot gases enter the preevaporator, as stated above. The overhead from the evaporator, leaving at about 250°F, contains not only the water vapor plus the hydrogen chloride and combustion products from the fluid-bed reactor, but also some hydrogen chloride that vaporizes directly from the entering plant liquor. To place the regenerated acid back to the pickle liquor circuit, the gas mixture from the preevaporator enters the bottom of an adiabatic absorption tower, where hydrogen chloride is absorbed by another bleed stream of the pickle liquor. The resulting liquid stream that is returned to the pickling plant contains 12 percent acid and about 70 g/l of iron. Meanwhile, unabsorbed gases go to a condenser, and the gases that do not condense are vented to the atmosphere.

Dravo Process

This process, also developed in Europe, is similar in most respects to the Lurgi process, except that the Dravo process uses a spray roaster instead of the fluid-bed regeneration step, and does not use a preevaporator. Although the Dravo system has been put into use in plants for Republic Steel Company and for McLouth Steel Company, the use of this and the Lurgi process proceeded slowly because of the high initial cost of the recovery units. It was estimated by experts in Pittsburgh that a typical generation plant would cost about $2 million, exclusive of acid storage tanks and the piping system. An installation of the Dravo system in Canada cost that amount for a plant to regenerate 30 gal/min of 13 percent hydrochloric acid. The proponents of the Lurgi system claim considerably lower costs; that is, a 26-gal/min plant would cost only $300,000 on shipboard at a German harbor. To this would be added 35 percent for assembly at the site, plus transportation charges and customs duties. This estimate is based on European labor rates, which of course would be considerably lower than such rates in the United States or Canada.

Haveg Process

This process, developed by the Hercules Powder Company, is a regeneration system for hydrochloric acid. This system, known as the Turbulator process, has the advantage of being very compact and capable of being installed completely inside the plant in an area of about 35 to 75 ft for a typical installation. The heart of the system is the Turbulator. See Figure 8-10. Short start-up and shutdown times permit batch and continuous pickling operation. Completed installations indicate substantial cost savings in initial construction, in purchases of hydrochloric acid, and in maintenance. The system has a decided advantage

Fig. 8-10   Turbulator unit. (*The Haveg Company.*)

over others in that it entirely eliminates stream pollution from pickling liquors and also the costly neutralization processes. It is said that ten installations of this method of regeneration are available, ranging in capacity from 1 to 44 gal/min of pickling solution. In this process the ferric chloride reacts with water and oxygen to form ferric oxide and hydrogen chloride. The recovered hydrogen chloride is returned to the pickling bath in 20 percent concentration. The system is closed with a loss of less than 1 percent of the hydrochloric liquor in the absorber. It is a development of Dr. C. Otto & Company of Germany, and the first installation went on stream in January 1966. It is stated that about seven other plants are in operation. The operational data given for a 17.6-gal/m plant are as follows:

Operating Cost per Hour

Fuel oil	2.1 gal/min at $0.12 per gal	= $15.12
Power	120 kWh at $0.015 per kWh	= 1.80
Absorbent water	18 gal/min at $0.20 per 1,000 gal	= 0.22
Other water	360 gal/min at $0.05 per 1,000 gal	= 1.08
Total		$18.22

This total represents a cost of $0.017 per gallon of pickling liquor processed, which, of course, will vary according to the locality. Figure 8-10 illustrates an actual installation. The estimated cost of an installation to regenerate 1060 gal/h was $415,000 in October 1966. The system is handled in the United States by the Haveg Corporation.

Japanese System

This system is for the neutralization of acid with ammonia gas for the recovery of marketable ammonium sulfate and magnetite iron oxide. The ammonia compound is used in fertilizer, and the iron is widely used in the manufacture of tape for tape recorders and magnetic cores. It is reported that a plant handling 1.5 million gal of this waste per day costs $500,000. This plant is said to consume 630 lb of ammonia per ton of ammonium sulfate plus 95 kWh of electric current, 3300 lb of steam, about 6.5 lb of diatomaceous earth, and 19,000 gal of cooling water. Four workers per shift are required. This plant is said to have replaced a former lime-neutralization system which cost $350,000 per year to operate. See Cost Index Chart in Chapter 11 for comparative current prices.

Singmaster & Breyer Process

This is a double-effect evaporator-crystallizer-system operation for the recovery of sulfuric acid. Low-pressure steam is used to concentrate acid to a point just below saturation. Slurry for a second-stage evaporator is centrifuged. The centrifuge cake, with 15 to 20 percent moisture, is fed to a roaster. A plant handling 100,000 gal/d is estimated to cost $2.10 million when treating waste pickle liquor with a concentration of 8.5 percent sulfuric acid and 13 percent ferrous sulfate. The regeneration cost is estimated to be about $15 per ton of acid. The estimated return of 10 percent on capital investment is claimed as profit on the operation.

PHENOLS AND PHENOLIC
COMPOUND DESTRUCTION

Phenols when combined with chlorides in the water produce a very disagreeable taste. An example of the nuisance caused by chlorine contact with phenol occurred in a large city in Ohio several years ago. A large iron and steel mill discharged its treated wastes into a large river. The state health officials had established a maximum of 75 ppb of phenol in the final effluent from the mill into the river, and this was being met by the mill. However, the river discharged into a large lake, the water of which was used as the supply for the same city in which the mill was located. The current from the river flowed close to the shoreline, past the intake of the water treatment plant of the city. Therefore, water from this plant was given a final protective dose of chlorine. This action resulted in many complaints about the taste of the water. The residual phenol, while below state limits, was above the 1 ppm concentration known to cause a disagreeable taste in the water when chlorine is present.

Ozone

Ozone is used effectively to reduce the concentration of phenols in the wastewater. Plastics industries, oil refineries, chemical companies, and coke plants are particularly concerned with the effluent phenol concentration. A refinery of the British-American Oil Company, at Bronte, Ontario, Canada, used 50 mg of ozone per liter of effluent. The waste flow was 300 gal/min, necessitating an ozone capacity of 190 lb/d. The phenol level in the discharge was less than 3 ppb.

The economy in the use of ozone lies in the operation of the ozone generation plant for the full 24-h day. If 120 lb of ozone is required and the waste is treated only during an 8-h operating period, the ozone generator must have a total daily capacity of 360 lb/d. This is of course much more expensive to install and operate than a unit producing 120 lb/24 h. Moreover, treatment over 24 h would require the use of holding tanks. But unless the flow is very large, the overall cost of the plant, including the holding tanks and pumps, would be less than the plant for an 8-h operation.

It is reported that the Soviet Union has used ozone to treat refinery and coke plant wastes. According to the report from the *Czech Technical Digest,* mercaptans and other sulfur-containing compounds virtually disappear after 3 min exposure to ozone. Phenols are also efficiently removed. One effluent had 30 to 50 mg/l of dissolved organics. After 15 min exposure to ozone, the final stream contained only 10 mg/l of organics. Ozone consumption was 1.5 to 2.7 mg/mg of contaminant. If oxygen is used instead of air in the production of ozone, the same amount of power will produce twice as much ozone. Air and oxygen must be dried to $-60°C$ and supplied to an ozonator at not over 70°F. Breathing-grade oxygen (99.5 percent) is recommended. Dosage rates that have been successful in treating various industrial waste ingredients are reported as phenol, 1.5 to 2.5 parts ozone to 1 part phenol; cyanide, 1.5 parts ozone to 1 part cyanide. In treating phenol in oil refinery wastes, ozone is used as a tertiary treatment to produce final effluents with not more than 3 ppb phenol. Reports from Philadelphia, Pennsylvania, where ozone has been used in the treatment of water for many years, indicate that the electric power requirements to produce 1 lb of ozone are 11.6 kWh, which also includes the current for lighting, heating, and the recording equipment.

The Welsbach Company of Philadelphia, pioneers in providing equipment for ozone production, also furnishes completely self-contained generating sets for smaller installations which will produce up to 78.6 lb of ozone per day. An entire unit is mounted on skids and is ready for connection to the utility supplies and treatment units. The largest of these units occupies a space of only 7 by 12 ft.

EVAPORATIVE STRIPPER

In some operations, the waste contaminants may be organic solvents. In order to discharge plant effluent, this contaminant must be removed. Since organic solvents have high vapor pressures and low boiling points, they lend themselves to evaporation at slightly elevated temperatures and depressed pressures. Such units are called evaporative strippers. They allow for the removal of the contaminants and their collection and possible reuse.

SEPTIC TANKS

The simplest of all sewage treatment plants is the standard septic tank. Such a unit is usually made of precast concrete and is delivered to the site as a complete tank and lid. In constructing these tanks, high-strength concrete is molded in shop-fabricated rigid steel forms which provide close tolerances on all dimensions. A controlled curing process assures attainment of maximum concrete strength. When properly installed and backfilled, the units will remain in sound structural condition indefinitely.

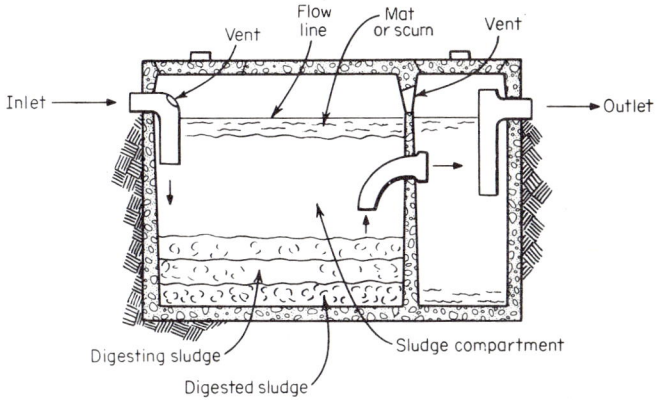

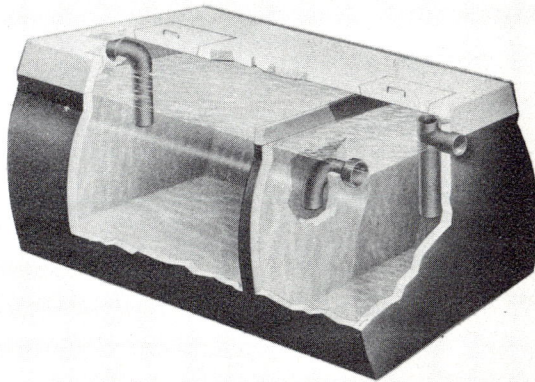

Fig. 8-11   Cross section of typical septic tank. (*M. C. Nottingham Company of California.*)

Septic tanks are designed to provide conditions in performing the following functions: (1) separation of floating and settling materials, (2) storage and digestion of separated matter, and (3) biological treatment of suspended particles, changing their character so that the discharged water with some solids will filter through the soil without clogging the soil pores.

Figure 8-11 shows the cross section of a typical septic tank. The two compartments are proportioned to carry the necessary sewage. The first or sludge chamber of the unit collects a mat of solids to promote rapid decomposition of the sewage. This mat or scum can be seen trapped on the surface of the liquid between the inlet elbow and the separation wall. The nonsettleable sewage materials contact the bioseptic environment in passing through the tank. At the bottom of this first compartment lie both the digesting and digested residues. The second compartment of the septic tank is where the bacterial treatment of the suspended particles occurs. The digested residue should be pumped out about once every 4 or 5 years. On certain facilities these tanks should be checked regularly and serviced more often. When too much material is allowed to accumulate in the tank, the flow velocity increases and will scour solids into the leach system, causing failure because of clogged soil pores.

Prefabricated septic tanks are available in sizes from 385 to 1500 gal. These units can also be used in batteries so that much larger capacities are achieved.

Leaching Systems

The effluent from a properly designed and properly operated septic tank may be disposed of in a very sanitary manner by leaching the effluent into the ground. The problem is first a hydraulic one of distributing the total volume of effluent over enough ground area so that it will be completely percolated away; otherwise the effluent will pond and most likely create a health hazard as well as a nuisance.

Septic tank effluent is not free of sewage materials and it is not free of intestinal bacteria which may include pathogenic bacteria (which cause disease) contained in the original sewage. Some of the sewage material and some of the bacteria are removed by the tank action, but enough remain to constitute a health hazard. A septic tank does not completely treat sewage; rather it should condition the sewage so that it may be percolated into the ground without clogging. The final treatment takes place in the ground.

Percolation into the ground is usually attained by one of two accepted methods. These are leaching trenches and leaching pits. A leaching trench consists of a long trench approximately 2 to 4 ft wide. Some may be wider. The length and depth of the trench depend upon the amount of surface area required for percolation. The trench is filled with crushed rock of ¾- to 1½-in size which is covered with a layer of soil. Within the trench is laid a clay pipe with open joints to carry the effluent to the full length of the trench and to distribute the flow equally.

The other method is the leaching pit. See Figure 8-12. Leaching or seepage pits are often used rather than horizontal trenches. They are useful when the

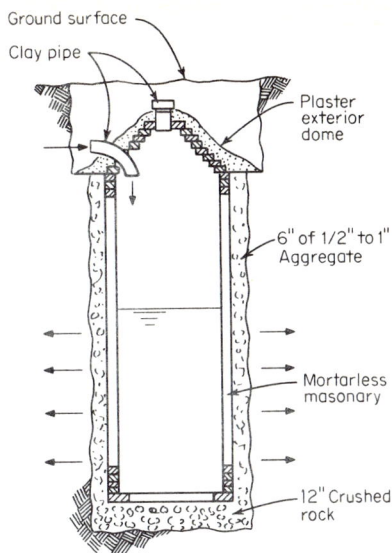

Fig. 8-12 Cross section of typical leaching pit.

Table 8–3 Soil Values

Type of soil	Required leaching area, ft²/100 gal	Max. absorp. capacity, gal/ft² of leaching area
Coarse sand or gravel	20	5
Fine sand	25	4
Sandy loam or sandy clay	40	2.5
Clay with considerable sand or gravel	60	1.66
Clay with small amount of sand or gravel	90	1.11

ground area available is limited. They are especially applicable where it is necessary to dig to a considerable depth to reach a suitable porous stratum. The vertical walls of the leaching pit are usually lined with masonry without mortar or precast concrete liners. The walls should be surrounded by 6 in of graded gravel to stabilize the structure. The pit bottom should be covered with 12 in of crushed rock. The factors influencing the design of the leaching pit are the same as for the horizontal trenches. The total area of surface of soil in contact with the liquid of the effluent is determined by its rate of percolation.

Table 8-3 shows the recommended soil values for leaching of effluent for various types of soils.

Factory-built Sewage Treatment Plants

There are many small and medium-size factory-built package sewage treatment systems available in the market. These units range in capacity from 300 gal/d to large units with capacities greater than 60,000 gal/d. The following are typical of some of the models that are being used by small industries throughout the country.

Nottingham Hygi-Aeration System

The Nottingham Hygi-Aeration sewage treatment system uses the extended aeration modification of the activated sludge process. These package treatment plants are fabricated of precast reinforced concrete, designed to withstand all normal loads imposed by the earth and the liquid pressure. High-strength, high-density concrete is used in these structures, which are reinforced with steel and well cured before installation.

Each model is rated in accordance with its daily capacity in gallons. There are three basic models which can be grouped into larger plants. These models, capacities, and BOD loading are shown in Table 8-4.

Based on an average of ten weekly determinations composited from daily grab samples, the performance data of the M. C. Nottingham Hygi-Aeration treatment system were found to be as in Table 8-5.

Diagrams of the various package plants are shown in Figures 8-13 to 8-16. The smallest unit is the HA-15-T, which consists of a small anaerobic chamber into which the sewage first enters through a pipe elbow. This chamber is approxi-

Table 8–4

Model	Plant capacity, gal	BOD loading
HA-15-T	1,500	3.2
HA-15-TR	2,500	6.2
HA-25	2,500	7
	5,000	14
	7,500	20
HA-25	10,000	27
	12,500	34
	15,000	40

Table 8–5

5-d	National Academy of Sciences criteria	Septic tank effluent	Effluent	Hygi-Aeration Influent	% removal
BOD	50 mg/l (max)	120–140 mg/l	275 mg/l	12.8 mg/l	95
COD	No standards	...	481 mg/l	61 mg/l	87
SS	150 mg/l (max)	150–350 mg/l	460 mg/l	44 mg/l	90
DO	No standards	None	None	4–6 mg/l	

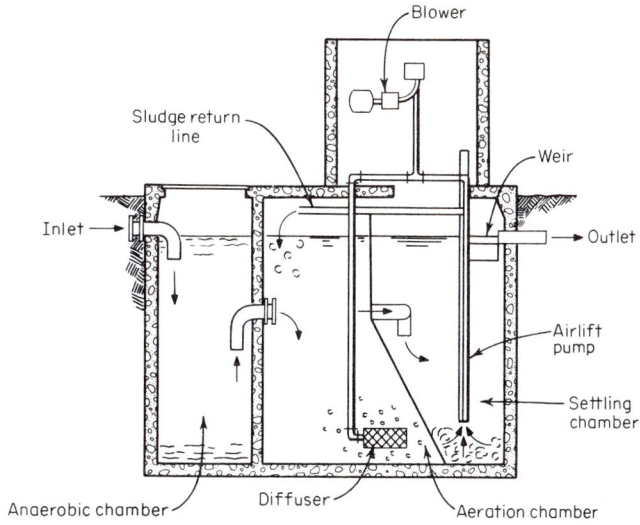

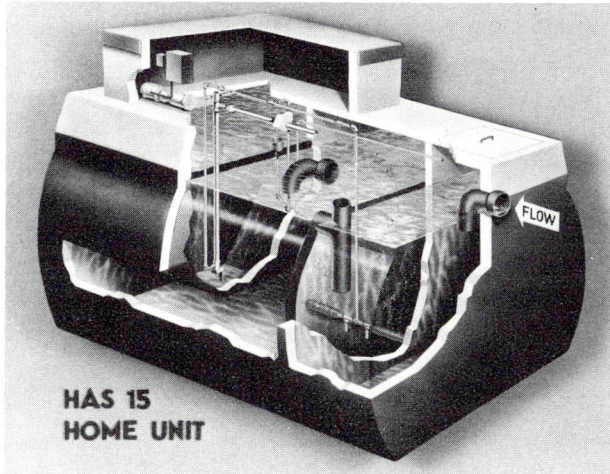

Fig. 8-13 Section of 1500-gal aeration plant. (*M. C. Nottingham Company of California.*)

mately 5 ft deep and has a manhole for access at the top. The flow out of this chamber is near mid-height of the water depth, thus preventing the floating material and the settled material from passing to the next stage, which is the aeration chamber. In this compartment is located a steel settling chamber. The settling chamber contains a vertical air lift tube and overflow pipes discharging through a weir opening. Air is pumped into the bottom of the aeration chamber through a diffuser. The same air pump operates the air lift, which returns the settled activated sludge to the aeration chamber. This returned material accelerates the bacterial activity in the aeration tank.

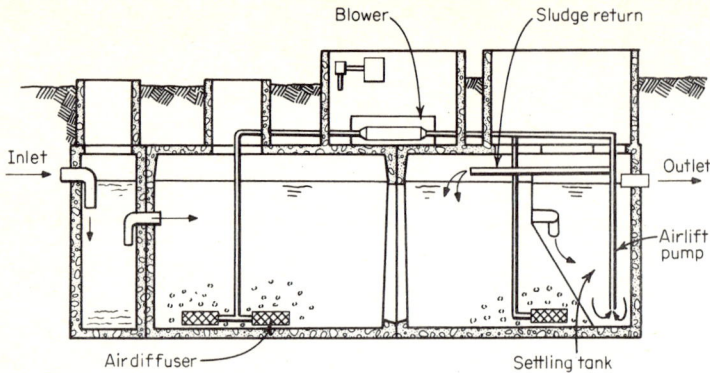

Fig. 8-14   Section of 2500-gal aeration plant. (*M. C. Nottingham Company of California.*)

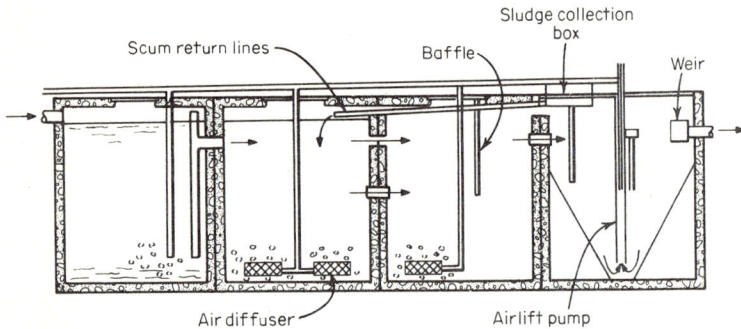

Fig. 8-15   Section of 5000-gal aeration plant. (*M. C. Nottingham Company of California.*)

Basically, the extended aeration system as used in these package plants consists of:

1. Primary screening, sedimentation, or comminution which is optional

2. 24-h or longer aeration period

3. Clarification by sedimentation

4. Disinfection by chlorination of the clear, odorless effluent where required

The design criterion of the plant is based on a loading of 20 lb of BOD per 1000 ft³ of sewage influent. The capacity of the unit is based on 24-hr sewage flow.

The aeration system is based on 2500 ft³ of air per pound of BOD. Diffusers are placed in the aeration chamber to provide complete mixing of the bank contents. The design of the clarifier is based on a surface overflow rate of less than 300 gal/ft² (d). There is a 4-h minimum detention time. A positive sludge return pump returns 100 percent of the daily flow. There is an orificed outlet

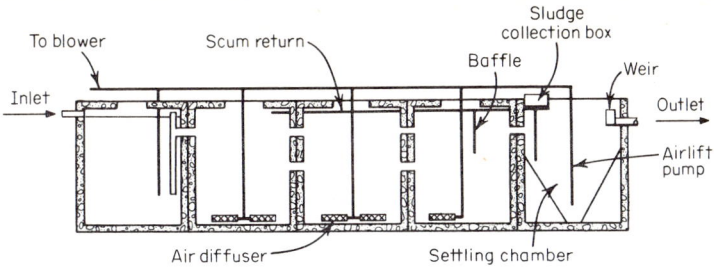

Fig. 8-16 Section of 7500-gal aeration plant. (*M. C. Nottingham Company of California.*)

weir to provide for peak flow control. For pretreatment there can be either a trash tank or a comminutor.

Domestic sewage quality is usually classified by four indexes. These are

1. 5-d BOD

2. COD

3. Suspended solids

4. Dissolved oxygen

The HA-15-TR 2500-gal aeration unit is similar in operation to the above unit except that the aeration chamber is considerably larger and there are additional air diffusers in the chamber. The HA-25 5000-gal unit is also different in that it has two separated aeration chambers, each with its own air diffusers. Each chamber is also fed by the return scum line which is air-lifted from the settling chamber. Flow from the first aeration chamber goes to the second chamber. From the second the flow is to the settling chamber, from which the sludge is returned to both aeration chambers and the overflow is discharged from the unit.

The HA-25, 10,000-gal unit and the HA-25 12,500-gal unit are both similar to the HA-25, 2500-gal except that there are four and five aeration chambers respectively.

All these systems are based on settling of the solids in the first chamber, mixing with activated sludge, aeration to induce bacterial action, and settlement of ash in the hopper bottom settling tank. The overflow leaves the unit in a 95 percent digested state. Cleaning the tanks of ash needs to be done only once in 2 or 3 years.

Tex-A-Robic

Another prefabricated sewage treatment plant is one constructed by Can-Tex Industries, a division of Harsco Corp. This unit, known by the name Tex-A-Robic, is a waste treatment plant with a mechanical clarifier. See Figure 8-17. The process that it uses is the contact stabilization treatment, which is a modified

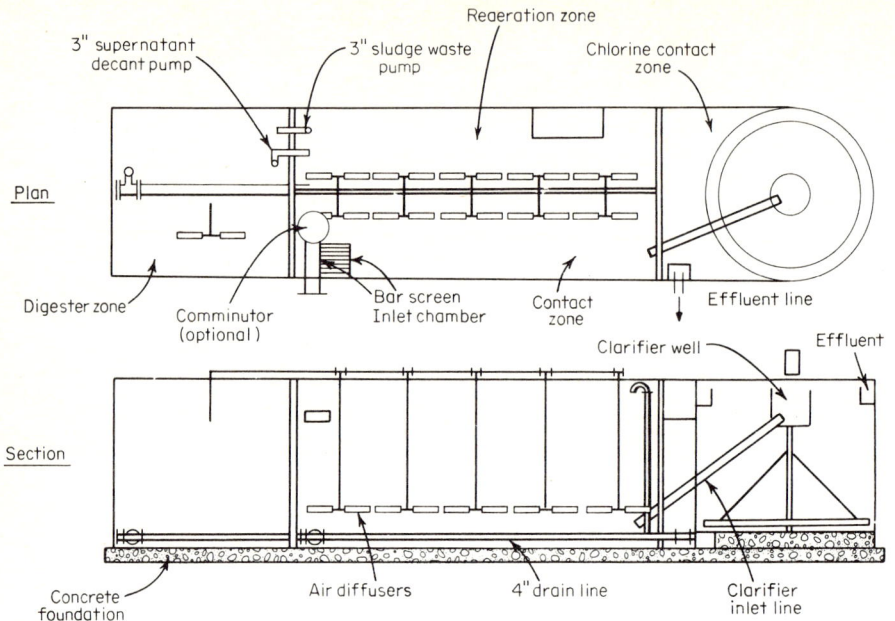

Fig. 8-17 Plan and section of contact stabilization treatment plant with clarifier and chlorinator. (*Cal-Tex Industries, a division of Harsco Corporation.*)

activated sludge process that separates the aeration volume into two zones. Sewage flows into the treatment plant through the influent piping, entering an outlet chamber containing a manually cleaned bar screen or, optionally, a comminutor to cut up or shred waste solids. From the inlet chamber the sewage flows to the "contact zone," where it is quickly and thoroughly mixed with the well-conditioned activated sludge from the "reaeration zone." The mixture then flows through a pipe to a loading well in the center of the clarifier or settling basin where the inlet velocity energy is dissipated and the flow is evenly dispersed. The sludge or solid material settles to the clarifier bottom, where it is collected and pumped back to the reaeration zone for recycling. In this zone the sludge is aerated for approximtely 7 h to provide a well-conditioned or "healthy" activated sludge to mix with the raw incoming waste in the contact zone.

The liquid in the clarifier, free of settleable solids, overflows the weir and passes through a chlorine solution to disinfect the effluent. This effluent can now be emptied into an intermittent stream or a dry water course without creating a nuisance.

There are many variations of sewage package treatment plants. The material of the shell of the unit may be steel plate or concrete. In the factory-built models concrete is precast and delivered in components. Some models include pretreatment such as screens or mechanical cutting mechanisms. The aeration chamber may be designed in many configurations to obtain maximum holding capacity, optimum circulation and aeration, and maximum mixing. The settling or stilling chamber is mainly in a variety of cone or hopper bottom shapes to provide efficient separation of solid and to permit complete digestion of the effluent. Some units are available with aftertreatment such as chlorinators or sludge holding tanks. The mechanical accessories to these units include blowers, air piping, diffusers, adjustable weirs, regulating valves, pumps, control panels, cathodic protection, electric wiring and conduits, mechanical skimmers, and other options. The complexity of the system depends on the method of discharge of effluent and on the nature of the sewage.

The key to all of these systems is the aerobic bacteria. These kinds of bacteria require air for life support. Agitation, settling, pH, and other controllables are carefully considered and employed as a means of maximizing the potential of bacterial reduction of organics in the wastewater.

TRICKLING FILTERS

In a report entitled "Trials of the Thermal Aerobic Filters at Coppy Clough" in March 1899, Naylor, in his book "Trades Wastes: Their Treatment and Utilisation," published in 1902 in London, discussed the results obtained on industrial wastes by trickling filters. It is evident from the photograph shown that these circular filters, with crushed-stone media and with revolving distributors, had been in operation for several years before the date of the report mentioned. Trickling, or percolating, filters have been very useful tools in the treatment of many types of industrial wastes.

Wastes which have been treated with appreciable success on trickling filters are

1. Aircraft
2. Beet sugar
3. Brewery
4. Cannery
   Fruit
   Vegetable
5. Carbohydrate
6. Chemical
7. Chemical cotton
8. Citrus processing
9. Corn chips
10. Cucumber canning
11. Cyanide
12. Dairy
    Milk processing
    Whey
    Cheese processing
13. Detergent reduction
14. Distillery
15. Fermentation industries
16. Food processing
17. Formaldehyde
18. Meat packing
19. Metal plating
20. Molasses distillery
21. Oil-containing
22. Organic chemical
23. Paper and pulp mill
24. Petrochemical
25. Petroleum (phenolic)
26. Pharmaceutical antibiotics
27. Pickle factory
28. Potato chips
29. Potato starch
30. Poultry processing
31. Radioactive laundry wastes

The advantages of trickling filters are

1. Ability to produce effluents of consistent quality even under shock load conditions

2. Ability to recover quickly from flash or shock loads

3. Elimination of need for constant manual attention

A few of the disadvantages of these units are

1. Loss of head necessary to operate the revolving distributors (usually 2.5 to 5 ft) above the centerline of the distributor arms

2. Ventilation ducts for the underdrain system

3. Odor and fly nuisance in intermittent operation (almost entirely eliminated by the continuous dosing employed on modern units with recirculation)

4. Need for final sedimentation units to catch the bacterial slime periodically unloaded from the bed media

The trickling filter requires a great deal of room when large volumes of wastes are to be treated. At the older rates of dosage of 2 million gal/d (acre), a flow of 2 million gal/d (acre) would require a filter with an area of 1 acre, or 43,560 ft².

Many industrial plants do not have much land to spare for nonproductive purposes. The late Harry Jenks, in the early 1930s, helped reestablish the trickling filter as a useful unit for treatment of large-volume flows by introducing his "biofilter" idea, which was to increase the capacity of trickling filters by recirculating the filter effluent back through the filter a number of times, according to the amount of initial BOD in the waste. This not only made it possible to apply waste liquors to these units at rates up to at least ten times the old rates, but materially reduced the amount of area required.

At the same time Jenks determined that since the bulk of the reduction of BOD in trickling filters was accomplished in the top 3 ft of the filter medium, there was no need to build units with 6 to 8 ft of stone, as had been the practice for low-rate units. This reduction in area requirements and in depth of the filters was such a saving in the cost of trickling filters that they have become competitive with the activated sludge process in overall cost and operation.

In spite of the improvement and economies developed by the ideas of Jenks and competitors, all using recirculation in one form or another, many industries with large-volume waste flows, especially those adjacent to municipalities where land was unavailable or costly, did not have the space available for even these smaller units.

About 1957 the Dow Chemical Company, manufacturers of plastics, developed what is called the Surfpac medium for trickling filters. This material was designed to distribute falling liquid wastes in thin films over large surface areas so that maximum efficiency of contact with aerobic organisms was attained. It provides a high percentage of void space for unimpeded draft ventilation and waste flow. It provides large-surface-area adherence of biological slimes. The material as produced by Dow Chemical Company consists of individual sheets of polystyrene or Saran plastic material, corrugated in two directions, having dimensions of 3 by 1.75 ft. The individual sheets are usually shipped stacked in bundles, and are then assembled into structually self-supporting modules at the point of use. In final assembly the modules occupy approximately five times the shipping volume, which materially reduces freight charges. In assembly the sheets provide approximately 1 in of free space between them. The modules are assembled in the filter structure in a grid pattern to give good distribution of flow of liquid, and also to assist in structural stability. The void space in an assembled filter bed is said to be 94 percent. The assembled weight of the individual modules is 4 to 6 lb/ft^3, a weight which enables them to be stacked in important depths, up to 30 and 40 ft, thus conserving the use of expensive land.

The availability of this material led to trials on various types of industrial wastes, and because of the variable character of different wastes, Dow has developed two types of plastic for this purpose, each suitable for certain waste components: (1) Dowpac 10, which has good resistance to alkalies, salts, dilute mineral acids, the lower alcohols, and water and is stated not to be suitable for some hydrocarbons, ketones, oxidizing acids, or vegetable fats and oils; (2) Saran, originally known as Dowpac 20, which is said to be extremely chemically resistant to all common acids and alkalies, with the exception of strong ammonium hydroxide, and is suitable for most alcohols, esters, ketones, nitroparaffins, benzene, xylene, and toluene which diffuse slowly through the interstices between the modules, but are said to have little effect on the material itself. The sheets of Dowpac 10 are assembled with a solvent adhesive supplied by the manufacturer. Dowpac 20 is heat-welded by special assembly machines supplied by Dow. In estimating the cost of plastic material for trickling filters, the cost of the assembly into modules must be included as an essential item.

Because of the light weight of this new material and its available void space, the development of small-diameter towers with great height has taken place. This has incorporated important savings in the use of this type of filter, since it materially reduced the amount of underdrain material required, a large item in the cost of the low-rate large-area filters. The enclosing structure for this type of filter may be made of aluminum or other light metal or wood, since no structural containment walls are necessary. The only protection needed is from wind and weather. The revolving distributors used are supported on the central riser pipe, and no crossbeams are necessary. In place of the vitrified underdrain tile used in ordinary trickling filters with stone media, the underdrains in the tower type may be made of pressure-treated lumber, concrete partition blocks, subway grating, etc. Since the assembled modules are rectangular in shape, to avoid expensive cutting and shaping of the material, the tower structures should be rectangular or hexagonal in shape.

As happens in the development of any new material or equipment, the advantages of this lightweight and resistant substance excited the interest of other manufacturers. Since that time similar plastic materials have been developed, and are marketed by the Ethyl Corporation, which is offering a polyvinyl chloride (PVC) packing named Flocor. This was developed in England by the Imperial Chemical Industries, Ltd., and consists of flat and corrugated sheets bonded into a module 2 ft in width and depth and 4 ft in length. The configuration of the sheets is a patented feature and is said to offer such advantages as the largest practical surface area with the lowest bulk density.

Another development in this field is the use of a polyvinyl chloride plastic called Koroseal, produced by the B. F. Goodrich Industrial Products Company. This material is shipped packaged, like the others mentioned, and assembled into modules in the field. The most outstanding demonstration of the use of this particular material has been at the Rome, Georgia, mill of a manufacturer of kraft paper. This filter, which handles a flow of 16 million gal/d, is 80 ft in diameter, with a medium depth of 20 ft and a total medium volume of 100,000 ft³. The surface area of the material is said to be 40 ft²/ft³, with a weight of 3.5 lb/ft³. The material is usually supported on epoxy-coated steel gratings in the tower, which has concrete block walls, with a total height of 30 ft. To fit the rectangular module shape, the tower is octagonal in form.

An additional development in the plastic-medium line is provided by American-Standard, New York. This is a cellulose-fiber sheet impregnated with plastic resin, made in a honeycomb design and suitable for stacking in a column.

Another variety of plastic material for trickling filters is offered by the Tex-Vit Company of Texas.

Table 8-6 illustrates the weight and surface-area advantages of these materials.

Since their development the various kinds of plastic media have been used successfully on a number of types of industrial wastes. In general, it may be said that they can be used on any waste that can be successfully treated by biological methods. Some examples of actual treatment of individual types of wastes are of interest. Trials on ammonia still wastes containing phenol cyanide, with high temperature and pH and with an initial phenol content of from 83.3

Table 8–6   Comparison of Plastic and Other Trickling Filter Media

Source	Brand name	Weight, lb/ft³	Surface area per ft³	Void space, %
Dow Chemical Co.	Surfpac	3.6	25	94
Goodrich Co.	Koroseal	2.7–3.5	40	94
Ethyl Corp.	Flocor	4.06	....	95
Raschig rings	...	30.3	22.7	74.9
Blast-furnace slag	...	68.0	20	49
Stone, granite	...	90.5	30	45

to 162 mg/l, resulted in a final effluent with 30.6 and 72 mg/l, or reduction values of 49.9 and 70 percent. The removal rate was established as 25 to 40 lb per 1,000 ft³ of filter material. On kraft paper mill wastes with a raw-waste BOD of 217 to 639 mg/l, the effluent was 67 to 319 mg/l, or removal percentages of 48 to 69 and removal rates of 35.7 to 75.4 lb per 1,000 ft³ of material. On rag mill wastes the average removal of BOD was 62.8 to 88.6 percent.

Another development in trickling filters with a new form of medium was reported by Ingram. He has developed a medium consisting of a honeycombed metal packing built up from corrugated expanded metal sheets, fastened together to form trays, which are alternately piled on each other at 90° angles. In this filter, air is admitted at several points in the depth of the filter. In treating whey wastes, it is reported that loadings of 460 to 690 lb of BOD per day per square foot of filter surface effected removals of 67 percent.

The types of industrial wastes on which filters with plastic media have been used are

1. Ammonia still
2. Antitoxins
3. Atomic
4. Beet sugar
5. Brewery
6. By-product coke
7. Cannery
8. Cellulose acetate
9. Chlorinated phenol
10. Coke plant
11. Corn cannery
12. Cyanide
13. Deinking
14. Distillery
15. Fermentation process
16. Food processing
17. Fiberboard reprocessing
18. Cane sugar
19. Frozen foods
20. Gas manufacturing
21. General organic chemical
22. Meat packing
23. Kraft process paper mills
24. Laundries
25. Petrochemical
26. Petroleum
27. Pharmaceutical (antibiotics)
28. Phenolic petroleum refinery
29. Pickle manufacture
30. Potato starch
31. Poultry processing
32. Pure phenol
33. Textile
34. 2,4-D
35. Vaccines
36. White water
37. Glycol
38. Tannery

Because of the higher loadings possible when a plastic medium is used, it has been suggested that the usual method of reporting results of trickling filter operation—pounds of BOD per unit of filter volume—be discarded and a new factor—pounds of BOD per unit of surface area of the medium—be used.

Where pretesting seems to be indicated before designing or specifying filters with plastic media, the Dow Chemical Company and the other manufacturers of this type of material have small pilot units, usually 3 ft in diameter and about 15 ft deep, which may be obtained for moderate rental fees.

In adapting the several progressive phases of industrial waste treatment, the following outline indicates those particular wastes which require treatment by certain operations.

Interception Methods

   1. *Fine screening:* Waste from
      a. Beet-sugar mills
      b. Glue factories
      c. Meat-packing
      d. Slaughterhouses
      e. Tanneries
      f. Textile mills (woolen)
      g. Vegetable canning

This method has practically no effect on reduction of biochemical oxygen demand (BOD).

Solids-separation Methods

   1. *Grit and dirt*
      a. Beet-sugar mills
      b. Poultry processing
      c. Vegetable processing

   2. *Gravitational settling.* Practically any waste which has an appreciable content of suspended or rapidly settleable particulate matter.

   3. *Precipitation.* With chemical additives, coagulants, and/or coagulant aids to cause the precipitation of fine, colloidal, or dissolved solids.

These methods will reduce BOD from 35 to 50 percent or more, depending upon the type of the wastes.

Oxidation Methods

   1. *Trickling filters.* To stabilize the organic matter in wastes by aerobic bacterial action, thus reducing BOD to a high degree, 90 to 95 percent.

   2. *Activated sludge method.* To stabilize organic matter, thus reducing the BOD of wastes by the assistance of oxygen provided by compressed air admitted to the wastes through ceramic or plastic diffusers; rotating mixers to inspire air; brushes or rotors to oxidize by spraying the wastes through the atmosphere. This method of which there are several variant forms, as described in Chapter

6, reduces BOD by 90 to 95 percent, depending upon the time the waste is exposed to oxidation.

The advantage of the activated sludge method lies in the ability to produce high-quality effluents in smaller areas and with less loss of head than with trickling filters.

The disadvantages of the activated sludge method are

   a. High cost of operation and maintenance
   b. Difficulty of dewatering sludge
   c. High power costs for air compression
   d. Greater sensibility to shock loads of toxic and organic substances
   e. Slow recovery from damage to the activated culture by shock loads
   f. Necessity for close control of volume and concentration of return sludge in the more conventional forms of this method

3. *Polishing phase.* This is a step or phase in treatment of industrial wastes made necessary by the increased emphasis on quality of effluents to prevent pollution of streams and the Clean Streams programs of the federal government. This final stage may be accomplished by lagoons, oxidation ponds, stabilization ponds, etc. Lagoons may also be considered as a conventional method since they have been used for many years as a tertiary step in waste treatment without many of the refinements in their design and operation that have been developed.

4. *Solids disposal.* Reduction of organic sludges by anaerobic digestion in closed tanks, with environmental control by temperature and pH, to reduce organic matter by gasification and liquefaction. Disposal of sludge ash by drying, natural atmospheric exposure, vacuum dewatering, thermal destruction, etc., as discussed in detail in Chapter 12 on sludge.

An anaerobic treatment process employs principles similar to the trickling filter and therefore deserves a place in this category. This process, developed by James C. Young and Perry L. McCarty at Stanford University at Palo Alto, California, employs an "anaerobic filter," which is a rock-filled bed similar to an aerobic trickling filter. Waste is distributed across the bottom of the filter, and the flow is upward through the rock bed, so that the filter is completely submerged. As the liquid passes through the filter, it comes into contact with a large active biological mass. A high degree of treatment is said to result even at nominal waste temperatures, and the effluent is essentially free of biological solids. Sulfides resulting from sulfate reduction, and the remaining BOD demand in the treated effluent, can be effectively removed by aeration or other suitable means. Low head loss through the filter indicates that power requirements for filter operation are very small. In addition, the methane released from decomposition of the wastes represents a usable power supply.

AERATION

The addition of oxygen to wastes by artificial means has long been practiced. The main developments have been in the introduction of oxygen into oxidation ponds or aeration lagoons by artificial means. In such large surface areas it has been very expensive to provide air because of the cost of the supporting structures to carry the mixers or other equipment. One development has been the Floating Aqua-Lator, which is simply a motor-driven agitator enclosed in a bowl-shaped container that floats on the surface of the liquid. It is movable, of course, and can be located anywhere in a basin. It is manufactured by Welles Products Corporation.

It is reported that the Standard Oil Company (Ohio) at Lima, Ohio, has equipped a 20-acre oxidation pond with these floating aerators for the purpose of biologically reducing oils, phenols, and other oxygen-consuming materials. A third stage follows a gravity separator and chemical flocculation and air flotation units.

A development for the supply of oxygen to oxidation ponds or lagoons is the Air-Aqua system which consists of a network of plastic tubes laid on the bottom of the pond and supplied with air from a compressor. Other types of surface aerators are stationary mixer-type units, with draft tubes and impellers of the turbine type, consisting of a horizontal plate with vertical blades welded to its periphery. Liquid is drawn up at the center and thrown out radially at the surface, the actual spray being very turbulent. Oxygen is absorbed from the air in the turbulent area. The action also causes currents through the liquid mass which diffuse the oxygen throughout the basin. The setting of the rotor with regard to its submergence governs the effectiveness of oxygen dispersion.

Another type of aeration device is the Kessner brush type, developed in the Netherlands before 1930, and its simulation in the rotor type used in oxidation-ditch aeration. The Inka system, developed by Dorr-Oliver, Inc., consists of rotary mixer units installed near the bottom of the basin.

An unusual aeration unit developed by Reeves of Dallas, Texas, consists of a V-shaped container or aeration tank, with an outside annular shell forming an interspace in which the air is admitted. The inner shell is perforated with many small holes that admit the air to the wastes in small bubbles at a great number of points and elevations. The unit is made of steel. It is claimed by the inventor that a unit 28 ft long by 21 ft wide with a total depth of 10 ft 4½ in with sixteen V-shaped channels could handle a flow of many millions. The unit is said to require about 70 ft³ of air per gallon of waste.

An unusual innovation in aeration was the system included in the sewage treatment plant installed to serve Expo 67 at Montreal, Canada. This system, developed by Aero-Hydraulics Corporation of Montreal, is installed in a basin divided into three zones, aeration and mixing, sludge settling, and polishing. The mixing tank is in three sections, separated by aluminum baffles, to provide flexibility in the flow. In the mixing tanks are aeration "mixing guns" of polyethylene pipe 12 in in diameter, standing vertically on the bank floor. Air is delivered to an air distribution chamber near the bottom of the stack pipes through polyethylene tubing. The air forces the water in the stack down, until breaking of a water seal at an opening creates an inverse siphon, allowing a bubble of air to pass up the stack. The action is regulated so that one bubble is rising in the stack at all times, creating an air piston effect. This pushes and pulls waste up from the bottom to provide complete circulation. The bubble of air bursts as it emerges from the stack, some 6 ft below the surface, and imparts oxygen, as well as agitation, to the contents of the tank. Effluent from this stage flows over a submerged weir into an aluminum-hoppered settling zone. No return sludge pumps are required. Return action is accomplished by a number of aeration stacks, similar to those described but without side ports. Organic matter from the hopper is drawn to the stacks through 8-in polyethylene pipes by bubble action. An additional set of guns, set at an angle, create directional flows to separate the aeration-mixing zone from the settling zone.

FILTRATION

Other methods for the removal of suspended solids and liquids include filtration. In this category reference is made only to those filtration methods which may be classed as "physical." Liquid under natural gravitational means or under pressure is made to pass through a natural or synthetic medium, leaving behind solids or a colloid of a very finely divided nature. In this discussion *filtration* is used as a loose term applied to filters in which biological action due to accumulation of bacterially active films or to slimes which coat the filter material is not included.

Pressure Filters

A pressure filter can be used to clarify the water by removing turbidity and suspended matter including dirt, rust, sand, metallic dust, oil, color, and other impurities. The filter operates in the following manner: As the water passes downward through the filter bed, suspended matter is removed and accumulates on the bed. As this foreign matter gradually accumulates, loss of head occurs, indicating the need for cleansing the bed by backwashing. This is accomplished by an upward flow of water which expands and loosens the bed, releasing the material for discharge. The filter bed is then down-rinsed through the filter to a waste sump, thus compacting the bed before returning to the normal filtering operation.

    The most common filter bed materials used for water clarification are sand and gravel. Where silica pickup from sand is objectionable, anthracite is often used. Activated carbon is used for removal of objectionable tastes and odors, organic material, chlorine residues, and color. Granular marble and calcite are used for neutralization of carbonic acid.

    Some of the accessories which are available with a typical pressure filter are

Alum feeder

Loss of head indicator

Backflow rate indicator

Air vent and sampling lines

Sight glass

Surface washer

Interconnecting lines

When considering the cost of operating, be sure to include replacement cost of the filter media, alum, labor, and volume of backwash water.

Gravity Filters

A gravity filter may be used for removal of suspended solids, both colloidal and granular. Frequently the unit is used as a last-stage polishing step before final discharge. Since the unit consists only of a container filled with the filtering media through which the waste permeates, its capacity is quite varied, ranging from 6 to 380 gal/min. The actual speed depends upon the filtering material and the size and consistency of the contaminants.

Influent water percolates through the filter bed to the false bottom, where it is collected by strainers. From the underdrain compartment filtered water flows to the backwash storage compartment and out to service. The water in the backwash storage compartment remains fixed, while water in the level control piping slowly rises as the filtering continues, and the bed collects impurities. This continues until the head loss reaches a predetermined level. At this point, the backwashing cycle begins. The backwash water is discharged separately for drying or further treatment.

Moving-Bed Sand Filters

The Newark College of Engineering, Newark, New Jersey, developed a type of filter that uses sand as the medium. This filter is known as the MBF, moving-bed filter. The operation of the unit consists of driving the filter medium through a pipe in one direction while simultaneously feeding the liquid to be filtered through the same pipe in the opposite direction. This produces a filtering action at the face, as well as in the depth, of the moving filter bed. The impurities removed are driven along the pipe with the filter medium and are removed from the face and depth as rapidly as is required by their buildup. Tests have indicated that the combined use of chemical treatment and the moving-bed filter would be able to produce effluents exceeding accepted turbidity standards set for potable water.

The filter operates as a continuously running system and does not have to be stopped for backwashing. Filter material is constantly being removed automatically from the system, cleaned, and returned to it. This action, combined with a variable rate of motion or of feed, gives the unit great flexibility to meet varying quality of influent feed. As the quantity and quality of influent change, the rate of filter-medium feed, its makeup, and its cleaning, can all be modified to maintain a constant output quality. The manufacturers of the equipment for the system, Scientific Developments, Inc., of New Jersey, state that cleaning of the filter medium can include heating and carbonizing steps, if required. Tests also indicate BOD reductions of 50 percent, phosphate removal from 30 mg/l to approximately 0.5 mg/l, and turbidities lowered from 60 to less than 1 Jackson unit. A 250,000-gal/d demonstration plant was built for use on an industrial waste. The Johns-Manville Company development department was actively engaged in the early stages of experimentation with this system. The system was described in more detail in *Industrial Water Engineering,* May 1967.

Coal as a Filtration Medium

In the endeavor to reduce the cost of treatment of industrial wastes, one effort has been directed to the discovery of alternative media for filtration. Experiments have been in progress for some time using anthracite or bituminous coal in a finely ground state as a medium. It is stated that filtration through graded beds of coal particles produces results superior to biological methods. Coal is cheap, as compared with the stone normally used in biological filters, or activated carbon, and has the added advantage of being usable as fuel after use as a filtering agent. This method also creates a potential new market for coal.

In operation, as developed by the Rand Development Company, of Cleveland, Ohio, finely crushed coal is proposed as a complete treatment method. The top layers of coal in the filters, with entrapped solids, are continuously removed and replaced with new layers. In this method a highly volatile bituminous coal is found to be most adsorbent. Coal adsorbs pollutants as well as detergents and phosphates and permits higher rates of application and longer filter runs.

Coated Steel-Mesh Filter Medium

One type of pressure-developed filter uses a stainless steel wire-mesh medium coated with diatomaceous earth.

Diatomaceous-Earth Medium

The use of diatomaceous earth as a filter medium was the subject of an extenive report published in *Industrial and Engineering Chemistry,* February 1953. This material is said to be very useful in removing emulsified oils and oil solids in large machine shops and in rolling mills, in the recovery of wool grease or lanolin, and in the filtration of waste effluents for reuse. It has also been used for the reduction of undissolved solids in cyanide streams destined for deep-well injection; also in wastes containing phenol, radioactive isotopes, toxic materials from pharmaceutical manufacture, and other materials preparatory to underground disposal. It has been used for filtration of effluents from secondary and tertiary treatments, particularly when a very high-grade water for reuse in certain industrial purposes is required.

Fiber Glass Screens

This is a conception for a tertiary phase in biological treatment of industrial wastes. It consists of screens of fiber glass immersed in a tank. It is claimed that it provides a high ratio of stationary surface to volume with minimum turbulence. Under conditions of continuous flow, and with retention periods of 10 to 36 h, a pilot plant was said to show reductions of 90 percent BOD, 92 percent turbidity, 94 to 95 percent suspended solids, and 96 percent coliform bacteria.

Electrodialysis or Dialysis

This is one of the oldest forms of membrane separation of colloidal matter from wastes by normal pressures acting on the membrane. It has been practiced on an industrial scale for more than 40 years in the rayon and pulping industries. In its simplest form, as dialysis, it consists in the passage of liquids of different concentrations through a selective membrane. For instance, it is effectively used in the desalination of brackish water by removing a large percentage of the salt from the water and permitting the water to pass through the membrane.

Electrodialysis, a later development, is the practical separation of ionic components of a solution in an electric field owing to the difference in transport number between the ions in the bulk solution and the interposed membrane. This system was developed in Europe about 1915. Electrodialysis has many uses, such as reduction of the citric acid content of orange juice; recovery of spent pickling liquor; demineralization of molasses and raw-sugar-cane juice; concentration of radioactive solutions; treatment of spent sulfite liquor to recover pulping liquors; and desalting of cheese whey into a useful by-product as the base of an infant-food formula. New membranes have been developed which greatly increase the usefulness in the recovery of mineral acids from synthetic organic compounds; treating stainless steel pickling liquor for recovery of nitric and hydrofluoric acids; and for purifying pharmaceuticals.

An electrodialysis cell is formed of two smooth, rigid plastic membranes with a spacer between which guides the water flow. One membrane carries a permanent positive charge, and the other membrane has a negative charge. Just as magnets of like charge repel each other, the positively charged membrane repels positive ions, yet permits negative ions to pass through. The membranes and electrodes are so arranged as to permit ions to leave an electrodialysis cell but not to return. Therefore water in one compartment of a cell is deprived of its salts, while the adjacent compartment is increased in concentration. Depending on the material to be handled, this system is expensive, since it requires electric energy, investment in plant, replacement of membranes and other parts, labor for operation, chemicals, etc. The solutions to be treated must be substantially free of all particulate matter, because the membrane apertures are so tiny as to be readily clogged. These systems are not filters in the true sense of the word.

It has been reported that single-pass electrodialysis will remove about 40 to 50 percent of the total dissolved solids. Power requirements are about 6 to 10 kWh per 1000 gal. Dissolved organics must be removed before introduction into the system to avoid fouling. Capital costs are said to be about 15 cents per 1000 gal of the volume of water to be treated in large plants. Operating costs are said to be less than 15 cents per 1000 gal. Laboratory dialysis test sets are available from several manufacturers, including Graver Water Conditioning Company. See Cost Index Chart in Chapter 11 for comparative current prices.

Microstraining

This method of fine straining was introduced in Great Britain in 1945, primarily as a polishing step in municipal water treatment. In a way, it is also membrane filtration, since the revolving drum of the unit is covered with a finely woven wire-mesh cloth to separate microscopic solids. The wire cloths used in this system average from 160,000 apertures, or 20 μm/in^2, down to 58,000 apertures, or 60 μm/in^2. In practice, the efficiency of this type of filtration depends upon the formation of a thin mat of the intercepted solids on the inner surface of the filtering fabric, which in turn intercepts solids finer than the apertures in the medium. Although microscopic suspended matter is filtered out, no appreciable reduction of true colloidal matter can be expected unless coagulants are used in the prior treatment phases.

The woven-wire-mesh cloth used on these units is made of stainless steel wire drawn down to diameters so small that they are barely visible to the naked eye, and is woven on special looms. The fabrics are unique in that they combine acceptable filtration efficiencies with high porosity, high flow ratings, low loss of head, and easy removal of intercepted solids by low-pressure water jets. Durability of fabrics is high, experience having demonstrated a possible life of 10 years or more. Although these units were developed for water treatment and are so used in numerous countries, including the United States, they offer possibilities for use in the treatment of industrial wastes as a final polishing step, particularly in those cases where the reuse of final effluents is planned. The process and the units were developed in England by Glenfield & Kennedy, and they are represented in the United States by Crane Company. Passavant-Werke of Germany has developed a similar idea.

To determine the possibilities of this unit in the sanitary engineering field, experiments on municipal sewage were carried out in Baltimore, Maryland, in 1954, and the results indicated that influent with a concentration of 525 to 589 mg/l of suspended solids was reduced to a final effluent with from 35 to 119 mg/l.

Reverse Flow

This system, also termed the "upside-down" filter, is a development of the DeLaval Separator Company. The trade name is the Immedium filter. The principal difference in this filter is that the wastewater flows up from the bottom; thus the entire depth of the filter bed is used for filtration and the storage of intercepted solids. The larger sand particles in the bottom of the bed function as a prefilter, and the fine particles at the top act as polishers. Thus the bulk of the turbidity and the fine solids is absorbed in the coarse-sand layers that have the greatest dirt-holding capacity, whereas in the normal filter operation the feed enters the fine-grain layer first, and this layer quickly becomes saturated with impurities which clog the interstitial space and create a loss of head, necessitating frequent backwashing. For oil and solids removal from steel mill waste waters, the filter is used following sedimentation.

In a case in Ohio for the treatment of acid wastes, a similar idea was used, with limestone as the filter medium and with upward flow of the acid wastewaters.

In-Depth Filters

This type of filter has multiple uses. It can remove phosphates and suspended solids and also those contaminants causing increases in BOD and COD. An in-depth filter is used immediately following treatment. It is capable of filtration to less than 0.5 ppm and is suitable for recovering nickel from plating rinse waters, copper oxides from hot mill cooling water, hexavalent chromium by treatment with barium carbonate, emulsified and floating oil, and phenolic resins.

The materials and utilities required are the replacement of filter media, backwash water, and electricity. The unit uses the same principles as gravity filtration and carbon adsorption. The filtering medium combines both properties. The effluent from the filter is recyclable. The backwash material must be further concentrated, dried, or disposed of.

MICRO-FLOC

A development of Neptune Micro-Floc, Inc., of Corvallis, Oregon, is known as the "mixed media" system. It was originally proposed for increasing the efficiency of water filtration plants. In theory it is a balancing of the media in a filter bed with respect to the specific gravity of the various types of materials and with respect to the size of the particles and their placement in the bed. By this system optimum results from the filtration step are obtained. The Micro-Floc mixed-media custom-designed filter beds are adapted to different capabilities to meet different conditions. These conditions include turbidity loading, operating rates, finished water usage, and chemical costs, as well as initial cost.

By using combinations of two, three, or more media of different specific gravities and different screen sizes, hundreds of combinations are available to match the requirements. In other words, this is a special sizing relationship between the high-density gravel in a filter bed and the adjacent filter medium which optimizes the performance of the gravel layer. Used on water treatment, savings of up to $5 per million gallons in coagulation chemicals have been reported when treating turbidities of up to 48 Jackson units.

This same company also manufactures a package for clarification which combines mechanical flocculation, settling, mixed-media filter bed, backwash storage, and clearwell, all in one steel tank. The settling unit is simply a group of shallow basins which tend to catch the floc at different levels in its downward passage. With sizes rated up to 750 gal/min, this unit may prove to be useful in industrial waste projects. For instance, a 750-gal/min unit, complete with steel tank, with two filter and two settling sections, occupies a space of only 32 by 11 ft and has a 10-ft-diameter filter outside. A complete installation for 750-gal/min water rating would occupy a total area of 30 by 32 ft.

Pilot studies to examine the feasibility of this type of unit on industrial wastes have been made in the following processes:

1. Treatment of metal-plating effluent

2. Tertiary treatment of effluent of wastes from electronics manufacturing

3. Treatment of steel mill effluent

4. Treatment of effluent from wallboard (gypsum) manufacture

5. Clarification of flue-gas scrubber streams

6. Reclamation of water from laundry wastes

7. Separation of cereal grains from hexane stream

8. Tertiary treatment of meat-packing effluent

9. Treatment of effluent from preservative processing of piling

10. Treatment of plywood mill (glue) effluent

11. Treatment of wastes from hydraulic-gravel classification

12. Treatment of milk processing effluent

13. Reclamation of water from soft-drink preparation rinse water

14. Treatment of combined municipal sewage and cannery wastes

The settler unit is unique in that the flow of coagulated material enters small tubes placed at different levels in the tank. The flow of liquid is passed lengthwise through the tubes, each of which acts as a shallow settling device. The tubes are slanted so that settleable material accumulates at the low end, while the liquid exists at the high end. This provides for gravity drainage of the settled material. Settled material in this system falls only a fraction of an inch, instead of feet, as in the conventional settling basin. It is claimed that excellent removal of settleable solids is achieved in 10-min detention time, instead of the several hours usually required in normal settling basins.

CENTRIFUGES

Another particularly effective method of suspended solids removal is the centrifuge. Both mineral suspended solids and biological suspended solids that reduce BOD and COD can be removed by the principle of masses accelerated in the direction of a force. Figure 8-18 shows a small centrifuge.

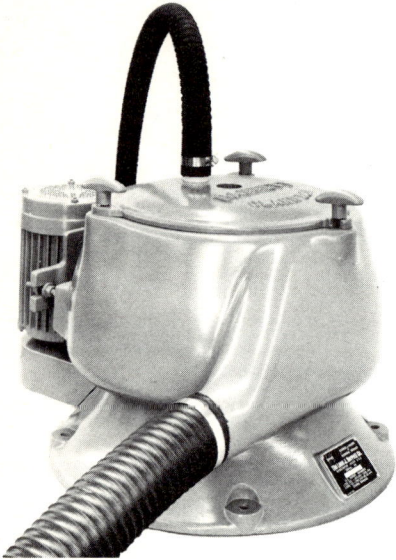

Fig. 8-18 A centrifuge. (*The Leon J. Barrett Co.*)

Vertical Basket Type

The vertical basket centrifuge is used to dewater chemical, industrial, and domestic waste. The basket centrifuge will produce a cake of 10 to 25 percent solid with essentially no polymer addition. The unit is best for maximum sludge concentration and should be considered when it is costly to burn, truck, or barge the dewatered waste.

Hydroxide sludges may first be treated with alum floc and then centrifuged with the basket unit. The solids as fed are floccy and are discharged as a very thick pudding that is 15 to 25 percent solid. No polymer addition (or less than 1 lb/ton of solid for 85 to 90 percent recovery) is required. Hydroxide sludges may be formed in any heavy-metal precipitation process. This type of centrifuge can therefore be used by metal-plating and finishing industries.

The basket is a spinning cylinder whose medium centrifugal forces settle solids against the wall and hold them there. It is the simplest centrifuge to build, operate, and maintain. Its undisturbed sedimentation gives high solids compaction and effluent clarity. The centrifugation process is interrupted in auto-mated operations to scrape and/or skim out the collected sediment. This sediment can be retained.

The basket capabilities and specifications are as follows:

Bowl diameter	12 to 60 in
Flow rate—feed	Up to 100 gal/min
Solids in feed	0.1 to 3%
Discharged solids	Up to 1050 lb
Temperature	Up to 450°F
Pressure range	Up to 5 lb/in²
Speed	Up to 2500 r/min
Gravity (G forces)	Up to 2100 G
Motor power	Up to 100 hp
Particles removed	1 μm and up

Nozzle Type

A nozzle type of centrifuge can separate, classify, or concentrate a wide variety of materials including alum floc, refinery wastes, and activated wastes from paper mills. The waste stream to the disk centrifuge bowl is diverted among many narrow channels. The particles are recovered at high throughput rates. Solids, less than 60 mesh, are continuously discharged from the bowl through peripheral orifices or nozzles which are limited in size and number. The nozzle-type centri-fuge is particularly suited for high-volume flows containing relatively low concen-trations of nonfibrous solids that do not pack to a hard cake. They must be fine enough to pass freely through the nozzles. These units are used by facilities to further reduce the suspended solids in the secondary aeration tanks or effluent from the first settling tanks.

Centrifuges are more fully discussed in Chapter 12, which deals with sludge handling.

CONCENTRATORS

A concentrator combines flotation with centrifugation to remove suspended solids, grease, foam, floatable solids, and settleable solids from the effluent stream. Indus-tries such as paper, textile, tobacco, meat-packing, potato, and poultry have found it very economical in contaminant reduction.

The unit contains a vertical cylindrical screen cage that rotates with ap-proximately 3 to 6 G of centrifugal force. The screens are of stainless steel or synthetic material and range from 10 to 100 μm in mesh size.

The incoming hydraulic flow, ranging from 100 to 2000 gal/d, enters the unit with approximately 10 ft of hydraulic head. The flow travels up through the center of the unit and onto two horizontal distribution domes. Flighted domes direct the influent against the inner top surface of the screens in a series of overlapping thin layers of liquid.

The particles, some even smaller than the wire openings, are caught in a continuous slough-off concentrate stream, washing down the inside of the screen and representing 5 to 20 percent of the hydraulic flow to the concentrator.

The effluent, highly aerated by the centrifugal screening action, is then discharged to the flotation cell. The flotation cell is operated solely from super-saturated air entrained in the concentrator effluent.

Soluble oils and grease that have passed through the centrifugal screens are lifted by the surface energies of the air bubbles, combined with the fine suspended solids in the foam, and build into particles which are many times the micron size of the screen openings. A skimmer removes the highly concentrated suspended solids materials at the end of the cell.

The concentrate may be dewatered, further concentrated on a smaller concentrator, or sent back to a settling tank or primary clarifier.

Table 8-7   Representative Test Data for Concentrator

Applications	Screen openings, μm	Aver. feed susp., mg/l	% removal	% BOD removal
Hydraulic debarker effluent	105	160	85–90	25
Cannery effluent	44	2004	83	Not avail.
Paper mill—white water	44	120	65	Not avail.
Potato processing	44	2250	89	62% COD
Poultry	25	5496	89	87

VIBRATING SCREENS

Vibrating screens are used to remove suspended solids from wastewater. The unit is particularly effective in large water-consuming industries since throughputs range from 1 to 2 million gal/d when 10- to 100-mesh screens are used. High throughputs are also achieved with even finer mesh. The separator is fed at the periphery of the top screen. Vibrator action and liquid flow cause the solid matter to travel toward a large center opening in the screen. The solids then fall to the lower screen deck and move to the outer periphery for final discharge from the solids spout. Most of the liquid passes through the top screen and is discharged out of the upper spout. Liquid contained in the solid material then falls to the lower deck, passes through the bottom screen, and is discharged out of the lower liquid spout.

The effluent is substantially free of suspended solids and is ready for discharge or reuse. The solid may be dewatered and hauled away or reused.

INCINERATION

Thermal waste incinerators are designed to eliminate the problem of disposal of liquid waste materials. They are readily adaptable to the disposal of virtually any liquid waste combustible, including streams which contain inorganic or other noncombustible materials present in combination with the oxidizable contaminants.

Solid and liquid wastes from filtration units, vibrating screens, oil-water separators, etc., are common influents for such units. Carbon from carbon adsorption filters is reactivated in thermal incinerators.

The fluid-bed incinerator converts most organic wastes by completely oxidizing combustible contaminants. It can be used on ordinary wastes, polymeric wastes, or shredded plastics. Its by-products are carbon dioxide, nitrogen gas, and water. If the unit is used in pyrolytic processes (burning in the absence of oxygen), carbon can be reactivated.

The unit is based on the principle of "fluidization" of finely divided particles on a bed of gas emerging under pressure. When lying on this bed, the particles become microscopically separated. The particles and gas behave like a liquid. The less dense particles rise while the more dense sink. The system has high heat transfer properties which can be exploited.

Solids are fed into the unit by gravity or through a screw or piston conveyor. The fluidizing gas is pumped into the chamber below a distribution plate. For incineration the unit is operated with a temperature level of 1500°F. Waste solids are rapidly combusted within the bed. The gaseous products of combustion, along with particulate matter leaving the top of the fluidized bed reactor, are passed through a solids separator and a scrubbing system for condensation of water and removal of potential air pollutants.

In pyrolysis operations, the off-gas system is designed for the recovery of the gaseous by-products. The 1500°F temperature assures complete combustion of organic material within the bed, permitting the use of high-temperature gases for preheating the chambers.

Liquid waste streams which contain 10,000 BTU/lb or more can usually be incinerated directly in a thermal burner without the requirement of an auxiliary combustion chamber. In some instances, where the content is less than 10,000 BTU/lb, it is possible to enrich the liquid waste with natural gas or other fuel. Aqueous wastes of other lean streams are usually sprayed into a hot combustion chamber fired by means of a thermal burner. A unit of this type now available is that of the Thermal Research and Engineering Corporation (Figure 8-19). This company now offers complete systems as package assemblies, including waste incineration and concentration, heat recovery, catalytic combustion, solvent vapor recovery, and other applicable liquid waste disposal techniques.

The installation shown in Figure 8-19 is that of a chemical company in Texas. The unit was installed in 1960 but is very similar to the present-day developments in this system. It was designed to handle 2,900 gal/d of various organic waste materials. The burner is fired at the bottom into the vertical refractory-lined incineration chamber, and the waste liquid is introduced directly into the burner as fuel. The burning of the liquid waste takes place within the combustion chamber of the burner. The secondary combustion chamber is only partially refractory-lined, to the inlet of the dilution air fan, which lowers

Fig. 8-19 Thermal incineration system. (*Thermal Research and Engineering Corporation.*)

the operating temperature and permits the use of carbon steel materials in the upstream end of the unit.

Another interesting installation of this system used on an exotic waste is that made at the Cape Kennedy Missile Test Center in Florida. This unit disposes of waste propellant of the Titan II missile. Since this fuel is toxic, it must be incinerated. This is also true of any of the propellant fuel found to be off specification in the storage tanks or as received at the Center.

This system is said to be applicable to the incineration of the following wastes:

1. Boron-caustic liquor

2. Carbon disulfite waste liquid

3. Chlorinated hydrocarbons

4. Diethylene triamine tar

5. Dimethyl sulfide in water

6. Ethyl alcohol and mixed aromatics

7. Liquid hydrocarbons

8. Magnesium chloride with miscellaneous solvents

9. Pharmaceutical wastes

10. Phenolic tars

11. Phenolic-tar-still bottoms

12. Sulfur- and phosphorus-bearing liquids

13. Toluene, diethyl ether, acetone mixtures

14. Wastes containing tars, ammonium sulfate, etc.

15. Water-pitch emulsions

Another process has been in use at the Dominion Rubber Company in Ontario, Canada, for several years. This operation consists of pumping the liquid wastes from feed tanks to the burner gun, where steam is introduced to atomize the waste. The atomized waste is ignited by a pilot flame fed by propane gas. The burner is capable of burning residues containing as much as 40 percent water. The advantages claimed for this system are low initial cost, low maintenance and operating costs, unattended operation, and smokeless burning.

Walker Process Equipment has developed a process called the Greaseburn incinerator system. Without auxiliary fuel, accumulated grease is prepared for smokeless, odorless combustion, with resultant minor ash residue.

Shell Chemical Company at Houston, Texas, incinerates liquid hydrocarbon wastes in a three-phase waste treatment system.

Another system of thermal incineration of wastes which has been used extensively is that of Prenco, Inc., Chicago. This method destroys both combustible and noncombustible industrial wastes completely, without smoke nuisance. By combining combustible wastes with others which have a sufficiently high BTU total per gallon, the combination can be effectively destroyed without the need for auxiliary fuel.

At the research center of the Marathon Oil Company at Littleton, Colorado, a high-temperature pyrodecomposition Prenco process for disposing of light and heavy ends from petroleum processes has been installed. This unit uses fuel oil as an auxiliary fuel. It takes advantage of the dissociation and ionization of waste substance constituents. The installation is said to dispose of 3,360 gal of waste oils per week. It operates at a temperature of 2000 to 2700°F, and the waste products are completely consumed. It has been suggested that this type of unit may be applicable as a complete waste treatment facility to prevent water pollution by waste oils and oily wastes.

A similar Prenco system for oil-bearing wastes has been installed at one of the automotive plants of the Chrysler Corporation at Twinsburg, Ohio. This installation handles all wastes from the plant except sanitary wastes. The cost of this installation was $168,000, which is said to be about half of that expended on previous methods at this plant. The installation handles 500 gal/h at 2200°F, with a total combustion and without smoke, odor, or visible ash. The cost of the plant was expected to be amortized in less than 2 years by halving the costs of waste disposal.

Another installation of the Prenco system is used by the Standard Oil Company at Lima, Ohio, to dispose of the wastes from an acrylonitrile plant which produces approximately 200 million lb of product per year. This installation is said to have cost $1.25 million and has an operating cost of about $1 million per year.

The Prenco Super E_3 Pyro-decomposition system is shown in Figure 8-20. It operates in the following manner. A mixture of auxiliary fuel (usually natural gas) and high-pressure air is first fed into the vertical retort to bring it up to the proper decomposition temperature. When the retort reaches the correct temperature, as determined by the temperature-measuring instruments, fuel flow is modulated and the waste is admitted to the air-waste entrainment compartment. From there the aerated waste is fed into the turbulence chamber, where it is mixed with high-pressure air and is injected into the high-temperature vertical retort. The waste is then broken down by molecular dissociation, oxidation, and ionization. The gases and any inert particles flow vertically through the air cone and out of the top of the retort in the form of an invisible, clear, pollution-free exhaust.

The Zimpro (Zimmermann) process, discussed in detail in Chapter 12, is reported to have been used in the treatment of combined municipal sewage treatment plant sludge and the wastes from ten textile mills in Kannapolis, North Carolina, and also for the treatment of glue factory wastes in Germany. This type of plant is also designed for recovery of minerals from wastes and to generate steam for use in the factory. Chromium has been recovered also.

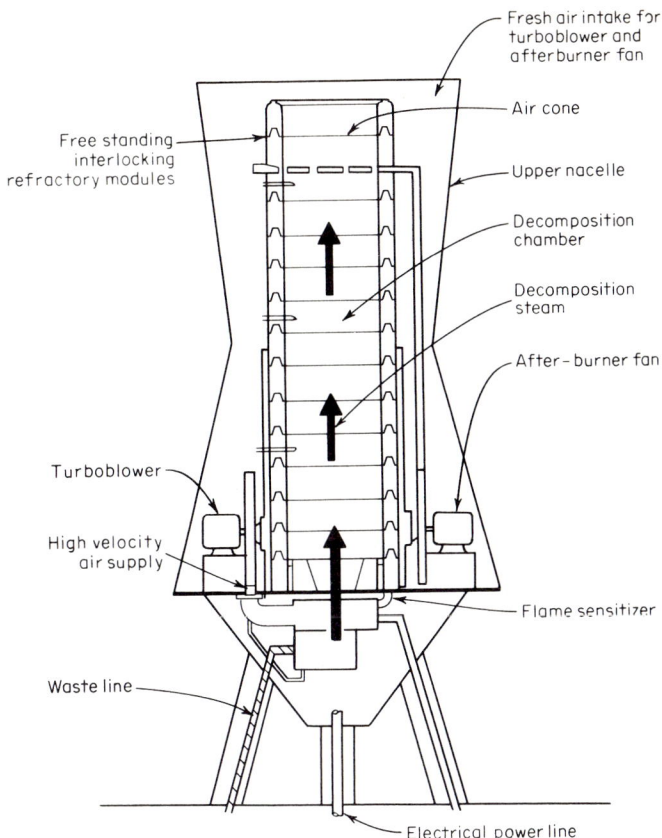

Fig. 8-20 Diagram of a pyro decomposition system. (*Prenco.*)

In 1966, a Zimpro installation was put into operation in Tasmania, Australia, for recovery of sodium from the pulping of eucalyptus wood. Although the wastes that had been dumped into the ocean were causing no pollution problem, the many tons of sodium in the waste represented an appreciable loss, and the reason for the recovery was economic.

Another Zimpro plant has been in operation since 1936, recovering vanillin from paper mill wastes. Experimental runs on various paper mill wastes indicate successful results on magnesium-base spent sulfite liquor; semichemical acid; neutral and alkaline sodium sulfite spent liquor; calcium-base spent sulfite liquor; sodium-base spent liquor; and kraft liquor.

The Zimpro process has also been applied to the destruction of cyanide in plating wastes. It is reported that 99 percent destruction of cyanide has been accomplished. In the treatment of wastes containing phenol the reduction has been from 275 mg/l to an effluent with only 0.3 mg/l.

At the plant of the Sohio Chemical Company, at Lima, Ohio, three types of harmful wastes are discharged from the manufacture of acrylonitrile. The most dangerous of these is the one containing hydrogen cyanide, which is toxic to animal life in concentrations as low as 10 ppm. The Lima plant also produces an organic effluent containing mixed nitriles, and dilute aqueous process wastes. As the production of the plant increased, a treatment system became inadequate, and the management decided to build two new waste incinerators to take care of the HCN vapors, the mixed nitriles, and the aqueous effluent. The burners composing the new system are John Zink thermal oxidizers provided by the John Zink Company of Tulsa, Oklahoma. These burners are fired by natural gas, and the various waste streams are admitted at different points. The mixed nitriles go through a steam-atomizing burner before entering the combustion chamber. The HCN vapors feed in through a specially designed raw-gas burner, and the aqueous waste is broken up into small droplets by commercial spray nozzles as it enters the chamber. The temperature inside the combustion chambers is 1600°F, at which the contaminants are quickly converted to CO_2, H_2O, and N_2.

These incinerators have been in service since September 1965 and are successfully handling all the plant wastes. Quantities of previously impounded liquid wastes have also been processed, and the land occupied by the impounding basins has been used as part of the site for new factory production facilities that were completed in 1966.

Cyanide concentrations in the oxidizer exhaust gases vary from 0 to 5 ppm, which is considered a safe level. Natural-gas consumption for the new incinerators amounts to about 4000 surface $ft^3/m(d)$. The waste streams are as follows: total waste water, 95,000 lb/h; mixed nitriles, 8500 lb/h; and HCN vapor, 5200 lb/h. The wastes and the organics are injected at 40 lb/in² gauge, and the HCN vapors at 2 lb/in² gauge. This installation was described in *Chemical Processing*, January 1967.

Although the cost of the thermal incinerators varies according to the material handled and the size of the installation, a very general approximation is given by the manufacturers. A system handling 5 to 10 gal/h of liquid waste, including the burner, combustion chamber, controls, and accessories, would cost a minimum of $5000. As the size increased, the cost of the plant would increase, at an approximate rate of $500 per gallon of material to be treated per hour of operation. See Cost Index Chart in Chapter 11 for comparative current prices.

ATOMIZED SUSPENSION TECHNIQUE

Another method of handling slurries or concentrated waste solutions from filtering units, clarifiers, or evaporators was developed by Dr. William H. Gauvin of Montreal, Canada. He calls the system the AST system. The reactor is an empty tower of stainless steel. The walls of the tower are kept at a high temperature. When a slurry or solution is atomized into the tower, the moisture in the droplets flashes into superheated steam. This acts as the carrier vapor as the evaporating droplets move downward in the tower. Partway down, all the moisture has been converted into steam, and the solid particles are suspended in an essentially nonturbulent vapor of their own creation. As this dispersion continues down the tower, various kinds of physical and chemical reactions can be introduced, one after another, all within a few seconds. For example, if the solids are organic material, a small amount of air can be introduced and they will burn. One unit using this process is said to be in operation in Canada.

OIL-WATER SEPARATOR

Oil must be separated from water before discharge of the waste. An oil-water separator is used for this purpose. It is based on the fact that most oil floats on water. Oily water is allowed to stand, and the oil forms a layer on the surface. The surface layer of oil is pumped off, or allowed to flow off, into a holding tank while the balance of the water is discharged to the sewer.

Only the floating oils and greases are removed, but they are frequently the only contaminant in some industrial wastes. The oil may be dried and reused or simply hauled away by a tank truck for reprocessing. A simple oil-water separator is shown in Figure 8-21.

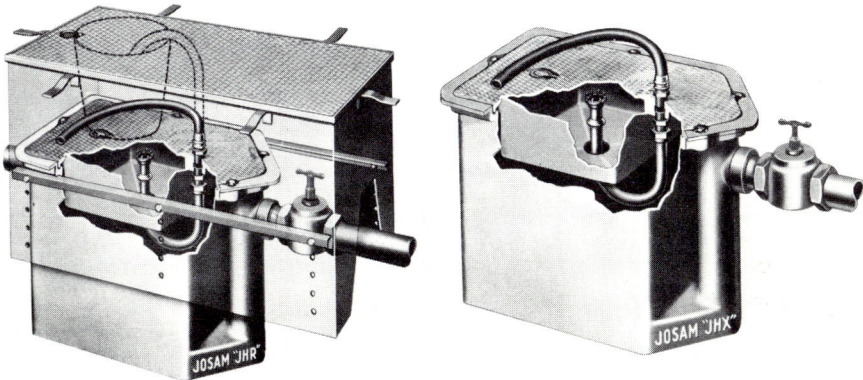

Fig. 8-21   Cast-iron grease traps. (*Josam Manufacturing Company.*)

OIL SKIMMER

This device is suitable for the removal of many kinds of oils from wastewater. These oils include fuel oil, kerosene, SAE 30 motor oil, and Bunker C oil. A $\frac{1}{2}$-hp motor drives a continuous polyester belt partially submerged in the water. Floating or stratified oils adhere to the belt as it enters the liquid. The recovered oil is lifted out of the wastewater and is removed from the belt by steel-supported hard rubber wipers. As it is scraped from the belt, the oil drops into a collection trough and exits from the unit. The trough can be positioned to discharge into a drum or tank for reclamation.

RADIOACTIVE WASTES

"At present there is no completely satisfactory answer to the problem of disposal of radioactive wastes." This statement, agreed to by many sanitary engineers experienced in industrial waste treatment, appears to epitomize the general feeling of the profession. Without doubt more engineers and scientists in all parts of the world have labored with this problem over recent years than any other problem in the field of industrial waste treatment. Many have worked longer with less tangible result on this than on any other type of waste. But the fault for the inconclusive results cannot all be placed on the researchers. Rather, the trouble is with the inherent nature of the contents of the wastes. They have a great tenacity for life, and there is no known method of shortening this life. In the treatment of normal types of organic industrial wastes, the entire cycle of treatment to an inoffensive, nonpolluting effluent is a matter of a few hours. But with several of the most dangerous radioactives, isotopes of cesium and strontium with half-lives of 28d to 30 yr and 61d to 19.9 yr, respectively, it can readily be appreciated that normal processes will have little or no effect. In fact, study of the very extensive literature relating to the treatment of radioactive wastes shows a principal interest in establishing practical and economical means for the disposal of the raw wastes rather than treatment to destroy the radioactive elements in the waste.

The half-life of the representative isotopes is found in wastes, is shown in Table 8-8.

The principal sources of radioactive wastes are from the various Atomic Energy Commission laboratories and development grounds, where studies in fission and fusion are being conducted; the manufacturing areas where extraction of the isotopes from uranium are being carried out; and the rapidly increasing number of power plants producing electric power by atomic means.

The problem of disposal of radioactive wastes is of increasing concern to the nuclear power industry, for two reasons: (1) the fear that has been planted

Table 8-8 The Half-Life of Significant Isotopes

Item	Half-life
Barium 140	12 d
Cesium 137	30 yr
Cesium 144	28 d
Iodine 131	8 d
Niobium 95	35 d
Strontium 89	53 d
Strontium 90	19.9 yr
Yttrium 91	61 d
Zirconium 95	65 d

in the minds of the people of the danger of leakage of radioactive radiation from these power plants, which are located in close proximity to residential and business areas; and (2) the present cost of disposing of the wastes, which is a definite and important item in the cost of producing power by nuclear means. As long as the established means of producing electric energy now existing are available, that is, by steam generation from existing fossil fuels or by hydroelectric means, nuclear power will not be salable unless the ultimate cost to the consumer is comparable with that produced by the older means.

The means for disposal of radioactive wastes that have been used or proposed up to this time are

1. Discharge into unused or abandoned salt mines or domes

2. Ion exchange

3. Dilution and storage in containers

4. Evaporation

5. Solvent extraction

6. Precipitation

7. Foam separation

8. Coagulation and flocculation

9. Underground storage with pretreatment

10. Disposal on deserts or arid areas

11. Burial in steel or concrete tanks

12. Absorption in clay pellets

13. Calcination in silicates

14. Disposal at sea in canisters

15. Deep injection wells

16. Oxidation ponds

17. Calcination in fluidized beds

18. Absorption in ceramic sponges

19. Mixture with municipal sewage in conventional treatment

20. Absorption into sodium silicate jelly and burial

21. Dilution

It must be understood that in none of these methods is the natural decay of the dangerous isotopes accelerated. They simply concentrate the isotopes into smaller volumes of liquid, but the ultimate disposal problem still remains, and will continue to exist during the natural life of the isotope. A brief survey of some of the above methods, proceeding numerically, produces the following observations.

   1. *Discharge into unused or abandoned salt mines or domes.* This method has been suggested by engineers of the Atomic Energy Commission. The difficulty with this method is that nuclear power plants are being established at many different places, and unused or abandoned salt mines are not always close enough to make this method economically feasible. This method would appear to have local possibilities only, because transportation costs to distant points would add materially to the cost of power.

   Abandoned salt mines have been suggested because they have considerable structural stability already and are usually to be found in geologically stable areas. This type of storage would be dangerous in earthquake-prone areas, where tremors could open passages into lower aquifers.

   2. *Ion exchange.* This method has been suggested by a number of scientists, but the high cost of equipment and operation is a deterrent.

   3. *Dilution and storage in containers.* One of the most common methods of disposing of radioactive wastes has been to place them in containers, which are then buried in the earth, stored in caverns, or dumped into the sea. The results of this practice and the effect on future generations are unknown; the storage environment and the containers must be of such structure that the radio-activity will be properly contained during its entire life, sometimes for hundreds of years. The containers themselves must be equally invulnerable to decay or damage. If the container of highly active wastes fails or deteriorates because of corrosion or attack by acid soils or underground waters, and releases the isotopes into the underground structure, the possibility of contamination of underground aquifers may persist for generations.

   Another danger peculiar to the tank or container storage of radioactive wastes is the heat continuously produced in the waste liquors by the beta and gamma radiation from the decay of the isotopes. This heat is a unique phenomenon, since it is not controllable by any known method. In a concentrated waste the heat generated is sufficient to cause the wastes to boil for long periods of time because of the decay of the long-lived isotopes cesium and strontium. It is quite possible that certain operating conditions could create a waste that would continue boiling for 100 years or more. Who can say what the effect of this action would have on metal or concrete containers?

   4. *Evaporation.* A method developed by General American Transportation Corporation reduces the volume of low-level wastes and would eliminate the need or use of large storage tanks. A similar process is being used at the Brookhaven National Laboratories on Long Island, New York, and at a nuclear power plant in Italy. The Atomic Energy Commission is said to be working on plans for a waste-fractionization process and in-tank solidification which would assist in reducing expensive tank storage. There are 145 storage tanks ranging in capacity from 50,000 to 1 million gal, each located in eleven tank farms up to 10 mi apart. Leakage from some of these tanks has already been detected in a number of instances. Evaporation is also a very expensive process, averaging from 5 to 12 cents per gallon of waste stored.

5. *Coagulation and flocculation.* High dosages of iron coagulants plus pH correction are said to effect 97 percent removal of strontium 90 from wastes. Slow sand filters fed at rates of 2 to 8 million gal/d (acre) accumulate soluble radionuclides in the *Schmutzdecke* (surface layer of solids). Accumulations of iodine 131, strontium 90, rutherium 106, cesium 144, and plutonium have been noted, with cesium 144 most prominent. This, of course, refers to the concentration of these isotopes, not their complete destruction.

Long periods of detection in oxidation ponds are economical in cases where land is cheap. These are most useful for treating partially decontaminated putrescible wastes containing concentrations of radionuclides slightly above the maximum permissible concentrations in water. Very long detention periods are essential. Because of the severe and differing restrictions placed on radionuclide concentration in final discharges in different areas, each case must be studied independently.

6. *Underground storage with pretreatment.* A French process has been developed in which radioactive wastes are entombed in tar and thus protected from future leaching by water and possible dispersion into underground areas after burial. The method, developed at the French Plutonium Production Center, provides for tar to replace entirely the 50 percent water content of the usual low-level waste concentrate, which resembles mud. The tar envelops each radioactive particle in an impermeable coat and forms a sludge of 65 percent solids, 35 percent tar that is stiffer than the original mud. Burial in the ground disposes of the sludge, and it is believed there is little risk of future contamination of the area. The plant is said to be capable of treating 1,400 lb/h of waste slurry.

It is noted that the Oak Ridge National Laboratory has developed a new process for incorporating aqueous intermediate and low-level radioactive wastes in asphalt for ultimate disposal by burial or storage. The asphalt solidifies the wastes and makes them insoluble. In this process aqueous wastes are introduced into an emulsified asphalt at any temperature below the solution's boiling point. The water is then boiled away, and the temperature of the mixture increased, until it flows freely, when it is drained into steel drums.

7. *Disposal on desert areas.* At the Hanford Works of the Atomic Energy Commission, at Hanford, Washington, liquid wastes containing light concentrations of radioactive material have been dumped on parched sands of the vicinity without apparent objectionable results (1957).

8. *Burial.* This method has been a common one, practiced for some time, principally for small volumes of radioactive wastes from the several laboratories and centers and productive establishments of the Atomic Energy Commission. It also has been employed for the wastes from some nuclear power developments. The cost is high, ranging from 35 cents to several dollars per gallon of waste contained and buried.

Russia is said to bury all radioactive wastes at special depots far from any population centers. Solid and liquid wastes are transported in special containers to forested grounds, where they are placed in deep bunkers, and sealed with turf-covered ferroconcrete slabs. The solid wastes are dumped into the bunkers, the containers are decontaminated with water, and the liquid wastes and decontamination filtrates are then slurried with cement. The slurry is poured onto the solid waste in the bunker and forms a monolithic block. Although the block is considered to be impervious and the local water table is below the buried waste, precautionary boreholes for water sampling surround the bunkers. In summer flowers and berries grow profusely on the slabs, apparently stimulated by radiation, which is too weak for detection. An experimental automated facility was constructed at this burial site. In this unit all types of radioactive wastes are delivered to conveyors and converted automatically into concrete cubes for burial.

9. *Absorption.* Absorption of radioactive wastes into clay pellets and subsequent vitrification has been suggested as an economic possibility for the concentration of radioactivity.

10. *Fixation in glass.* Fixation of radioactive wastes in glass seems to have promise for long-term storage, as proposed in Great Britain and in the United States.

With the rapid expansion of atomic power plants, the Soviet Union and Czechoslovakia are engaged in intensive research into radioactive waste disposal. Czechoslovakian research follows the research and development line already investigated in Canada, where an attempt was made to bury waste material contained in glass blocks. Another similar possibility is to use basalt instead of glass. A representative of Canada has advised that glass blocks containing 1000 Ci buried in the ground 7 years had leaching rates so low that it was possible to bury the material without protection in restricted areas.

An engineer of the Soviet Union has discussed the possibility of injecting radioactive wastes deep into the earth's crust where geological conditions permit. A testing ground for this purpose has already been established at the Soviet Nuclear Research Institute in Melekess. The first stage of the research was industrial disposal to a depth of 1550 m (4700 ft). It was reported that experiments for dealing with liquid waste in this way were successful.

Another engineer from the Moscow Disposal Center of the Soviet Union reported that the safety factor of wastes solidified in cement blocks was ten times lower than blocks buried without waterproofing. Still another Russian engineer reported that heat emitted by radioactive materials could be used to achieve solidification of clay soil and provide insulation against water.

Calcination reduces high-level radioactive wastes to safer, more easily stored solid masses which occupy very much less space. A number of pilot-plant tests have been carried out at the National Reactor Test Station of the Atomic Energy Commission in Idaho.

As an example of the effectiveness of calcination in locking in dangerous materials, the results of work done at an abandoned plant of an extractor of beryllium from Brazilian ore is worth mentioning. This plant, in operation for 9 years, had accumulated the final sludge from the extraction process in a series of lagoons covering an area of 5.7 acres. When the operation was moved to another location and the land offered for sale, it was believed that this accumulation of sludge, which still had a beryllium content of about 1.0 percent, although not radioactive, was toxic, and that it should be removed and the lagoons leveled. A contract was given to an engineering firm to determine the best way to dispose of the sludge, which amounted to about 20,980 yd^3.

In carrying out this contract twenty-six different disposal possibilities were investigated. The most effective solution from the standpoint of reduction of the waste volume was calcination. Since the sludge had a silica content of 67 percent as SiO_2, this was an indication that calcination might prove satisfactory. To prove this, a small rotary gas-fired kiln was built. Operated at 2000°F, the resultant calcined material represented only 25 percent of the original mass. Immersion of some of the calcined material in distilled water for 19 d indicated that only about 0.5 percent of the fused material would go into solution. Although the results of the calcination demonstration were indicative of the value of this method, the estimated cost of treating this one-shot mass of sludge in this way was not considered economically feasible. This work was done in 1956, showing the interest at that time in this method.

It is reported that at the Brookhaven, Hanford, and Oak Ridge AEC Laboratories the fusion-into-glass method has been studied, resulting in a final product, a homogeneous dense material less subject to leaching than powders or granules.

11. *Calcination in fluidized beds.* This method has been successfully used at Atomic Energy Laboratories at Idaho Falls, Idaho.

12. *Formation of jellylike sodium silicate mass.* One idea in radioactive waste disposal is reported to have been developed by the General Electric Company at the Hanford Works of the Atomic Energy Commission. It is said to be a process to eliminate the use of steel tanks for burial. The process as described uses a milky-white jelly of sodium silicate and a solution containing aluminum. It is hoped that this substance will entrap dangerous atomic by-products, such as cesium 137 and strontium 90, and keep them safely buried. As discussed earlier, heat is generated in stored radioactive wastes, and with this new method it is hoped that this heat would turn the sodium silicate–aluminum solution into a hard glasslike material which would imprison the "hot" wastes until their final decay.

Cost of Radioactive Waste Treatment

A comprehensive study of the cost of handling and treating radioactive wastes resulted in the following information:*

Cost of off-site disposal of solids wastes
$2.65 per 100 lb
$1.02 per cubic foot

* See Cost Index Chart in Chapter 11.

$0.08 per ton-mile (1 ton carried 1 mi)
if transported to distant points from the point of origin

Estimated cost of burial of solid radionuclide wastes
$9 to $11 per cubic yard in 1965

Estimated cost of burial of concentrates wastes in trenches
$0.37 to $2.00 per gallon

Cost of disposal by total evaporation
$0.12 to $0.138 per gallon

Costs at Los Alamos, New Mexico, Atomic Energy Laboratories
Annual operating cost of the waste treatment plant, $5.86 per 1000 gal; initial cost of plant, including laboratory facilities, $350,000

SUMMARY

Almost every wastewater problem has been investigated by some industry or another. Many have found solutions. In many cases package units now exist to treat some of the most hazardous of pollution problems. Individual problems may require a sequence of treatment units. Regardless of the problem, diligent investigation of the alternatives will result in a treatment method satisfactory to industry and the environment.

Chapter Nine

The role of the federal government

In 1972 Congress amended the Federal Water Pollution Control Act (Public Law 92-500), referred to as "1972 amendments." This legislation established a truly national program to combat water pollution. For the first time, there was established a system of national effluent limitations and national performance standards for industries and publicly owned waste treatment plants. The program extends to all navigable and underground waters of the United States, both interstate and intrastate.

The objective of this program is "to restore and maintain the chemical, physical, and biological integrity of the Nation's waters." Pursuant to this objective the act proclaims several goals for the nation:

1. That a major research and demonstration effort be made to develop technology necessary to eliminate the discharge of pollutants into navigable waters, waters of the contiguous zone, and the oceans

2. That areawide waste treatment management planning processes be developed and implemented to assure adequate control of sources of pollutants in each state

3. That federal financial assistance be provided to construct publicly owned treatment works

4. That the discharge of toxic pollutants in toxic amounts be prohibited

5. That wherever attainable, an interim goal of water quality which provides for the protection and propagation of fish, shellfish, and wildlife and provides for recreation in and on the water be achieved by July 1, 1983

6. That the discharge of pollutants into the navigable waters be eliminated by 1985

Congress placed the responsiblility of administering the broad scope of the act on the Environmental Protection Agency. In order to establish the necessary national standards for the act, there was created within the EPA a Water Pollution Control Advisory Board. This board consists of the Administrator of the EPA or his designee, who shall be chairman, and nine members appointed by the President of the United States. It is the duty of this board in cooperation with the states to develop and publish regulations specifying minimum guidelines for effluent discharge and to develop and publish relevant information and procedures for compliance with such limitations.

EFFLUENT GUIDELINES

The guidelines will provide a criterion for water quality. This will be accomplished by identification of discharge constituents for classes of point sources and thereafter setting effluent limitations.

A source is any building, structure, facility, or installation from which there is or may be a discharge of pollutants. A point source is any source discharging to navigable or underground waters of the United States.

The effluent limitations will take into consideration:

1. Control measures and practices applicable to point sources
2. "Best practicable" treatment technology
3. Process and procedure innovations
4. Total cost of application of technology in relation to the effluent reduction benefits to be achieved from such application
5. The age of equipment and facilities involved
6. The process employed
7. The engineering aspects of the application of various types of control techniques
8. Process changes
9. Nonwater quality environmental impact including energy requirements

These guidelines apply to "point sources," but what about those dischargers that do not qualify as point sources? These consist of the typical industrial wastewater discharge into a publicly owned treatment plant.

Indirectly applicable to these nonpoint sources are the federal guidelines for publicly owned treatment plants. All sewage treatment plants in operation on July 1, 1977, must provide a minimum of secondary treatment and must comply with any effluent limitations imposed by the EPA or by a state.

In order that treatment of those pollutants which are not susceptible to treatment by publicly owned treatment plants shall be dealt with, the EPA will establish pretreatment guidelines for such pollutants. These guidelines apply only to those dischargers as explicitly designated by the EPA.

Pretreatment standards are designed to achieve two purposes: to protect the operation of publicly owned treatment works and to prevent the discharge of pollutants which pass through such works inadequately treated. The pretreatment standards are intended to be national in scope. In many cases it will be necessary for a state or municipality to impose more stringent pretreatment standards than are specified by the EPA to enable the publicly owned treatment works to comply with the effluent limitations applicable to its operation as a point source.

The general pretreatment standards to be met will be in accordance with the policy that

No discharge to publicly-owned treatment works shall contain the following described materials:

1. Wastes which create a fire or explosion hazard in the publicly-owned treatment works.

2. Wastes which will cause corrosive structural damage to treatment works, but in no case wastes with a pH lower than 5.0.

3. Solid or viscous substances in amounts which would cause obstruction to the flow in sewers, or other interference with the proper operation of the publicly-owned treatment works.

4. Wastewaters at a flow rate and/or pollutant discharge rate which is excessive over relatively short time periods so that there is a treatment process upset and subsequent loss of treatment efficiency.

The EPA has stated that "any owner or operator or any source to which the pretreatment standards . . . are applicable, shall be in compliance with such standards" by July 1976 for existing industrial facilities and by May 1974 for new industrial sources of pollution.

The EPA has published information on pretreatment unit operations. These guidelines are established to assist municipalities, states, and federal agencies in developing requirements for the pretreatment of wastewaters which are discharged to publicly owned treatment plants. These guidelines also explain the relationship between pretreatment and the effluent limitations for a publicly owned treatment plant.

The industries thus far denoted by the EPA as requiring pretreatment methods are: paper and allied products, dairy products, textiles, seafoods, pharmaceuticals, leather tanning and finishing, sugar, petroleum refining, meat products, grain milling, fruit and vegetable, beverages, plastic and synthetic materials, blast furnaces, steel works—rolling and finishing, organic chemicals, metal finishing, inorganic fertilizer, electric and steam generation, aluminum, flat glass, cement, lime, concrete products, gypsum, asbestos, inorganic chemicals, and industrial gas products.

In order to illustrate the manner in which pretreatment standards are established for various industries, the examination of the metal-finishing industry by the EPA is described below.

1. *Industry Description*

This industry includes Standard Industrial Classification (SIC) 3471. The industries covered under this classification include those primarily engaged in various types of plating, anodizing, coloring, forming, and finishing of metals. The metal-finishing industry operations are related closely to those of many other industries, including transportation (automobile parts and accessories), electrical, and jewelry.

The metal-finishing operation involves cleaning, conversion coating, organic coating, plating, anodizing, coloring, and case hardening. Acid pickling is the most common type of cleaning of metal being prepared for plating. Sulfuric acid is the most commonly used pickling agent, but phosphoric, hydrochloric, hydrofluoric, and other acids are used as well. Alkalies, dichromates, and numerous proprietary compounds are also used in various combinations for descaling, degreasing, desmudging, stripping, brightening, or otherwise preparing different metals (zinc, brass, steel, copper, etc.) for plating or anodizing.

The plating solutions for nickel, chromium, copper, cadmium, zinc, tin, and silver may be basically cyanide, acid, or alkaline. Anodizing is done either in sulfuric acid or in chromate solutions. Coloring is accomplished with dyes, nickel acetate, and chromates. Cyanides are used in case hardening.

2. *Wastewater Characteristics*

The characteristics of the process wastewaters from the industry are shown in Table 9-1.

The metal-finishing industry usually generates a continuous stream of rinse waters containing dilute concentrations of heavy metals and cyanides and intermittent batch dumpings of spent acid and cleaning solutions. The nature of metal-finishing operations and the consequent fluctuating (cyclic) characteristics of the wastewater should be taken into consideration in the design of treatment facilities.

Water is used extensively in metal-finishing processes to clean, strip, pickle, and rinse the metal products before and after plating operations. The rinse waters constitute the major volume of wastewaters, while spent solutions discharged intermittently add major pollutants to the total effluent. The wastewaters contain, in general, spent acids, alkalies, oil and grease, detergents, cyanides, and various heavy metals (Cr, Ni, Cu, Ag, Fe, Zn, and Sn). The metal-finishing plants differ from one another with respect to their processes, metals and chemicals, and the characteristics of wastewater may vary widely from one to another. However, their wastewaters all contain primarily inorganic pollutants, particularly heavy metals. In addition, the wastewaters frequently are highly toxic due to the presence of cyanides and heavy metals.

In general, the types of wastewaters from metal-finishing industries are:

1. Acid wastes
2. Alkaline wastes
3. Heavy metals wastes
4. Cyanide-bearing wastes
5. Miscellaneous wastes (dyes, soluble and floating oils, etc.)

Table 9-1   Wastewater Characteristics for the Metal-finishing Industry

Characteristics	
Industrial operation	Year-round (BATCH)*
Flow	Continuous–VARIABLE
BOD	Low
TSS	Average–high
TDS	HIGH
COD	Low
Grit	Present
Cyanide	HIGH
Chlorine demand	HIGH
pH	ACIDIC
Color	Present
Turbidity	Present
Explosives	Absent
Dissolved gases	Present
Detergents	Present
Foaming	Absent
Heavy metals	HIGH
Colloidal solids	Absent
Volatile organics	Present
Pesticides	Absent
Phosphorus	Present
Nitrogen	Present
Temperature	Normal†
Phenol	Low
Sulfides	Absent
Oil and grease	Present
Coliform (total)	Absent

* Characteristics which may require pretreatment or are significant to joint treatment plant design are shown in Uppercase.

† Temperature similar to domestic wastewater.

SOURCE: Federal Guidelines, "Pretreatment of Discharges to Publicly Owned Treatment Works," U.S. Environmental Protection Agency, Office of Water Programs Operations, Washington, D.C. 20460 (U.S. Government Printing Office, 1973).

Any of these wastewaters may occur as either dilute rinse waters or concentrated baths. Except for the cyanide-bearing wastes, the wastewaters are generally connected to a common sewerage system for treatment and disposal. The cyanide wastes usually are collected in a segregated sewer system in order to prevent the release of toxic hydrogen cyanide gas under acidic conditions. However, the cyanide wastes can be mixed with other waste streams provided that any acid streams are neutralized prior to mixing with the cyanide waste stream.

The major constituents in the wastewater generated from metal-finishing operations are cyanides, metal ions (Cr^{6+}, Ni, Fe, Cu, and Sn), oil and grease, organic solvents, acids, and alkalies. The wastewaters characteristically are so toxic and corrosive to sewers and equipment that they require pretreatment before discharge to municipal sewers.

A wide variety of processes are used in metal finishing operations, resulting in widely varying wastewater characteristics. Typically, these wastewaters have poor treatability characteristics without adequate pretreatment.

3. Pretreatment

The pretreatment unit operations for various types of joint treatment facilities are shown in Table 9-2.

The pretreatment processes generally involve separate treatment of cyanide wastes and other acid wastes containing metal ions. The cyanide wastes can be treated with ferrous sulfate and lime to convert highly toxic cyanides to less toxic cyanates or cyanide complexes, or can be oxidized to CO_2 and N_2 with chlorine under alkaline conditions. The acid waste streams are treated first to reduce hexavalent chromium to trivalent chromium, using ferrous sulfate, scrap iron, or sulfur dioxide, and then precipitating the metal ions (Cr^{3+}) as metal hydroxides.

In addition to the effluent limitations and the processes shown in Table 9-2, the degree of reduction in heavy metals waste loadings should consider the sludge handling and disposal methods used for the combined domestic and metal-finishing wastewaters. Some processes (e.g. anaerobic digestion) concentrate these metals, and this can lead to process failure unless adequate pretreatment is provided.

Although no definite limitations have been set forth specific to those discharges to publicly owned treatment plants, those that will be forthcoming must be at least as stringent as those limitations for the treatment plant's effluent.

The EPA has published general effluent guidance for all industrial categories and has specifically listed maximum effluent concentrations for specific parameters applicable to particular industrial classification industries. The specific guidelines are directed at the major or more significant dischargers. The general limitations consist of the following major guidelines for discharge into the national waters:

Table 9-2   Pretreatment Unit Operations for the Metal-finishing Industry

Suspended biological system	Fixed biological system	Independent physical chemical system*
Equalization + neutralization + cyanide removal + chromium reduction + chemical precipitation (heavy metals) + solids separation	Equalization + neutralization + cyanide removal + chromium reduction + chemical precipitation (heavy metals) + solids separation	Equalization + cyanide removal + chemical precipitation + neutralization

* Chemical precipitation may not be needed, depending on the processes used in the independent physical chemical joint treatment plant.

SOURCE: Federal Guidelines, "Pretreatment of Discharges to Publicly Owned Treatment Works," U.S. Environmental Protection Agency, Office of Water Programs Operations, Washington, D.C. 20460 (U.S. Government Printing Office, 1973).

pH. The normal range in which all discharges should fall is generally accepted to be between a pH of 6.0 and a pH of 9.0. Any discharge outside of this range should be well justified and should consider such things as the addition of total dissolved solids from neutralization and buffering capacity of the receiving water. If the pH of the discharge is significantly different from the pH of the receiving water within the 6.0 to 9.0 range and the discharge is a significant portion of the receiving water, toxic effects due to pH fluctuations within this range can be shown. Therefore, it is imperative that the discharge be evaluated carefully even within the range of 6.0 to 9.0 (e.g., 7.0 ± 0.5).

Coliform. If sewage is present, the following effluent limit should be imposed: Organisms isolated in the fecal coliform test shall not exceed 1000 organisms per 100 ml.

Total Dissolved Solids. No limitations have yet been established.

Oil and Grease. Vegetable- or animal-based oil and grease are generally less significant than petroleum-based oil and grease at the same levels. Conditions for petroleum-based oil and grease are to reflect an achievable concentration of 10 mg/l as the basis for determining limitations.

Toxicity. National standards for toxicity limitations via bioassay procedures have not been set.

Heavy Metals. A general limitation of 1 mg/l maximum total dissolved heavy metals should not be imposed across the board. The figure is, however, a good point of reference when establishing limits for a discharge involving a number of dissolved heavy metals which could present a hazard to water quality owing to synergistic toxic effects.

Table 9-3 shows the EPA effluent limitations for the metal-finishing industry. It is included here as an example of the federal requirements for a NPDES discharge permit. The limitations stated herein are to be met as a minimum requirement by July 1, 1977, for all discharges within the standard industrial classifications 3471 and 3479.

DISCHARGE PERMITS

The 1972 amendments created the National Pollutant Discharge Elimination System (NPDES). This system is a nationwide program regulating the issuance of discharge permits for discharges of pollutants into the nation's waters. The system replaces the permit program established under the 1899 Refuse Act. The system is as comprehensive as it is complicated. There is, however, one basic rule: No discharge of any pollutant is allowed without a permit.

The National Pollutant Discharge Elimination System is the key to applying the national effluent limitations to specific polluters. Publicly owned treatment

**Table 9-3   Effluent Limitations for the
Metal-finishing Industry**

Effluent parameter	Concentration	
	mg/l	*lb/1000 gal*
COD*	—	—
TSS	10	0.0834
Cyanide (dest. by Cl_2)	0.03	0.00025
Fluoride	18	0.150
Aluminum	0.2	0.00167
Barium	1.0	0.00834
Cadmium	0.1	0.000834
Chromium, CR^{6+}	0.05	0.000417
Chromium. CR^{3+}	0.25	0.00209
Copper	0.2	0.00167
Iron	0.5	0.00417
Lead	0.05	0.000417
Manganese	1.0	0.00834
Nickel	1.0	0.00834
Silver	0.05	0.000417
Zinc	0.5	0.00417

* Limited significance in this industry; should be considered on a case-by-case basis.

SOURCE: "Interim Effluent Guidance for NPDES Permits," U.S. Environmental Protection Agency, Office of Permit Programs, Washington, D.C. 20460.

plants as well as industrial dischargers must obtain permits. The issuance of a permit informs the discharger of the effluent quality required of his operation. If a polluter cannot immediately comply with the effluent limitations, the permit will set target dates for installing needed pollution abatement equipment. Even though the discharger has been given a "period of grace," the permit will state precise limits for the interim period which will be strictly enforced.

It is within the discretion of the EPA, the regulating agency for NPDES permits, to authorize a state, which has the capability of administering a permit program, to issue permits consistent with the limitations set forth in the NPDES program. Such state-issued permits must meet with the approval of the EPA and must be at least as stringent as the federal permits but may include such additional limitations as the state deems justified. For further discussion of the EPA requirements for such authorization and the states' ability to increase the limitations of discharge, see Chapter 10.

In creating the form for the NPDES permits, the EPA has stated that the general information required of an applicant for permit shall be the same regardless of whether the applicant is attempting certification by the EPA or by a state agency. All applicants are required to submit to the appropriate agency the contents of existing or proposed effluent. The federal effluent limitations shall be the basis for the issuance of permits for discharge, and the EPA requires a particular method of reporting such information in the permit application.

Permits are to be written with limitations in terms of kilograms (pounds) of each pollutant per day excepting in certain industries (such as mining operations and utility discharges) where wastewater generation is not related to production levels. The monthly operating average effluent limitation, in kilograms (pounds) per day, should be calculated using the production basis multiplied by the appropriate effluent level as specified in the guideline limitations for specific industrial categories. Once a permit has been issued based on the use of effluent volume, production levels, and attainable abatement technology, the discharger may meet his requirement via any combination of effluent volume and concentration providing he does not violate water quality standards in doing so. If a discharger drastically reduces his water usage he may augment receiving water diffusion capacity to prevent localized toxic effects or take advantage of improved treatability to reduce pollutant discharge, at his option, within the life of the permit.

A manufacturer should not be limited to less than full utilization of actual installed production capability.

It is important to note that production capability estimates by the applicant will be carefully scrutinized so as to prevent the circumvention of the effluent limitations by gross overstatements of this capability. The manufacturer is required to meet the permit requirements without exception. Seasonal variations in treatment efficiency, production levels, or product mix are the inherent difficulties of the particular operation and are not considered to be reasonable explanations for failure to comply with permit requirements.

The duration for which a permit will be issued is dependent upon the level of effluent limitation achievable by the particular applicant. If a discharger is expected to achieve the 1977 effluent levels on time, then a permit will be issued for an extended period of time. Short-term permits will be issued only in those cases where a specific water quality problem may not be resolved or where the technology for pollution abatement is unavailable. In such cases a plan for eliminating the pollution problem should be submitted and a permit to discharge issued with a termination date shortly after completion if the discharger has effectively complied with the interim limitations and has the ability to comply with the national goals.

The 1972 act also regulates the disposal of sludge, the disposal of dredged or fill material, and ocean dumping of any material. The law prohibits the disposal of sludge from sewage treatment plants into the navigable or underground waters of the nation without a permit from the EPA. Like the industrial discharge permits, the permit issuing and regulating program may be taken over by a state if authority to do so is granted by the EPA.

Under preexisting law the issuance of permits for the disposal of dredged or fill material was under the authority of the U.S. Army Corps of Engineers. The 1972 act reaffirms this authority with one exception. Dredged or fill material may be dumped only in specified disposal sites, and the EPA has the authority to veto a selected site for such disposal. The veto power rests in the EPA's prevention position with respect to adverse effects on national waters from the disposal of such materials.

The EPA will establish, pursuant to the 1972 act, guidelines for the disposal of pollutants to ocean waters. The guidelines will cover the effects of pollutants on human health and welfare, on marine life, shorelines, and beaches and cover alternatives to ocean disposal of pollutants.

ENFORCEMENT

Various alternatives are available to EPA officials by which the standards and limitations previously discussed may be enforced. These may be combined into five categories.

Monitoring

The EPA may require the owner or operator of any point source to: (1) establish and maintain records; (2) make reports; (3) install, use, and maintain monitoring equipment or methods; (4) sample operation effluent in a prescribed manner; and (5) provide such other information as may be reasonably required.

Inspection

An authorized representative upon the presentation of his credentials (1) shall have a right of entry to, upon, or through any premises in which an effluent source is located or in which any records required to be maintained are located; and (2) may at reasonable times have access to and copy any records, inspect any required monitoring equipment or method, and sample any effluents which the owner or operator is required to sample.

Any information obtained by the EPA by either of these methods shall be available to the public except where such disclosure would violate the Trade Secrets Act. Even in the latter case the information may be disclosed to officers, employees, or other representatives of the United States concerned with carrying out the 1972 amendments. In conjunction with the EPA's monitoring and inspecting operations there may also be authorized a state monitoring and inspection program. Such a program is authorized only for the inspection and monitoring of facilities within the jurisdiction of the state. The state programs will enable the EPA to establish a nationwide water quality surveillance system to monitor water quality so that the proposed 1977, 1983, and 1985 goals may be reached.

Order to Comply

Whenever any person is found to be in violation of any condition or limitation imposed by a discharge permit, after appropriate notice, an order to comply with such condition or limitation shall be served on the violator. Such orders to comply are administrative in nature and as such are enforceable in court. In addition, violations of such orders are subject to a civil penalty not to exceed $10,000 per day of such violation.

Civil Actions

The Administrator of the EPA is authorized under the 1972 act to commence a civil action for appropriate relief, including a permanent or temporary injunction, for any violation for which a compliance order would issue. Such action would be brought in the federal district court having jurisdiction where the polluter resides or is doing business.

Criminal Penalties

Any person who willfully or negligently violates any permit condition or permit limitation contained in a permit issued by the EPA or any state shall be punished by a fine of not less than $2500 or more than $25,000 per day of violation, or by imprisonment for not more than 1 ycar, or by both. Subsequent violations are punishable by a fine of not more than $50,000 per day of violation, or by imprisonment for not more than 2 years, or by both.

Any person who knowingly makes any false statement, representation, or certification in any application, record, report, plan, or other document filed or required to be maintained under the 1972 amendment, or who falsifies, tampers with, or knowingly renders inaccurate any monitoring device or method required shall, upon conviction, be punished by a fine of not more than $10,000, or by imprisonment for not more than 6 months.

These enforcement procedures are seen by concerned officials as "fair but firm." The full implementation of the effluent limitations and performance standards cannot be accomplished without providing the legal muscle to back them up. Although the EPA may lack the sufficient manpower to watch over the nation's waters and protect them to the extent that the 1972 amendments have expressed, the proposed national water quality surveillance system should provide the EPA with the assistance that it needs to achieve the expressed goals.

SUMMARY

The federal government's program for setting and enforcing standards to prevent, reduce, and eliminate water pollution is now in its crawling stages. The interim effluent limitations for NPDES permits are the cornerstone for dealing with the problem that our technology and progress have created and that we affectionately call the "pollution situation." The crucial date of the program is 1985. By that time the research, guidance, and leadership are to culminate into an adult program capable of claiming the achievement of "zero discharge." The future for industry in light of this burden in conjunction with the current energy crisis seems awesome. The pretreatment requirements that are now imposed and will become even more stringent are going to economically burden some industries almost to the breaking point. There is little consolation that can be offered here for those who must face this problem in the wallet. There is the thought that the nation whose political philosophy has enabled industry to grow and prosper is making an effort to prolong the physical conditions to sustain that growth.

Chapter Ten

Ordinances and regulations of states, counties, and cities

All dischargers of wastewater are subject to the scrutiny of one or more governing agencies. Each of these has formulated water quality programs in the form of enforceable regulations.

There is a network of these regulations (Table 10-1) which involves all levels of governing bodies. The apex of this network is the federal government.

Table 10-1   Regulations Network

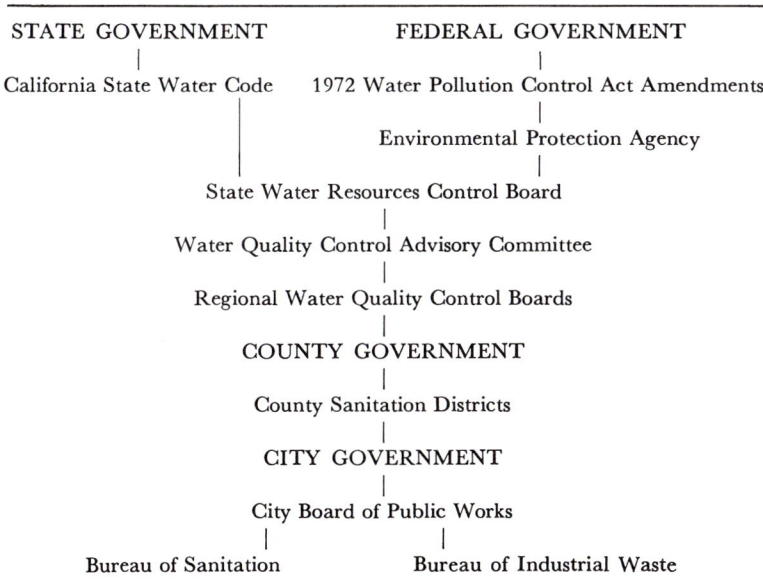

STATE GOVERNMENT	FEDERAL GOVERNMENT
|	|
California State Water Code	1972 Water Pollution Control Act Amendments
	|
	Environmental Protection Agency
	|
State Water Resources Control Board	
|	
Water Quality Control Advisory Committee	
|	
Regional Water Quality Control Boards	
|	
COUNTY GOVERNMENT	
|	
County Sanitation Districts	
|	
CITY GOVERNMENT	
|	
City Board of Public Works	
|	|
Bureau of Sanitation	Bureau of Industrial Waste

In the field of wastewater Chapter 9 revealed that the 1972 amendments are the primary regulation from which the Environmental Protection Agency was authorized to set forth guidelines for wastewater discharge. The federal regulations encourage state governments to develop water quality programs, and the state regulations affect all governing agencies down the line within their jurisdictions. The end result is that when a city needs to develop a program of its own, it looks up the network and incorporates the portions of the existing regulations that are mandatory and those that the city's particular situation require.

GETTING TO THE STATES

In 1965 the federal government issued guidelines directed at the states for the establishment of water quality standards for *interstate* waters. Each state was to decide how it wanted to use the interstate waters within its jurisdiction. The federal criteria demanded that the quality of water for a specific use has to be protected under the state program. If a state wanted to use interstate waters as a source of drinking water, it had to enact protective measures to safeguard the quality of the water for such a use. Such water quality programs were not required for use of *intrastate* waters.

In 1972 the Federal Water Pollution Control Act Amendment expanded the old water quality standards program. Both *interstate* and *intrastate* waters must meet federally established criteria. The interstate standards established under the 1965 law are to remain in effect unless they are inconsistent with the standards set by the Environmental Protection Agency (see Chapter 9). All state programs concerning interstate waters must obtain EPA approval. The states must also adopt and submit for approval a program for protecting the quality of intrastate waters.

The states are not limited to those criteria established by the federal government. As previously discussed, the federal criteria are guidelines. Any state, finding that the effluent limitations set forth by the EPA are not sufficient to meet its water quality control needs, may impose more stringent standards on pollution sources.

In order for any state to enforce those standards for water quality that it deems necessary, the state must undertake a program to control the discharge of pollutants. The most practical way to impose such controls is by requiring every discharger to control his effluent and submit to an appropriate agency proof that there is compliance with quality standards. This is the same method as proposed under the National Pollutant Discharge Elimination System permit program. Rather than have a state discharge permit in addition to the national one, a state may adopt a permit program incorporating the basic federal requirements and its own particular limitations.

The Environmental Protection Agency must approve any such permit program. EPA approval requires that a state permit program must assure compliance with the federal law and must include:

1. Requirements for monitoring and reporting discharges

2. Adequate provisions for entry by an authorized official of the state onto dischargers' premises to inspect polluting facilities and any required monitoring and sampling equipment therein

3. Procedures for making information concerning discharges public

4. Enforcement machinery

5. Adequate funds and a staff of qualified personnel to operate the program

A state permit program must also contain provisions for public notice of all permit applications and must provide an opportunity for a public hearing before a permit is granted. Also, no permit issued under such a program may extend permission to discharge for a period of more than 5 years.

After a state permit program is approved and goes into effect, the EPA may in its discretion review, demand revision, or revoke such a program. The basis for such EPA action would be failure to comply with the federal regulations. Within the administration of any state program the EPA has the power to review any application for permit previously issued to determine whether it meets the requirements of federal law.

In lieu of the federal limitations on discharge into the interstate waters, there is a high probability that many states will simply adopt these limitations into their intrastate programs. The federal government has given the funds to the EPA to research water quality. Few states have the funds to conduct a similar program directed at the state's particular water pollution problems. Thus the states are likely to be content with the conclusions reached by the EPA.

What this means to the industrial discharger (to a state-owned sewage treatment plant) is that the standards now demanded by such a plant will probably be changed so as to be consistent with the federal requirements. Also, there is little chance that publicly owned treatment plants which must obtain federal permits to discharge to the oceans, underground waters, or interstate waters will take it upon themselves to treat other than the biological contaminants of sewage. Thus the cost of bringing the treatment plant's effluent up to federal requirements will be borne by each discharger to such plant. Each discharger whose operation produces an effluent below the federal standards will be required to pretreat his effluent.

The EPA has enunciated pretreatment procedures in such a manner that the states can say to their industrial dischargers, "Follow the federal guidelines for pretreatment." Chapter 9 describes the way the EPA has suggested a metal-finishing operation must conduct itself so as to meet pretreatment standards.

State Water Quality Control Programs

Although all of the fifty states have some type of water pollution control program, California is known to be the leader in this area of environmental protection.

For this reason, and because the author is most familiar with the intricacies of California regulations, California will serve here as the example of the manner in which a state regulates the use of, and discharge into, the waters within its jurisdiction.

The California State Water Code, Division 7, is known as the Porter-Cologne Water Quality Control Act. This act contains the California plan for water pollution control and abatement. The policy of the state as stated in the act is

. . . that the water resources of the State be put to beneficial use to the fullest extent of which they are capable and that the waste or unreasonable use or unreasonable method of use of water be prevented. Moreover, because of the widespread demand and need for the full utilization of the water resources of the State for beneficial uses, it is the policy of the State that the disposal of wastes into the waters of the State shall be so regulated as to achieve highest water quality consistent with maximum benefit to the people of the state and shall be controlled so as to promote the peace, health, safety and welfare of the people of the State.

The State Water Resources Control Board was established by the Porter-Cologne Act and therein designated as the state water pollution control agency for all purposes stated in the 1972 amendments. This agency works in conjunction with the Water Quality Control Advisory Committee and the Regional Water Quality Control Boards to administer statewide programs for water quality control. Such administration includes research programs, programs delegated to the state under federal law, dispersal of federal funds for any such water quality control program, and programs limiting the discharge of wastes into the waters of the state.

In March 1972 the State Water Resources Control Board proposed a water quality plan for the ocean waters of California. This plan was adopted in December 1972 and acted to supersede all less restrictive provisions of existing water quality control plans and policies of the state board and the regional boards. The EPA has approved this plan.

The general requirements of this plan are that wastewater discharge must be essentially free of:

1. Material which is floatable

2. Settleable material which adversely affects aquatic life

3. Substances toxic to marine life

4. Substances which decrease natural light to aquatic life

5. Discoloration which is undesirably unaesthetic

The exacting limits of water quality constituents are as follows:

Physical Characteristics. Dissolved oxygen concentration shall not be depressed more than 10 percent from natural conditions. The pH shall not be changed more than 0.2 units from natural conditions. The range of 6.0 to 9.0 is permissible. Other limitations are

	More than 50% of the time	*More than 10% of the time*
Settleable solids, mg/l	0.0	0.2
Suspended solids, mg/l	50.0	75.0
Turbidity, JTU	50.0	75.0
Total grease and oil, mg/l	25.0	40.0
Petroleum grease and oil, mg/l	15.0	25.0
Floating particles, dry weight, mg/l	1.0	2.0

Chemical Characteristics. All chemical constituents are measured in mg/l.

	More than 50% of the time	*More than 10% of the time*
Cyanide	0.1	0.2
True phenolic compounds	0.5	1.0
Total chlorine residual	1.0	2.0
Ammonia (N)	40.0	60.0
Conservative substances:		
Arsenic	0.05	0.10
Cadmium	0.02	0.04
Total chromium	0.2	0.4
Copper	0.2	0.4
Lead	0.05	0.10
Mercury	0.005	0.01
Nickel	0.2	0.4
Silver	0.02	0.04
Zinc	0.5	1.0
Total identifiable chlorinated hydrocarbons	0.002	0.004

REGIONAL WATER QUALITY CONTROL

The state of California is divided into nine regions, devised through demarcation of state basin areas (see Figure 10-1), each with its own regional water quality control board. It is the duty of each board to adopt a water quality control plan for the region to "ensure the reasonable protection of beneficial uses and the prevention of nuisance."

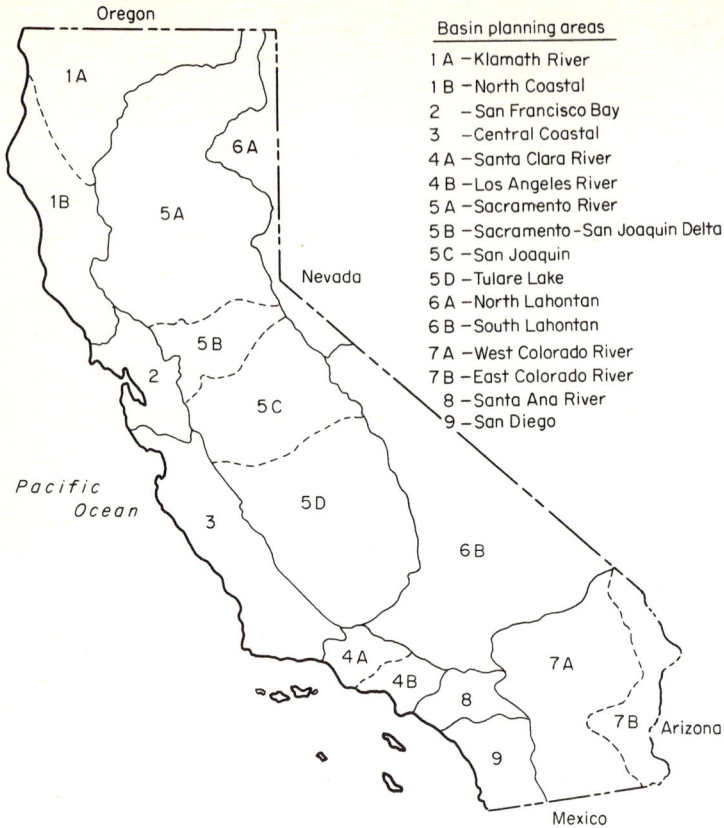

Basin planning areas

1 A — Klamath River
1 B — North Coastal
2 — San Francisco Bay
3 — Central Coastal
4 A — Santa Clara River
4 B — Los Angeles River
5 A — Sacramento River
5 B — Sacramento-San Joaquin Delta
5 C — San Joaquin
5 D — Tulare Lake
6 A — North Lahontan
6 B — South Lahontan
7 A — West Colorado River
7 B — East Colorado River
8 — Santa Ana River
9 — San Diego

Fig. 10-1   Basin location map. (*California State Water Resources Control Board.*)

The points which are considered by each regional board in establishing water quality objectives include:

1. Past, present, and probable future beneficial uses of water

2. Environmental characteristics of the hydrographic unit under consideration, including the quality of water available thereto

3. Water quality conditions that could reasonably be achieved through the coordinated control of all factors which affect water quality in the area

4. Economic considerations

A regional board may specify certain conditions or areas where the discharge of waste, or certain types of waste, will not be permitted. No such determination or specification is to be made without there first being a public hearing on the proposed prohibition.

The authority of the Regional Water Quality Control system, as set forth in the Porter-Cologne Act, extends to:

Any person discharging or proposing to discharge waste within any region that could affect the quality of the waters of the state, other than into a community sewer system, and any person who is a citizen, domiciliary, or political agency or entity of this state discharging waste or proposing to discharge outside the boundaries of the state in a manner that could affect the quality of the waters of the state within any region.

All such dischargers are required to file, with the appropriate regional board, a report of their discharge and from then on to report any material change or proposed change in the character, location, or volume of the discharge. Any discharger failing to file such a report when required to do so is guilty of a misdemeanor. The regional board may also require a particular discharger to furnish such technical or monitoring program reports as the board may specify. Failure to comply with such board requests is considered a misdemeanor.

Regional Control and Planning

The EPA has required each state to prepare and approve water quality control plans for drainage basins as a condition for future receipt of construction grants by communities. Under the present federal-state construction grant programs a community may receive up to 55 percent of the capital cost of a wastewater treatment project from the EPA and an additional 25 percent from the California State Water Resources Control Board, leaving as little as 20 percent of the cost to be met by local funding. Under such a program, federal and state officials must be assured that the investment will purchase protection of the waters from the effects of wastes and make maximum use of wastewater as a resource.

One such water quality control plan which operates within the regional water quality control program is that devised by the California Regional Water Quality Control Board—Los Angeles Region. The plan now in operation is an interim plan adopted in 1971. The final plan is expected in the near future. The area which this interim plan governs consists of the Los Angeles and Santa Clara River basins (see Figure 10-2).

The overall objective of the interim plan is to set forth a definitive program of actions designed to preserve and enhance water quality and protect beneficial water uses in a manner which will result in maximum social and economic benefits to the people of the state of California.

Actions will be directed toward implementing the following goals:

1. Protect and enhance all state waters, surface and underground, fresh and saline, for present and anticipated beneficial uses including aquatic environmental values.

2. The quality of all surface waters shall be such as to permit maximum recreational use where this use is otherwise practical.

3. Achieve maximum practical use of freshwater through reclamation and reuse by industries, municipalities, and agriculture.

4. Continually upgrade the quality of waste treatment systems to assure consistently high-quality effluents.

Fig. 10-2   Santa Clara River Basin and Los Angeles River Basin. (*California State Water Resources Control Board.*)

5. Develop a planned system for water use and waste discharge to assure protection of the aquatic resource for future uses and achieve harmony with the natural environment.

Regional Pretreatment Standards

The cessation of discharge of environmentally hazardous substances into the sewer systems is the most economical method of eliminating them from the community treatment plant's effluent. Pretreatment of industrial wastes is essential in those areas where effective source control cannot be achieved.

The Los Angeles Area regional board has set general guidelines for the pretreatment of industrial wastes discharged to treatment plants within the region.

A.   Industrial and municipal effluents shall be so treated as to assure essentially complete removal of the following substances:

1. Chlorinated hydrocarbons

2. Toxic substances

3. Harmful substances that may enter food webs

4. Excessive heat

5. Radioactive substances

6. Grease, oil, and phenolic compounds

7. Excessively acidic and basic substances

8. Heavy metals such as lead, copper, zinc, mercury, or mercuric compounds

9. Other deleterious substances

B. Sewering entities are encouraged to prohibit discharge of such substances by limiting them at their source.

Tables 10-2 and 10-3 contain the discharge limitations applicable to the publicly owned treatment plants in this region which discharge to the Pacific Ocean. The pretreatment standards promulgated by the municipalities controlling these plants will be geared so that the treatment plants are able to comply with the regional board requirements.

Table 10-2 Effluent Quality Requirements for Discharges to Ocean Waters, Physical Characteristics

	Unit of measure	Concentration not to be exceeded more than:	
		50% of the time	*10% of the time*
Grease and oil (hexane extractables)	mg/l	10.0	15.0
Floating particulates (dry weight)	mg/l	1.0	2.0
Suspended solids	mg/l	50.0	75.0
Settleable solids	ml/l	0.1	0.2
Turbidity	JTU	50.0	75.0
pH	Units	Within limits of 6.0 to 9.0 at all times	

Table 10-3 Effluent Quality Requirements for Discharges to Ocean Waters, Chemical Characteristics

	Unit of measure	Concentration not to be exceeded more than:	
		50% of the time	*10% of the time*
Arsenic	mg/l	0.01	0.02
Cadmium	mg/l	0.02	0.3
Total chromium	mg/l	0.005	0.01
Copper	mg/l	0.2	0.3
Lead	mg/l	0.1	0.2
Mercury	mg/l	0.001	0.002
Nickel	mg/l	0.1	0.2
Silver	mg/l	0.02	0.04
Zinc	mg/l	0.3	0.5
Cyanide	mg/l	0.1	0.2
Phenolic compounds	mg/l	0.5	1.0
Total chlorine residual	mg/l	1.0	2.0
Ammonia (expressed as N)	mg/l	40.0	60.0
Total identifiable chlorinated hydrocarbons	mg/l	0.002	0.004
Toxicity concentration	tu	1.5	2.0

Regional Enforcement Procedures

The Porter-Cologne Act empowers the regional boards with various administrative enforcement procedures and remedies with which it can ensure implementation of water quality control plans.

When a regional board finds a discharge of waste is taking place in violation of discharge prohibitions prescribed by the regional board or the state board, the board may issue an order to cease and desist and direct those persons not complying with the requirements or discharge prohibitions to (1) comply forthwith; (2) comply in accordance with a time schedule set by the board; or (3) in the event of a threatened violation, take appropriate remedial or preventive action. In the event of an existing or threatened violation of waste discharge requirements in the operation of a community sewer system, cease and desist orders may restrict or prohibit the volume, type, or concentration of waste that might be added to such system by dischargers who did not discharge into the system prior to the issuance of the cease and desist order.

Any person who discharges waste into the waters of the state in violation of any waste discharge requirement or other issued by a regional board or the state board, or who intentionally or negligently causes or permits any waste to be discharged or deposited where it is, or probably will be, discharged into the waters of the state and creates, or threatens to create, a condition of pollution or nuisance, shall upon order of the regional board clean up such waste or abate the effects thereof, or in the case of threatened pollution or nuisance take other necessary remedial action.

Upon failure of any person to comply with such cleanup or abatement order, the attorney general, at the request of the board, shall petition the superior court for that county for the issuance of an injunction requiring such person to comply therewith. In any such suit, the court shall have jurisdiction to grant a prohibitory or mandatory injunction, either preliminary or permanent as the facts may warrant.

Like the federal water quality regulations, these state and regional regulations permit more localized governmental authorities (counties and cities) to enforce additional limitations on the disposal of waste or any other activity which might degrade the quality of the waters within its jurisdiction. Thus these state regulations are viewed by local authorities as the bottom-line restrictions. They can and have enacted regulations of their own in an effort to control effluent contents which have deleterious effects on facilities and the environment peculiar to their jurisdiction.

COUNTY CONTROL

Los Angeles County was chosen to illustrate this level of governmental regulation of industrial waste discharge.

The sanitation districts of Los Angeles County were formed in 1923 to construct, operate, and maintain facilities to collect, treat, and dispose of sewage and industrial wastes. Today there are twenty-six sanitation districts and combined they are the largest in the world.

A sanitation district is any agency of the county government which operates the system of trunk sewers serving homes, commercial establishments, and industry. The district builds and operates only trunk sewers and sewage disposal systems. Local sewage collecting systems or lateral sewers are built and operated by the various cities under their engineering department's control. All liquid wastes originating from the district's boundaries are removed by the district's sewerage systems provided that the wastes do not (1) damage structures, (2) create nuisances such as odors, (3) menace public health, (4) interfere with sewerage treatment processes, or (5) impose abnormal costs on the district for their collection and disposal. Storm waters or uncontaminated waters of any origin are not permitted to be discharged into trunk sewers.

The sanitation districts require all persons who discharge or cause to be discharged any industrial wastewaters directly or indirectly to sewerage facilities owned by the districts to first obtain a district's permit for industrial wastewater discharge. The discharge of industrial wastewater in excess of the quantity or quality limitations set by the permit is prohibited and punishable by fines.

The permit for industrial wastewater discharge may require pretreatment of industrial wastewater before discharge, restriction of peak flow discharges, discharge of certain wastewaters only to specified sewers of the district, relocation of point of discharge, prohibition of discharge of certain wastewater components, restriction of discharge to certain hours of the day, payment of additional charges to defray increased costs of the district created by the wastewater discharge, and such other conditions as may be required.

The Permit Form

The permit application form used by the sanitation districts of Los Angeles County is typical of other wastewater permits including the NPDES permit (see Figure 10-3).

First, identification of the discharger is required by firm name and address as well as the property owner's name and address. Second, the legal description of the property concerned and the point at which the plant sewer line will connect to the public sewerage system must be submitted in detail. Third, a description of the type of industry is required as well as the Standard Industrial Classification (SIC) number, which is the federal government code number for the particular type of industry. This classification establishes the business category, whether manufacturing or nonmanufacturing.

Also required is the number of full-time and part-time employees in order to estimate the nonindustrial wastewater rate of consumption and discharge.

The remainder of the permit application form requires a detailed explanation and description of the process(es) to be employed by the plant. These will be discussed in detail.

The raw material used in production, as well as the products produced,

```
APPLICATION NO.                    EXISTING INDUSTRY                         PERMIT NO.
                           PERMIT FOR INDUSTRIAL WASTEWATER DISCHARGE
  E 31545                  SANITATION DISTRICTS OF LOS ANGELES COUNTY
                               2020 Beverly Blvd., Los Angeles, Calif. 90057
                           John D. Parkhurst, Chief Engineer and General Manager
```

01 _____ , Calif.* ____/____/____
MO.   DAY   YR.

APPLICATION IS HEREBY MADE BY _____ A
                              PRINT   (FIRM NAME)
03 (Mailing Address) _____ C05 _____ E
                    (STREET)   (CITY)   (STATE)   (ZIP)

07 _____ G of the property located at:
        (OWNER, TENANT, ETC.)
09 (Street) _____ 111 (City) _____ (Zip) _____
   PRINT   (ADDRESS OF PROPERTY PRODUCING WASTEWATER DISCHARGE)

Assessors Map Book No. _____ Page No.* _____ Parcel No.* _____ K
                              (LEGAL ADDRESS OF PROPERTY PRODUCING WASTEWATER DISCHARGE)

   PRINT   (LOCATION OF POINT OF WASTEWATER DISCHARGE TO SEWERAGE SYSTEM)
for a Permit for Industrial Wastewater Discharge to the sewerage system.

13 Type of Industry* _____ M. 17 _____ Q
                  (GENERAL DESCRIPTION)   (FEDERAL SIC NOS.)

19 Number of Employees (Full Time)* _____ (Part Time)* _____ S

21 Raw Materials Used* _____ U
                        (GENERAL DESCRIPTION – ADD ADDITIONAL SHEETS AS NEEDED)

   Products Produced _____
                        (GENERAL DESCRIPTION – ADD ADDITIONAL SHEETS AS NEEDED)

   Wastewater Producing Operations _____

   _____
                        (FULL DESCRIPTION – ADD ADDITIONAL SHEETS AS NEEDED)
31 Time of Discharge – * _____ AM/PM to _____ AM/PM,   Days per Week* M T W Th F Sa Su
                  (WORKING DAY – CROSS OUT AM OR PM)   (CIRCLE DAYS)

* Wastewater Flow Rate* _____ AA (Gallons Per Day) _____

   Constituents of Wastewater Discharge _____

   _____
              (GENERAL DESCRIPTION – ATTACH CHEMICAL ANALYSES RESULTS TO THIS APPLICATION)

Person in company responsible for industrial wastewater discharge:

41 _____ BB
   PRINT   (NAME)   (POSITION)   (TELEPHONE NUMBER)

I affirm that all information furnished is true and correct and that the applicant will comply with the conditions stated on the back of this permit form.

Date _____ , 19 _____

Signature for Applicant _____
(COMPANY ADMINISTRATIVE OFFICIAL)   (NAME)   (POSITION)
Approved by City or County Official   Approved by Sanitation Districts of Los Angeles County

Date _____   Date _____
For Dept. of County Engineers .... ☐   John D. Parkhurst, Chief Engineer and General Manager
   City of _____ ☐
Name _____   by _____
Position _____   Position _____

Note: A permit fee may be required by the local City or County Agency.
       This form when properly signed shall be a valid permit unless suspended or revoked.
                    RETURN THIS COPY TO APPLICANT WHEN APPROVED

Fig. 10-3 Application form for permit for industrial wastewater discharge. (*Sanitation Districts of Los Angeles County.*)

must be described. The wastewater-producing operation is then described. This is the most important part of the application. A description including a flow diagram should be included. The time of discharge given in the hours of the day is necessary since the sanitation district is concerned with the peak flow period into their treatment plant. They are more liberal with amount of flow discharged during off-hours than daylight hours.

The wastewater flow rate in gallons per day is to be given. This may be determined from water meter readings when domestic waste and industrial

SANITATION DISTRICTS OF LOS ANGELES COUNTY
INDUSTRIAL WASTEWATER
CRITICAL PARAMETER REPORT FORM

PERMIT NO.

Ident. Code	PARAMETER 1/	2/	QUANTITY VALUES	Ident. Code	PARAMETER 1/	2/	QUANTITY VALUES
A	Flow (Total)		gals/day	V	Manganese - Total		mg/l
B	Flow (Peak)		gals/min.	W	Mercury - Total		mg/l
C	COD		mg/l	X	Molybdenum - Total		mg/l
D	SS (Suspended Solids)		mg/l	Y	Nickel - Total		mg/l
E	pH		Units	Z	Selenium - Total		mg/l
F	Total Dissolved Solids		mg/l	AA	Silver - Total		mg/l
G	Ammonia (N)		mg/l	BB	Sodium - Total		mg/l
H	Sulfide		mg/l	CC	Thallium - Total		mg/l
I	Cyanide		mg/l	DD	Tin - Total		mg/l
J	Fluoride		mg/l	EE	Titanium - Total		mg/l
K	Aluminum - Total		mg/l	FF	Zinc - Total		mg/l
L	Antimony - Total		mg/l	GG	Oil & Grease (HEXANE EXTRACT)		mg/l
M	Arsenic - Total		mg/l	HH	Phenols		mg/l
N	Beryllium - Total		mg/l	II	Surfactants (MBAS)		mg/l
O	Boron - Total		mg/l	JJ	Chlorinated Hydrocarbons (EXCEPT PESTICIDES)		mg/l
P	Cadmium - Total		mg/l	KK	Pesticides (CHLOR. HYCARB.)		mg/l
Q	Chromium - Total		mg/l	LL	Radioactivity (ALPHA, BETA, GAMMA)		pCi/l
R	Cobalt - Total		mg/l	MM	Temperature		Degrees °F
S	Copper - Total		mg/l	NN	Color		Units
T	Iron - Total		mg/l	OO	Thiosulfate (S)		mg/l
U	Lead - Total		mg/l				

	NON-CRITICAL PARAMETERS (Report When Available)				OTHER PARAMETERS (Report When Requested)		
PP	Calcium		mg/l	A1	Raw Material or Product Quantities		per day
OO	Magnesium		mg/l	A2			
RR	Potassium		mg/l	A3			
SS	Barium		mg/l	A4			
TT	Nitrate		mg/l	A5			
UU	Chloride		mg/l	A6			
VV	Bromide		mg/l	A7			
WW	Sulfate		mg/l	A8			
XX	Phosphorus-Ortho		mg/l				

NOTES:

1/ Report all critical parameters required by the Sanitation Districts and any other critical parameter known to be present in the wastewater. Those parameters required by the Districts but known to be absent from the wastewater may be reported by placing the word absent in the appropriate space.

2/ If values are obtained by measurements or analyses write A in this column. Analysis values must be determined, using representative 24-hour composite samples, by a State Certified or Districts Approved Laboratory. If values are obtained by estimate, write E in this column. Estimated values are acceptable for new plants only.

(PRINT)   NAME AND ADDRESS OF LABORATORY PERFORMING ANALYSES AND FLOW MEASUREMENTS

(PRINT)   NAME OF COMPANY HAVING WASTEWATER DISCHARGE   SIC NUMBER(S)

(PRINT)   ADDRESS OF WASTEWATER DISCHARGE

(PRINT)   ADDITIONAL LOCATION DATA (DATA ABOVE SHOULD BE FOR ONLY ONE DISCHARGE POINT TO THE SEWERAGE SYSTEM)
Statement of Accuracy of Data

I hereby affirm that the above data comprise a true and correct representation of the wastewater discharged from the stated discharge point.

Date: _____ Location: _____, California _____

(SIGNED) NAME   POSITION (ADMINISTRATIVE OFFICER OF COMPANY WITH WASTEWATER DISCHARGE)

Fig. 10-4   Critical parameter report form. (*Sanitation Districts of Los Angeles County.*)

wastewater are mixed in the plant. When industrial wastewater is separate, it can be measured by a flowmeter.

A general description of the constituents of the wastewater must be given. For an existing plant a full laboratory chemical analysis should be attached. For new plants an estimate of flow and constituents is expected.

For each SIC number there is an Industrial Wastewater Critical Parameter Report Form which lists all constituents required to be reported (see example form, Figure 10-4).

The applicant is required to furnish data on items listed or affirm that they are absent from the wastewater discharge.

If values were obtained by measurement or analysis, the name of the laboratory must be given. This laboratory must be state-certified or district-approved.

If the values are obtained by estimate, as in a proposed plant, it should be so noted. This estimate should be verified by suitable measurements made within 6 months after the start of operations.

Finally, the name, position, and telephone number of the person in the company responsible for industrial wastewater discharge must be given. An administrative official of the company must affirm and sign that all information given on the form is true and correct.

When a plant has more than one point of discharge, a separate permit is required for each.

Together with the permit form the discharger should furnish data and descriptive plans of the company's facilities for examination and approval. The plans should include:

1. A plot plan showing disposal of rainwater.

2. A floor plan showing the origin of the wastewater, identifying processes creating the wastewater and listing accurately for each wastewater connection to sewerage, the total daily flow in gallons and the peak flow in gallons per minute. For companies having extensively complicated processes, in addition to the above, a schematic mass balance flow diagram of the wastewater sources from water supply to wastewater discharge should be included showing water and wastewater total daily flows, evaporation losses, losses to product, etc.

3. The source of all water used in the plant and the average total volume of water used per month from each source.

4. The location of all sewers on company property as well as the location of the connection of the outfall sewer from company property to the public sewerage system.

5. Detail plans to be used for the pretreatment of the wastewater before discharge to the sewer.

6. An automatic full-time total flow measurement system incorporating flow indication, totalization, and recording equipment for any company having a peak 30-min flow rate over 100 gal/min or a total daily sewer discharge of 75,000 gal/d or more.

7. An automatic effluent pH measurement and control system to regulate neutralization of acid wastes for any company proposing to discharge substantial amounts of acid wastes.

8. A description of liquid waste materials leaving the plant other than in the industrial wastewater discharged into the sewerage system, indicating types and materials and daily quantities.

Industrial Wastewater Treatment Surcharge

An industrial wastewater surcharge must be paid to the district annually, following the fiscal year in which charges accrue, by those industrial wastewater dischargers whose contribution of flow, chemical oxygen demand, suspended solids, or peak flow creates costs in excess of the value of dischargers' ad valorem taxes. The treatment surcharge is based on the appropriate district's sewerage system's total maintenance, operation, and capital expenditures for providing industrial wastewater collection, treatment, and disposal services.

Periodic measurements of flow rates, flow volumes, chemical oxygen demand, and suspended solids for use in determining the annual industrial wastewater treatment surcharge and such measurements of other constituents believed necessary by the chief engineer are required of all industrial wastewater dischargers.

All wastewater analysis must be conducted in accordance with appropriate procedure contained in "Standard Methods." There are two standard analytical methods for analysis of industrial wastewater: (1) *Standard Methods for the Examination of Water & Wastewater,* 13th edition, 1971, American Public Health Association, N.Y., N.Y. 10019. (2) ASTM Standards, Part 23, *Water; Atmospheric Analysis,* 1972, American Society for Testing and Materials, Philadelphia, Pa. 19103.

The following is an explanation of how to use the analytical data to complete the surcharge form. Table 10-4 gives some typical data for plating companies. Companies 1 and 2 discharge less than 6 million gal/year; hence they have the option of using either the short form or the long form of the Industrial Wastewater Treatment Surcharge Statement. If they choose the short form, the only data they need for the calculation of the surcharge are annual flow volume. Companies 3, 4, and 5 discharge more than 6 million gal annually; therefore, they must use the long form. Companies using the long form need all the kinds of data shown in Table 10-3 in order to calculate the surcharge. In addition, the quantities of COD and SS must be calculated in thousands of pounds per

Table 10-4 Typical Plating Company Data Needed to Calculate the Industrial Wastewater Treatment Surcharge

	Company				
	1	*2*	*3*	*4*	*5*
Annual flow*					
ft³/year	121,672	674,387	1,092,470	9,660,027	33,603,709
gal/year	912,540	5,057,903	8,193,525	72,450,203	252,027,818
Peak flow rate, gal/min	9	43	85	295	583
COD, mg/l	8	11	80	164	266
Suspended solids, m/l	38	157	316	247	157
Working days per year	250	260	260	312	360
Avg. working hours per day	8	16	8	16	24
Ad valorem tax	$28.05	$505.73	$1517.00	$7609.63	$2475.96

* 1 ft³ = 7.5 gal.

year. These quantities are calculated in Tables 10-5 and 10-6. Table 10-7 summarizes the surcharge payable by the five companies.

Pretreatment Standards

Los Angeles County sanitation districts have pronounced interim industrial effluent limits for various wastewater constituents. The ultimate limitations are expected before January 1977. These limitations are

Table 10-5   Calculation of Quantity of COD

$$COD \ (1000 \ lb/year) = flow \ (million \ gal/year) \times COD \ (mg/l) \times \frac{8.34}{1000}$$

	Company				
	1	2	3	4	5
Flow, million gal/year	0.91	5.06	8.19	72.45	252.03
COD, mg/l	8	11	80	164	266
COD, 1000 lb/year	0.06	0.46	5.46	99.09	559.11

Table 10-6   Calculation of Quantity of Suspended Solids

Suspended solids (1000 lb/year) = flow (million gal/year)

$$\times \ \text{suspended solids} \ (mg/l) \times \frac{8.34}{1000}$$

	Company				
	1	2	3	4	5
Flow, million gal/year	0.91	5.06	8.19	72.45	252.03
Suspended solids, mg/l	38	157	316	247	157
Suspended solids, 1000 lb/year	0.29	6.63	21.58	149.25	330.00

Limitations for arsenic, cadmium, lead, silver, phenolic compounds, total identifiable chlorinated hydrocarbons, ammonia, and radioactive substances are as yet undetermined.

CITY CONTROL

The city of Los Angeles, although differing from most other cities because of its great size and population, faces typical problems of controlling industrial wastewater discharges to the waters within its jurisdiction. In those areas where a city is able to utilize the treatment facilities of neighboring or surrounding

Table 10-7 Surcharge Summary

	Company						
	1		2		3	4	5
	Short form	*Long form*	*Short form*	*Long form*	*Long form*	*Long form*	*Long form*
(1) Flow volume charge	165.51	75.76	915.86	421.25	681.82	6,031.46	20,981.50
(2) COD charge	......	0.29	......	2.19	25.94	470.68	2,655.77
(3) Suspended solids charge	......	3.19	......	72.93	237.38	1,641.75	3,630.00
(4) Peak flow charge	......	0	......	785.18	523.60	1,427.80	2,565.20
(5) Calculated payment (1 + 2 + 3 + 4)	165.51	79.24	915.86	1,281.55	1,468.74	9,571.69	29,832.47
(6) Minimum payment	200.00	200.00	200.00	200.00	200.00	....	200.00
(7) Total payment (higher of 5 or 6)	200.00	200.00	915.86	1,281.55	1,468.74	9,571.69	29,832.47
(8) Ad valorem taxes	28.05	28.05	505.73	505.73	1,517.00	7,609.63	2,475.96
(9) Net surcharge (7 − 8)	171.95	171.95	410.13	775.82	−48.26	1,962.06	27,356.51
(10) Surcharge payable (9 not less than zero)	171.95	171.95	410.13	775.82	0	1,962.06	27,356.51

Total chromium, mg/l	5.0
Copper, mg/l	20.0
Mercury, mg/l	2.0
Nickel, mg/l	10.0
Zinc, mg/l	30.0
Cyanide, mg/l	10.0
Highly acid, alkaline, or corrosive substances, mg/l	0.0
Toxic substances, mg/l	0.0

counties, the city may simply adopt county regulations as to the criteria for water quality. Los Angeles, cities like it, and those without available county facilities must construct and operate their own treatment facilities.

According to the Los Angeles Municipal Code,

It is the policy of the City of Los Angeles to assure that the highest and best use of the sanitary sewer system is for the collection, treatment and disposal of domestic sewage, and that the highest and best use of the storm drain system is for the collection and disposal of storm water. The use of either of these systems for industrial waste discharge is a privilege which is subject to the requirements of this section.

The Board of Public Works, through its Bureau of Sanitation, is responsible for operating and maintaining the sanitary sewers, storm drains and watercourses, sewage and storm water pumping plants, and the sewage treatment and disposal facilities of the city of Los Angeles.

The regulation of these wastes is necessary in the eyes of the City Council to:

Protect the facilities from damage by deleterious wastes; protect the treatment processes; protect the operating and maintenance personnel; preserve capacity in the sanitary sewer for sanitary wastes and appropriate industrial wastes; insure the safety and welfare of the public; prevent contamination, pollution and nuisance, and protect the established beneficial uses of receiving waters as required by the California Regional Water Quality Control Board, Los Angeles; and, conserve a reclaimable resource.

Accordingly, the Board of Public Works adopted the following criteria for disposal of wastes:

1. Storm waters may not be collected and discharged to the sanitary sewer system.

2. Single-pass cooling water may not be discharged to the sanitary sewer system. The blowdown or bleed-off from the cooling towers or other evaporative coolers may be accepted in the sanitary sewer system.

3. Material which will readily settle such as sand, glass, metal fillings, diatomaceous earth, etc., or floatable material which is readily removable must be removed from the wastewater prior to discharge to either the sanitary sewer system or the storm drain system.

4. Petroleum products or any other product which by reason of its nature or quantity may cause a fire or explosion or in any way be injurious to persons must be removed from the wastewater prior to discharge to either the sanitary sewer system or the storm drain system.

All discharges to the sanitary sewer must meet the following general requirements:

pH, allowable range	5.5 to 11.0
Temperature, maximum	140°F
Dispersed oil, grease, and fats, maximum	600 mg/l
Floatable oil, etc., maximum	None visible
Hexavalent chromium, maximum	5.0 mg/l
Cyanide, maximum	2.0 mg/l
Chlorinated hydrocarbons	Essentially none
Mercury	Essentially none

All discharges to the storm drain system must comply with the appropriate Water Quality Objectives adopted by the California Regional Water Quality Control Board—Los Angeles Region. For specific conditions and locations, more restrictive limits may be prescribed by the Board of Public Works.

City Industrial Wastewater Permit

All industrial wastewater dischargers within the jurisdiction of the city of Los Angeles must file for an Industrial Wastewater Permit with the Bureau of Sanitation. Accompanying such application should be a completed set of plans showing the proposed facility, or descriptions or specifications thereof, indicating the methods of compliance with the requirements imposed on the waste discharge.

A permit is required for any discharge to other than the sanitary sewer system. A permit is not required for the discharge to the sanitary sewer system when the discharge is less than 100 gal/d and pretreatment is not required, or from cooling towers, evaporative condensers, or other recirculating water heat exchange devices with a rated capacity of 5 tons or less.

The Board of Public Works announced that a quality surcharge fee will be assessed annually for commercial and industrial waste discharge to the sanitary sewer system. This fee is imposed in order that the city of Los Angeles might qualify for federal and state construction grants to provide the higher degree of sewage treatment demanded by federal and state water quality control legislation.

The surcharge fee is assessed against dischargers of wastewater stronger than residential sewage. Sewage strength is measured by the amount of solids which must be removed and treated, and the amount of oxygen required to stabilize the wastewater before discharge. The stronger the sewage, the more treatment is required, and higher costs are incurred.

SUMMARY

The exacting limitations contained in the discharge regulations are better understood in terms of the regulations network (Table 10-1). It can be easily seen that the limitations set by the city of Los Angeles do not directly correspond to those of the Regional Water Quality Control Board—Los Angeles Region. For example, the Los Angeles city limit for total chromium is 5.0 mg/l, and the regional limit is 0.005 mg/l (50 percent of the time). The discrepancy does not mean, however, that Los Angeles is not conforming to regional standards. It must be remembered that the limitations are set for the discharge of specific *concentrations* of constituents. Thus, dilution factors become an important consideration. Los Angeles can calculate, knowing the flow rate from city-owned treatment facilities to regional waters, the maximum concentration of chromium in its receiving waters from industry so that the ultimate effluent concentration will be within the regional limits.

The net result is that the regulations are complied with and that the network operates so that a comprehensive and effective water quality control program is attained.

Chapter Eleven

Designing economy into industrial waste treatment

At the national level, the water cleanup program is practically as costly as the space program. Some EPA officials estimate a cost exceeding $100 billion to meet the 1983 target standards. This would be about twice the total cost of all public sewage work for the period of the past 118 years. The cost to California alone is $6 billion. This does not include the cost to be borne by private industry. Just to meet the 1977 standards, the nation's industry will have to spend $13.5 billion. And under the 1972 Federal Water Pollution Prevention and Control Act amendments the federal government will carry 75 percent of the cleanup bill.

In a day of soaring construction costs, the question, "How much will it cost?" is not to be taken lightly. Because of the enormous amounts of money required in this cleanup program, there must be a constant effort to provide the best engineering know-how combined with a search for economy.

As Edmund Burke stated, "Economy is a distributive virtue and consists not in saving but in selection." That quotation strikes at the heart of the subject. True economy is not merely saving money. It rests in the careful selection of demonstrated facts that will introduce lasting financial benefits into the design, construction, and operation of industrial waste treatment facilities.

To design and install a truly economical facility requires a great deal of thought, research, and experience and a little bit of luck. This chapter endeavors to acquaint the reader with two sides of an economical design. One side involves cost comparisons. As an example, consider this question: Should one use gravity flow lines that require saw-cutting of existing concrete slabs, excavation, pipe laying, backfill, and new concrete slab, or should one use a pressure line requiring overhead piping? Which is more economical? The answer of course depends upon the situation and requires careful cost comparisons.

The other side is less tangible. It requires creativity. What innovations can be made, given the situation, to provide the most economical design possible? Perhaps an abandoned building on plant property can be restored to house a treatment facility. Perhaps the wastes from various parts of the plant can be combined for a neutralizing effect.

In considering the true cost of various treatment facilities, many items should be investigated in addition to the so-called list price of the equipment. The real price of an item is difficult to ascertain. There are many imprecise terms such as the list price, retail price, wholesale price, markup, discounts, original equipment manufacturer's price, resale price, and others. In general, there are three prices on most equipment: (1) owner's or user's price, (2) contractor's price, and (3) original equipment manufacturer's price (OEM). The price directly to the owner is the highest and the OEM is the lowest.

The so-called list price of most equipment may be far less than the in-place cost. Price lists printed in catalogs do not include discounts, sales tax, freight, or installation. They also do not include the overhead and profit of the general contractor and subcontractor. Their extra costs can vary greatly. If an installation is awarded to a general contractor, it is very likely that he will in turn contract the work to various subcontractors. Each subcontractor, a specialist in a single trade, is responsible for the purchase of particular equipment and its installation. For example, an electric motor might cost an electrical subcontractor $260, but the final cost to the general contractor might be $505.63. This is explained by a typical summary sheet by the electrical contractor which includes all the burden costs as follows:

Material (motor)	$260.00	
Sales tax, 5%	13.00	
Freight, 5%	13.00	
Total material cost		$286.00
Labor to install	121.00	
Payroll tax, 24%	29.00	
Tools, 3%	3.63	
Prime cost		$439.63
Overhead and profit, 15%		66.00
Total charge		$505.63

The general contractor in turn may add 15 percent for his overhead and profit, or $75.84, making cost to the owner a total of $581.47. But despite the overhead and profit factor, the contractor can frequently install an item of equipment for the owner at a lower cost than the owner could because of the contractor's discounts. This is an important consideration. Many plants prefer to have their plant managers run the construction of the facility, negotiating with each subcontractor individually for labor only. But buying the equipment directly from the manufacturer is not necessarily the most economical method of constructing the facility. To determine which is more economical, the OEM prices must be determined and comparisons made.

Regardless of the type of wastewater treatment considered, the full cost implications must be analyzed. These costs are generally divided into capitalization costs and operating costs. Amortization of equipment over its useful life at prevailing interest rates must be calculated. Operating expenses include cost of water, electricity, and chemicals. Labor costs should include manpower required for operation and maintenance.

As a suggested tabulation of costs the following outline is recommended.

1. Capital costs—equipment, foundation, and utilities

2. Operating costs—treatment cost per 1000 gal of wastewater

    a. _____ gal/h _____ gal/24 h

    b. _____ operators per shift (surveillance and adding chemicals)

    c. _____ laboratory technicians

    d. Chemicals: _____ lb/1000 gal @ $ _____/lb

    e. Summary:
       Operator $_____/h _____ h/week
       Technician $_____/h _____ h/week
       Chemicals $_____/1000 gal _____1000 gal/week
       Power $_____/_____gal _____gal/week

3. Maintenance costs

    a. Labor

    b. Materials

There are other general considerations one must be aware of. Some machines are portable and can rest on the floor; they require only a simple 115-V receptacle. Other machines, designed to perform the same general function, may require a foundation, a fused switch, and any number of extras. An example of this situation can be found when one is ordering a centrifuge.

Other factors that could affect the overall cost of the project include changes during construction. It goes without saying that when dealing with any type of construction, any changes made during the progress of construction could have disastrous financial affects.

ESTIMATING THE COST OF A TREATMENT FACILITY

Examination of a particular industry's solution to their wastewater treatment problem should be helpful in understanding the many kinds of costs incurred. A Los Angeles–based industry required the construction of a wastewater treatment facility.

The project involved the collection of wastes from several buildings spread out over the plant. The wastes were to be collected at a single treatment facility. To accomplish the installation of the system, the new facility had to be constructed, pipelines from the various buildings had to be installed, and a decision for disposal of the concentrated wastes had to be made. Recycling of the treated waters was considered for the future.

In order to transport the wastes from one building to the treatment facility, a drain line was extended from the inside of the building through an existing parking lot to a new neutralizing tank. Transporting wastes over this long distance involved many details of demolition work. The installation of the new facility itself involved demolition work.

Demolition Costs

In both new and existing facilities, the removal of paving, clearing work sites, removing existing structures, and removal and relocation of pipelines, conduits, and machinery precede all work. In this particular project concrete saw-cutting was required for a short distance outside the building.

The subcontract price of saw-cutting is based upon the depth of the cut in inches and the total lineal footage of each cut. There is also a minimum cost for mobilizing the equipment at the jobsite. For example, to single saw-cut a 4-in concrete slab, up to 50 ft in length, would cost $34. The cost for a cut 50 to 100 ft in length would be $50. For each foot over 100 ft, 50 cents per lineal foot should be added. A similar situation for a 6-in slab would run $53, $76, and 76 cents respectively. The removal of the concrete after saw-cutting is not included in these prices, nor is the initial mobilization cost.

At the site of the central treatment facility, concrete demolition was required. Such work is broken down into two parts, labor and equipment. The cost for breaking up a concrete pavement is $2.02 per square yard when done by a pneumatic breaker. If the work must be done by hand, the cost is $11.55 per square yard. The cost of renting an air compressor for the pneumatic breaker is $26 per day.

In order to run the pipeline through the existing asphaltic parking lot, some demolition work was required. The cost of removal of concrete pavement is about 6 cents per square foot.

Trenching Costs

Once the concrete and asphalt had been cut and removed, pipe trenches were excavated. There are various machines used for trench excavation. These include trenching machines and backhoes. Trenching machines are available as vertical boom or wheel type. They can dig a trench from 4 ft deep and 12 in wide to one $8\frac{1}{2}$ ft deep and 36 in wide. The backhoe is commonly available in $\frac{1}{2}$-in and $\frac{3}{4}$-yd^3 capacity.

The daily rental cost of a diesel trenching machine for a trench 6 ft deep and 12 in wide is $138. The daily rental cost for the $8\frac{1}{2}$-ft by 36-in machine is \$198.* Using a ¾-yd³ backhoe, the rate of excavation and the cost of labor with one operator and two laborers varies with the type of soil. For light soil the rate is 53 yd³/h, for an average soil the rate is 21 yd³/h, and for heavy soil the rate is 20 yd³/h. The labor costs are 31, 81, and 84 cents,* respectively, per cubic yard excavated.

In summary, the major cost in some pipelines is the cost of trenching, which can include saw-cutting, concrete demolition, hauling away, earth excavation, and backfill. All this must be added to the cost of the pipe material and installation.

Pipeline Costs

The arteries of any treatment facility are its pipelines. Pipelines can be a major economic consideration. A facility remote from the source and discharge point of the wastewater can be very costly; therefore careful attention must be given to the piping costs. The cost of piping can be subdivided into (1) material cost of the pipe, (2) fittings, (3) valves, (4) labor for installation, and (5) contractor's overhead and profit. In the case of underground piping the cost of field survey, excavation, and backfill must be added. For aboveground piping the cost of the pipe supports, pipe hangers, and wall penetrations must be considered.

Unit Prices of Piping Installation. Tables 11-1 to 11-7 show costs based on the prevailing early 1974 cost index. They were furnished by Richardson Engi-

Table 11-1   Pipe Erected on Pipe Racks outside of Buildings
Service Piping Using A120 Steel Pipe, Cast and Malleable Iron Fittings,
and Bronze and Iron Valves

Pipe size & weight, lb/lin. ft	Cost/100 lin. ft	Erection crew @ $11.40/h		Overhead @ 67%	Total cost	Total incl. 10% profit	Standard unit price range/lin. ft
½" 0.85	$ 22.50	(3.0)	$34.20	$22.91	$ 79.61	$ 87.57	$0.78 to $0.96
¾" 1.13	29.80	(3.3)	37.62	25.21	92.63	101.89	0.92 to 1.10
1" 1.68	41.00	(3.6)	41.04	27.50	109.54	120.49	1.10 to 1.30
1¼" 2.28	54.60	(3.9)	44.46	29.79	128.85	141.74	1.30 to 1.60
1½" 2.73	65.10	(4.1)	46.74	31.32	143.16	157.48	1.40 to 1.70
2" 3.68	87.50	(4.8)	54.72	36.66	178.88	196.77	1.80 to 2.20
2½" 5.82	136.00	(5.4)	61.56	41.25	238.81	262.69	2.40 to 2.90

SOURCE: Richardson Engineering Services, Inc., Solana Beach, Calif.

* See Cost Index Chart for current comparative prices.

Table 11-2   Pipe in or on Building Basement Walls or First Floor Walls
Service Piping Using A120 Steel Pipe, Cast and Malleable Iron Fittings,
and Bronze and Iron Valves

Pipe size & weight, lb/ lin. ft	Cost/ 100 lin. ft	Erection crew @ $11.40/h		Overhead @ 67%	Total cost	Total incl. 10% profit	Standard unit price range/lin. ft
½″ 0.85	$ 22.50	(9.2)	$104.88	$ 70.27	$197.65	$217.42	$2.00 to $2.40
¾″ 1.13	29.80	(11.5)	131.10	87.84	248.74	273.61	2.50 to 3.00
1″ 1.68	41.00	(12.4)	141.36	94.71	277.07	304.78	2.80 to 3.40
1¼″ 2.28	54.60	(13.5)	153.90	103.11	311.61	342.77	3.10 to 3.80
1½″ 2.73	65.10	(14.5)	165.30	110.75	341.15	375.27	3.40 to 4.10
2″ 3.68	87.50	(16.6)	189.92	127.25	404.67	445.14	4.00 to 4.90
2½″ 5.82	136.00	(18.6)	212.04	142.07	490.11	539.12	4.90 to 5.90

SOURCE: Richardson Engineering Services, Inc., Solana Beach, Calif.

Table 11-3   Pipe Laid under Building Ground Floor Slab
Service Piping Using A120 Steel Pipe, Cast and Malleable Iron Fittings,
and Bronze and Iron Valves

Pipe size & weight, lb/ lin. ft	Cost/ 100 lin. ft	Erection crew @ $11.40/h		Overhead @ 67%	Total cost	Total incl. 10% profit	Standard unit price range/lin. ft
½″ 0.85	$ 22.50	(2.8)	$31.92	$21.39	$ 75.81	$ 83.39	$0.75 to $0.91
¾″ 1.13	29.80	(3.0)	34.20	22.91	86.91	95.60	0.86 to 1.10
1″ 1.68	41.00	(3.3)	37.62	25.21	103.83	114.21	1.00 to 1.30
1¼″ 2.28	54.60	(3.6)	41.04	27.50	123.14	135.45	1.20 to 1.50
1½″ 2.73	65.10	(3.9)	44.46	29.74	139.35	153.29	1.40 to 1.70
2″ 3.68	87.50	(4.4)	50.16	33.61	171.27	188.40	1.70 to 2.10
2½″ 5.82	136.00	(4.9)	55.86	37.43	229.29	252.22	2.30 to 2.80

SOURCE: Richardson Engineering Services, Inc., Solana Beach, Calif.

neering Services, Inc., and appear in their publication "Process Plant Construction Estimating & Engineering Standards." These figures should be adjusted to the cost index at the time the analysis is made. The first five tables are based on service piping using A120 steel galvanized pipe, cast and malleable iron fittings, and bronze and iron valves. The work is performed by a mechanical contractor. Only threaded pipe is considered in sizes from ½ to 2½ in.

Table 11-4   Pipe Laid Underground outside of Building
Service Piping Using A120 Steel Pipe, Cast Malleable Iron Fittings,
and Bronze and Iron Valves

Pipe size & weight, lb/ lin. ft		Cost/ 100 lin. ft	Erection crew @ $11.40/h		Overhead @ 67%	Total cost	Total incl. 10% profit	Standard unit price range/lin. ft
½"	0.85	$ 22.50	(2.3)	$26.22	$17.57	$ 66.29	$ 72.92	$0.66 to $0.80
¾"	1.13	29.80	(2.5)	28.50	19.10	77.40	85.14	0.77 to 0.94
1"	1.68	41.00	(2.8)	31.92	21.39	94.31	103.74	0.94 to 1.10
1¼"	2.28	54.60	(3.0)	34.20	22.91	111.71	122.88	1.10 to 1.40
1½"	2.73	65.10	(3.2)	36.48	24.44	126.02	138.62	1.30 to 1.50
2"	3.68	87.50	(3.7)	42.18	28.26	157.94	173.73	1.60 to 1.90
2½"	5.82	136.00	(4.1)	46.74	31.32	214.06	235.47	2.10 to 2.60

SOURCE: Richardson Engineering Services, Inc., Solana Beach, Calif.

Table 11-5   Pipe Interconnecting Equipment outside of Building
Service Piping Using A120 Steel Pipe, Cast and Malleable Iron Fittings,
and Bronze and Iron Valves

Pipe size & weight, lb/ lin. ft		Cost/ 100 lin. ft	Erection crew @ $11.40/h		Overhead @ 67%	Total cost	Total incl. 10% profit	Standard unit price range/lin. ft
½"	0.85	$ 22.50	(11.5)	$131.10	$ 87.84	$241.44	$265.58	$2.40 to $2.90
¾"	1.13	29.80	(12.7)	144.78	97.00	271.58	298.74	2.70 to 3.30
1"	1.68	41.00	(13.8)	157.32	105.40	303.72	334.09	3.00 to 3.70
1¼"	2.28	54.60	(15.0)	171.00	114.57	340.17	374.19	3.40 to 4.10
1½"	2.73	65.10	(16.1)	183.54	122.97	371.61	408.77	3.70 to 4.50
2"	3.68	87.50	(18.4)	209.76	140.54	437.80	481.58	4.30 to 5.30
2½"	5.82	136.00	(20.7)	235.98	158.11	530.09	583.10	5.30 to 6.40

SOURCE: Richardson Engineering Services, Inc., Solana Beach, Calif.

To use Tables 11-1 to 11-5:

1. Select the proper table according to the condition under which the pipe is to be installed and price the pipe accordingly.

2. The labor hours shown in parentheses were used to calculate the cost of labor for installing the pipe only.

3. Add cost of hangers and supports.

4. Add cost of fittings and valves.

5. Testing is included in the cost.

Table 11-6
Carbon Steel Pipe, 2 Inches and Larger, Steel Butt Weld Fittings, and Steel Valves, All Field-fabricated and Installed

Pipe size, schedule & weight, lb/lin. ft	Cost/100 lin. ft	Erection crew @ $11.40/h	Overhead @ 67%	Equipment @ $14.00/h	Total cost	Total cost incl. 10% profit	Standard unit price range/ lin. ft
2-inch:							
40 Std.	3.65	$ 96.80	(27.6) $ 314.64	$210.81	$ 60.30	$ 682.55	$ 750.81	$ 6.80 to $ 8.30
80X-Stg	5.03	136.00	(32.4) 369.36	247.47	70.75	823.58	905.94	8.20 to 10.00
160	7.44	226.15	(40.1) 457.14	306.28	87.60	1077.17	1184.89	10.70 to 13.00
3-inch:							
40 Std.	7.58	178.60	(40.7) 463.98	310.87	88.90	1042.35	1146.59	10.30 to 12.60
80X-Stg	5.03	249.00	(47.6) 542.64	363.57	103.95	1259.16	1385.08	12.50 to 15.20
160	14.32	370.50	(56.6) 645.24	432.31	123.60	1571.65	1728.82	15.60 to 19.00
4-inch:							
40 Std.	10.79	223.50	(48.3) 550.62	368.92	105.50	1248.54	1373.39	12.40 to 15.10
80X-Stg	14.98	348.50	(58.0) 661.20	443.00	126.70	1579.40	1737.34	15.60 to 19.10
160	22.51	619.50	(71.8) 818.52	548.41	156.80	2143.23	2357.55	21.20 to 25.90
6-inch:							
40 Std.	18.97	369.60	(66.9) 762.66	510.98	146.10	1789.34	1968.27	17.70 to 21.70
80X-Stg	28.57	664.40	(81.4) 927.96	621.73	177.75	2391.84	2631.02	23.70 to 28.90
160	45.34	1263.00	(104.9) 1195.86	801.23	229.10	3489.19	3838.11	34.50 to 42.20

SOURCE: Richardson Engineering Services, Inc., Solana Beach, Calif.

Table 11-7 100'–0' of Type 304L Welded Stainless Steel Pipe, Erected in a Combination of Horizontal and Vertical Runs Between Tanks, Pumps, and Other Treatment Equipment. All Work Performed *outside* of Building.

Pipe size, schedule & weight, lb/lin. ft		Cost/100 lin. ft	Erection crew @ $11.40/h		Overhead @ 67%	Equipment @ $14.00/h	Total cost	Total cost incl. 10% profit	Standard unit price range/lin. ft	
½-inch:										
10S	0.67	$ 172.72	(15.0)	$171.00	$114.57	......	$ 458.29	$ 504.12	$ 4.50 to	$ 5.50
40S	0.85	227.37	(15.0)	171.00	114.57	......	512.94	564.23	5.10 to	6.20
¾-inch:										
10S	0.86	209.59	(15.0)	171.00	114.57	......	495.16	544.68	4.90 to	6.00
40S	1.31	278.72	(15.0)	171.00	114.57	......	564.29	620.72	5.60 to	6.80
1-inch:										
10S	1.40	304.28	(15.0)	171.00	114.57	......	589.85	648.84	5.80 to	7.10
40S	1.68	328.70	(15.0)	171.00	114.57	......	614.27	675.70	6.10 to	7.40
1½-inch:										
10S	2.09	396.68	(17.3)	197.22	132.14	......	726.04	798.64	7.20 to	8.80
40S	2.72	497.10	(19.6)	223.44	149.70	......	870.24	957.26	8.60 to	10.50
2-inch:										
10S	2.64	478.42	(18.4)	209.76	140.54	$40.20	868.92	955.81	8.60 to	10.50
40S	3.65	579.16	(23.0)	262.20	175.67	50.25	1067.28	1174.01	10.60 to	12.90
3-inch:										
10S	4.33	677.75	(25.3)	288.42	193.24	55.25	1214.66	1336.13	12.00 to	14.70
40S	7.58	1070.03	(33.9)	386.46	258.93	74.05	1789.47	1968.42	17.70 to	21.70
4-inch:										
10S	5.61	772.66	(27.6)	314.64	210.81	60.30	1358.41	1494.25	13.50 to	16.40
40S	10.79	1429.49	(40.3)	459.42	307.81	88.00	2284.72	2513.19	22.60 to	27.60

SOURCE: Richardson Engineering Services, Inc., Solana Beach, Calif.

266

Table 11-6 indicates the cost of carbon steel pipe 2 in and larger with steel butt weld fittings and steel valves. All costs are for field fabrication and installation. The cost of welded pipe is based on:

1. The equipment charge of $14 per hour for erecting pipe spools is for a 5-ton hydraulic truck crane and a flatbed truck, both maintained but not operated. The operator's time is included in the erection crew's man-hours. The equipment for field butt welds at $1.80 is for a welding machine.

2. To the total cost of the pipe add the cost of fittings, flanges, and valves.

3. Add field erection welds. The number of welds must be determined from the configuration of the spools.

4. Cost of labor is included.

The straight-run pipe may be estimated as follows:

1. For material cost of pipe use cost shown on Table 11-6.

2. Reduce cost of erection, overhead, and erection equipment by 60 percent.

3. Add 10 percent profit.

4. Add material cost of butt-welded fittings and flanges and add 10 percent profit.

5. Add welding cost of all joints of pipes and fittings.

6. Add cost of valves.

Table 11-7 shows the cost of furnishing and installing stainless steel pipe.

Cast-Iron Pipe Costs. Cast-iron pipe is used in installations where the corrosive properties of the wastes are not too severe. Installation costs of cast-iron pipe in the following table assume fittings spaced every 50 ft but do not include the cost of excavation or backfill. The tabulated prices are for centrifugal cast 150 CI pipe cement lines, tar and coated. If the order is less than a carload (LCL), 15 percent must be added to the material cost. The costs include also

Size, in	Slip-on joint, cost per lineal ft			Mechanical joint, cost per lineal ft			Bell and spigot, cost per lineal ft		
	Mat.	Inst.	Total	Mat.	Inst.	Total	Mat.	Inst.	Total
4	$1.70	$1.90	$ 3.60	$1.80	$2.30	$ 4.10	$1.65	$2.85	$ 4.50
6	2.55	2.80	5.35	2.65	2.85	5.50	2.50	3.85	6.35
8	3.65	3.45	7.10	3.80	3.85	7.65	3.50	4.20	7.70
10	4.80	4.70	9.50	5.00	4.95	9.95	4.65	5.75	10.40
12	6.15	5.75	11.90	6.35	6.25	12.60	6.00	7.25	13.25
14	8.00	6.25	14.25	8.00	6.80	14.80	7.50	8.05	15.55
16	9.15	7.45	16.60	9.50	8.05	17.55	8.90	9.20	18.10

subsistence, overhead, and profit (Subs O & P). Slip-on joint costs include plain rubber gaskets and joint lubricant. Mechanical joint listing includes bolts, nuts, glands, and rubber gaskets. For open site work which requires few fittings, and little installation by laborers, deduct 10 percent from material cost and 60 percent from the installation price.

Tank Costs

In our example of the industrial plant it was more economical to treat the wastes at a central facility than to provide separate treatment units. At the treatment station, holding tanks were needed along with reaction tanks, mixing units, feeder pumps, and pH meters. When selecting the tanks to be used in the system, the initial cost should not be the only consideration. The material and life-span of the tank also should be considered. In our example, acid and other corrosive liquids were often handled. Nonferrous material was therefore recommended. Some other common tank materials are fiber glass–reinforced polyester (FRP); polypropylene (PP); conventional polyethylene (CPE); linear polyethylene (LPE); and cross-linked linear polyethylene (XLPE). The various descriptive terms refer to the molecular structure of the synthetic materials.

Typical costs for 500- and 2000-gal conical-bottom tanks with fiber glass casing and conventional polyethylene linear are as follows (other sizes may be interpolated):

500-gal capacity	52-in diameter × 64 in	$ 702
2000-gal capacity	78-in diameter × 120 in	3408

For ordinary aboveground steel storage tanks, the following budget prices and weights can be used:

Height	Cost	Weight
15-ft-diameter cone-roof tank:		
16 ft	$ 8,380	12,100 lb
24 ft	9,518	15,410 lb
32 ft	11,038	18,720 lb
20-ft-diameter cone-roof tank:		
16 ft	$ 9,673	
24 ft	11,385	
32 ft	13,081	

The above prices include a 24-in shell manway, a 1-ft 8-in opening, a 1-ft 6-in opening, 1-ft 2-in draw-off, and a 1-ft ladder. An approximate price per unit capacity where 1 bbl equals 42 gal is as follows:

$14.90–$18.00/bbl for 16' high tank
$11.40–$13.90/bbl for 24' high tank
$ 9.90–$12.10/bbl for 32' high tank

Pump Costs

Pumps are another crucial item in a treatment facility. Adequate sizing of all pumps, especially those sump pumps which supply the treatment facility with its wastes, will eliminate many hours of costly maintenance. Sump pumps should be sized to run only one-quarter of the total operating time. It is often economical to provide two pumps or a duplex sump pump. If one pump is in need of repair, the entire plant operation need not stop. If the extra pump saves 4 h of process time, it is well worth its initial investment. Vertical sump pumps and sewage pumps are provided for various heads of pressure. It is not necessary to oversize a pump. Cautious calculation of the head of pressure necessary for a particular facility is sufficient. See Chapter 6 on design.

The following are typical pump costs:

Vertical sump pumps:

50 gal/min	20-ft head	$440
	60-ft head	580
100 gal/min	20-ft head	460
	60-ft head	590

Vertical nonclog sewage pumps:

50 gal/min	17-ft head	$545
100 gal/min	20-ft head	545

In the treatment facility itself metering pumps must also be considered. Some common pumps are listed below.

Chemical metering pumps:

Stainless steel	$0 - 7.5$ gal/h $- 8000$ lb/in^2		$345
Carpenter 20	$0 - 7.5$ gal/h		410
Labor cost for installation, 4 h @ $14/h			56

Chemical feed package:

Includes pump	7.5 gal/h	860 lb/in^2	316 SS

Mounted under tank, interconnected piping, tank and stainless steel agitator

Tank material

Polyethylene	50 gal	$ 750
	150 gal	1100
Stainless steel	50 gal	1035
	150 gal	1370

Total Cost of Equipment Installation

In general, any piece of equipment required in an installation should be completely investigated. A checklist for analyzing the cost of such treatment equipment is shown below.

CHECKLIST FOR ANALYZING COST OF TREATMENT EQUIPMENT

1. Capacity in gallons per minute or gallons per hour
2. Weight, shipping and in place

3. Size, for floor area and height clearance

4. Electrical requirements, voltage, phase, full-load amperage

5. Construction material, corrosion resistance

6. Maintenance requirements, manual, schedule, spare parts

7. Installation time, changeover

8. Chemical requirements, reagents, cleaning

9. Air requirements, ft^3/m, pressure

10. Water requirements, volume, pressure

11. Prices

12. FOB point

13. Original chemical requirements

14. Operating costs

15. Pipe sizes, materials, fittings

16. Installation costs

   a. Erection time in man-hours

   b. Installed cost

   c. Hourly rate for hauling and lifting of equipment

Sewer Costs

After treatment, the wastes must enter the sewer system. The noncorrosive wastes may now travel in vitrified clay pipe. A typical breakdown of subcontractor prices for installation of a sewer line is shown below. These prices include the trenching to 4 ft, installation of pipe, and backfill. Various types of pipe material are listed below. These include cement asbestos pipe (CAP), reinforced concrete pipe (RCP), and vitrified clay pipe (VCP).

6" diameter	CAP	$6.00/lin. ft
8"	CAP	$7.00/lin. ft
12"	RCP	$9.00/lin. ft
18"	RCP	$13.00/lin. ft
6"	VCP	$7.00/lin. ft
8"	VCP	$7.77/lin. ft
10"	VCP	$10.00/lin. ft
12"	VCP	$11.75/lin. ft
18"	VCP	$14.50/lin. ft

Since these prices are from subcontractors, the overhead and profit of a general contractor must be considered.

Manholes

Sewer lines require manholes. Many governmental agencies require that manholes be "control manholes" equipped with sampling equipment, flow metering equipment, and a pH meter. Manholes require covers of various types, depending upon their location and the traffic over them. The cost of the system must also include the price of "connection to the main sewer line."

ECONOMIC ANALYSIS OF A FACILITY

Armed with a general idea of the costs associated with the installation of a wastewater treatment facility, the designer is ready to sit down and think, read, ponder, and discuss. In designing an economical waste treatment, the engineer should investigate the following items:

1. The number and kinds of liquid wastes requiring treatment
2. Characteristics of wastes:
   a. Periodicity of flow
   b. Volume
   c. Changes in concentration
3. Requirements of local authority
4. Possibility of combining wastes for "neutralizing" effect
5. Possibility of combining distant plant wastes into one treatment plant
6. Possibility of combining wastes of a neighboring plant
7. Changes in plant processes to reduce volume and/or concentration
8. Feasibility of recovery of contaminants
9. Treatment of wastewater to remove contaminants and recycle water
10. Types and quantities of chemicals used in plant processes
11. Survey of allotted site for treatment facility to determine suitability for purpose; to ascertain any obstructions, restrictions, etc.
12. How plant management intends to pay for facility:
    a. Reserve funds
    b. Borrow funds
    c. Lease equipment
    d. Governmental assistance
13. Laws that enable industry to claim local relief from property and other taxes for facility

14. Possibility of local municipal treatment plant obtaining governmental assistance and treating industry waste; industry would:

   a. Pay plant operation costs

   b. Pay bond service charges on amount chargeable to that industry

   c. Pay costs above grants or loans

   d. Lease entire facility from municipality

Recycling Water

Recycling should be an important consideration; if it is not to be implemented initially, then it should be in the near future. The most obvious advantage in treating industrial wastewater is being able to reuse the effluent. This has the economic advantage of reducing the cost of purchased fresh water as well as reducing any well pump operation. It also provides an inexpensive storage of firefighting water.

The treatment of the wastewater is only one of the objectives to be dealt with. Recycling wastewater effluent usually calls for a separate in-plant system. This water system, also known as "plant water," "industrial water," or "nonpotable water," must be kept separate from the potable or domestic water supply. The recycled water cannot be used at drinking fountains, lavatories, showers, emergency eyewashes, or other personal use by the plant's occupants. But the recycled water can be used for cooling, boiler makeup, washing, rinsing, and processing as well as for irrigation of the landscape areas. The last must be done with government approval. Typical equipment using single-pass water or recycled water includes:

Spot welders

Hydraulic oil cooling

Refrigerators

Drop hammers

Power brakes

Degreasers and stills

Air compressors

Evaporative condensers

Stretch machines

Boiler feed pumps

Pumps and fans—bearing cooling

Heat exchange equipment

Process tanks—rinse water

Condensate from process tank coils

Paint booth water curtains

Heat-treat quench pits

Vibrator deburrers

Whenever any plant water is being reused, the separation of the two water systems is controlled by local, state, and federal agencies. The method of separation depends upon the degree of hazard in accidental human use of the industrial water. In some cases a simple check valve is adequate. In cases where the industrial water could contain toxic material, only a positive air gap is permitted. The toxic condition of the water could be caused by an accidental malfunction of the wastewater treatment facility of improper piping connections.

All water pipes and outlets should be clearly identified, indicating the nature of the water. Each outlet on the nonpotable water line which could be used for drinking or domestic purposes should be posted DANGER—UNSAFE WATER.

Despite all the plant precautions required when using recycled water, recycling goes a long way in meeting the EPA's long-range goals of zero discharge. A treatment facility designed for recycling will be flexible enough to meet the future requirements and economical enough to justify its existence.

In the balance of the chapter the author discusses various problems as they occurred in actual projects. The results were economies in design, construction, and operation of the wastewater treatment facility.

Comparative Cost Index

Year	Average cost
1940	100
1945	140
1950	200
1955	270
1960	320
1965	350
1970	420
1975	650

Throughout this text references are made to costs of various types of construction and equipment. As noted in the Introduction many of the examples contained in this revised edition are the unaltered work of Edmund B. Besselievre. In order that the reader may better understand these examples, the costs related thereto may be updated by use of this Comparative Cost Index.

For example, if it is stated that the cost of a particular construction project or piece of equipment in 1955 was $10,000, to find the relative cost in 1975 simply multiply the 1955 cost by the ratio of 1975 cost to 1955 cost:

$$\$10,000 \times 650/310 = \$20,967$$

Costs for any particular item may vary from that computed but, in general, the use of this index will facilitate a more accurate picture of the examples contained herein.

Case 1: A Steel Company in Ohio

The single waste flow was a daily volume of spent pickling liquor. The State Department of Health had demanded that the management provide facilities to so treat these wastes that the final effluent could be discharged into the river. When the engineer was called in, he learned that the management had already had a report and plans made for the treatment by a firm of engineers from a distant city. The method of treatment recommended in the report was in accord with recognized practice at that time, but the estimated cost was high. The management said, "We just don't have that kind of money." Showing proof that the engineering firm which had made the report had been released from any further connection with the project, they requested a new survey.

A complete survey of the mill and mill property was then made, and it was discovered that there was a large building near the river edge of the property which did not appear to be in regular use. This building was adjacent to a concrete tank, also unused, and alongside the building was a railroad spur. With these items in mind and the consent of the management to use them, the engineer prepared plans and specifications; they were submitted to the State Department of Health and approved by them in writing, and proposals were made for the construction of the plant. The original plans had not taken into account these available facilities, or had not considered them worthwhile. Instead, a new fine control building, etc., had been proposed, at a total estimated cost of $275,000. The firm proposal for the plant, using the existing structures mentioned, was $86,000.

At the time this project was under consideration a pilot plant was in operation at a city about 10 mi away, experimenting on a German process of recovery of sulfuric acid and iron from spent pickling liquors. The management of the steel mill knew of this plant. During the discussions with the state engineers it was suggested that the steel company might be well advised to consider trucking their spent liquors to this experimental plant. The mill management had considered this, but had ruled it out as entirely beyond their financial capacity. The total volume of spent pickle liquor discharged from this mill was about 10,000 gal/d. The cost of hauling to the experimental plant had been estimated as 2 cents per gallon of waste. The daily cost for hauling the wastes would be $200; hauling the reclaimed sulfuric acid back to the plant would add another $100, or a total of around $300 per day. Since this was more than the cost of new acid per day plus the estimated cost of conventional treatment of the waste liquor, it was given no further thought.

Case 2: A Textile Mill in Massachusetts

A small family-owned mill was engaged in the custom dyeing of dacron. Their wastes were being discharged through a lagoon into a small stream which was a tributary of a larger river that passed into another state. This put the mill under the jurisdiction of a state compact, the regional New England Interstate Water Pollution Control Commission, which had very strict regulations on waste discharges into any waters under its control. The final overflow from the lagoon at this mill was free of solids but was highly colored—a fine purple hue. An engineering firm was called in. The cost factor was an urgent matter to the owners of the mill. At that time little had been done on treatment of wastes from the dyeing of dacron, but extensive tests indicated that chemical treatment could eliminate the disturbing color and other polluting components from the wastes.

The available company property was small, located on a main highway and providing very little space for any new structures. It was found that on the property there was a steel warehouse building, used only for storage of barrels. This was taken over to house the treatment units, which could readily be installed in it, away from the notice of neighbors. The effluent from the treatment flowed finally through the lagoon. The total estimated cost of the plant was $36,000.

Case 3: An Automobile Manufacturing Plant in Michigan

This problem involved multiple wastes from one of the largest manufacturers of automobiles in the world, the plant of the Pontiac Motor Division of the General Motors Corporation, at Pontiac, Michigan. Although it was designed, constructed, and equipped by a somewhat unorthodox method (Figures 11-1, 11-2), the results being produced by this plant, as attested to by the superintendent of the waste treatment plant, show that the methods, described here, have proved their worth.

The order for the treatment of the wastes from this plant arose from complaints that cyanide had been detected in the water supply of a municipality downstream. The wastes from the automobile plant were discharged into the sewerage system of the city of Pontiac. The limit of cyanide concentration in the waters of the river had been set at 1 mg/l. In ordering the motor company to treat the wastes, the city incorporated this state limitation without any consideration for the dilution in the sewerage system, which was, roughly, 10 to 1. The motor company called upon an engineering firm, which, after careful survey, presented a proposal to study, test, design, construct, equip, and supervise the initial operation of the complete waste treatment facility. The contract was signed in early 1955, and the plant was put into operation in 1957. As a site for the treatment plant the company allotted an area of ground immediately adjacent to the plating department. This plot was about 2.71 acres, between two railroad spurs which supplied the plating department and other factory sections with material for the cars. The area at the time of allotment was cluttered with a test track which ran through it; innumerable underground sewer, water, and drain lines which "could not be moved"; and a high-water tank and two steel high-tension electric utility towers which also "could not be moved."

The first step of the engineers was to inspect the operations in the plating department to determine the types of waste that would require treatment to satisfy the demands of the city and state officials. The result showed that there were six different classes of wastes which would require treatment to some degree or other:

1. Acid wastes and alkaline wastes

2. Plating wastes containing cyanide

3. Plating wastes containing chromium

4. Oily wastes containing free oil and containing emulsified oil

It was noted that all these wastes were produced in one section of the plating department, on one side of a wide roadway paralleling the building. The allotted treatment plant site was on the opposite side of this road. The first restriction established by the plant management was that this road should not be closed at any time by ditches built to carry pipes across it. Therefore all waste lines from the plating plant had to be carried overhead, across this road. This introduced an unforeseen pumping operation inside the plating room to elevate the wastes to the overhead lines, but actually worked to the advantage of the design, permitting gravity discharge into the main holding tanks of the waste treatment plant. Figure 11-1 shows this construction. These elevated lines, completely exposed to the heavy weather of the northern climate, were provided with steam tracer lines to prevent freezing.

From study of the different types of wastes it was concluded that combining them would not be practical or economical. Therefore the final design resulted in five different and separate treatment sections, as follows:

1. *To treat the acid and alkaline wastes together to get the benefit of mutual reactions.* The plant records and previous study by another engineering firm showed that the average daily volume of flow of alkaline wastes was about 200,000 gal/d, and the acid waste flow about 60,000 gal/d. This indicated that the alkaline wastes would always be in sufficient volume to correct the pH of the acid to a level within the bounds of the city regulations, but the resultant combination might still be above the limit of 6.0 to 10.5 for the final effluent to the city system. To provide for a remedy, if this emergency arose at any time, a storage tank for 10,000 gal of sulfuric acid was provided, with

Fig. 11-1 Overhead pipeways carrying separated wastes to treatment tanks, waste treatment plant, Pontiac Motor Division, General Motors Corporation, Pontiac, Mich. (*K & H Engineering Company.*)

two Milton Roy acid pumps, controlled by a pH meter, so that any high-alkaline condition above the limit would be automatically neutralized to the prescribed point.

The sulfuric acid tank was located immediately adjacent to the private railroad spur leading to the operating building, so that acid could be purchased in tank carlots for price advantage.

2. *To treat the wastes containing cyanide.* These wastes were in two categories: the daily overflows of spent plating liquors from the plating vats, which contained on an average about 436 lb of cyanide per day; and the periodic dumps of sludge when the plating tanks were completely cleaned. These dumps, which occurred at odd intervals, perhaps a fortnight to a month apart, were found to contain an average of 4769 lb of cyanide in a small volume of about 20,000 gal of sludge.

After studying all accepted methods of destroying or reducing cyanide, it was decided that the method of treatment for this large plant under the conditions mentioned was the gaseous-chlorine method, which breaks down the cyanide into nitrogen and carbon dioxide. The stoichiometric amount of chlorine needed to break down 1 lb of cyanide is 6.83 lb, but since this is an ideal condition, rarely found and not safe to count upon, it was decided to design the chlorine supply on the basis of 8 lb/lb of cyanide.

Computing the amount of chlorine that would be required to treat the total amount of cyanide in the normal wastes of one day's operation of the plating plant, plus a cleanout dump on one day, and treating this combination in one 16-h operating day of the treatment plant, showed that a possible total amount of 5205 lb of cyanide would require breakdown.

The largest commercial size of automatic chlorinator at the time of design having a maximum capacity of 6000 lb/d, it developed that to supply the chlorine necessary to break down 5205 lb at a rate of 8 lb/lb of cyanide, a total of 41,640 lb of chlorine would be required in 16 h. To supply this would require eleven chlorine-dispensing units, each with a capacity of 6000 lb/d. The approximate cost of this array of units, with their accessory evaporator units, control equipment, etc., would have been over $128,000. To accommodate this collection of equipment in the operating building would require approximately 2592 ft². On reflection, this was considered a very high cost for equipment that might be needed only 1 d per month. An alternative was sought and found.

The alternative was to provide a dump sump at ground level for the cyanide sludge, which was brought out from the plating plant in motorized buggies, and could be dumped into the sump. From the sump the sludge would be pumped into one of the two strong-cyanide storage or holding tanks.

After mechanical mixing to homogenize the wastes and prevent separation, the sludge would be pumped by another set of pumps to one or other of the two large storage or holding tanks provided for the daily flow of weak-cyanide-bearing wastes. In order to prevent overloading of the cyanide treatment system, the pumps which transferred strong-cyanide wastes to the weak holding tanks were, deliberately, of such low capacity that the contents of one strong-cyanide holding tank could not be pumped into one of the weak-cyanide holding tanks in less than 1 week. This eliminated any chance of error in the operation. By this method of control it was estimated that the maximum daily amount of cyanide that would require treatment would be about 780 lb. At the rate of 8 lb of chlorine per pound of cyanide, this amount of cyanide would require only 6248 lb of chlorine per 16-h day. This reduction, in turn, resulted in the need for only three chlorinators, each with a 24-h capacity of 6000 lb, instead of eleven units and accessory equipment. These three units, with all the needed adjuncts, actually cost only $35,000, and the space required in the operating building was reduced to a total of 720 ft².

The study of the most economical means of supplying the chlorine for this large establishment resulted in the decision to use the single-tank-car system, in which chlorine is delivered in a railroad tank car to a special siding or spur directly outside the chlorination area, where the car is parked and used as the supply facility. This system provided the most favorable price for chlorine. Since these tank cars have capacities of 16, 30, and 55 tons of chlorine, the plant purchasing department was able to call for the car of the right size to meet their requirements, using the car as the supply point for the full 30 d of the allowable time for the car to remain on the siding. The cost of a special siding and protective devices and piping to convey chlorine into the plant in this case was $11,183, but this cost would soon be amortized in the saving in cost of chlorine.

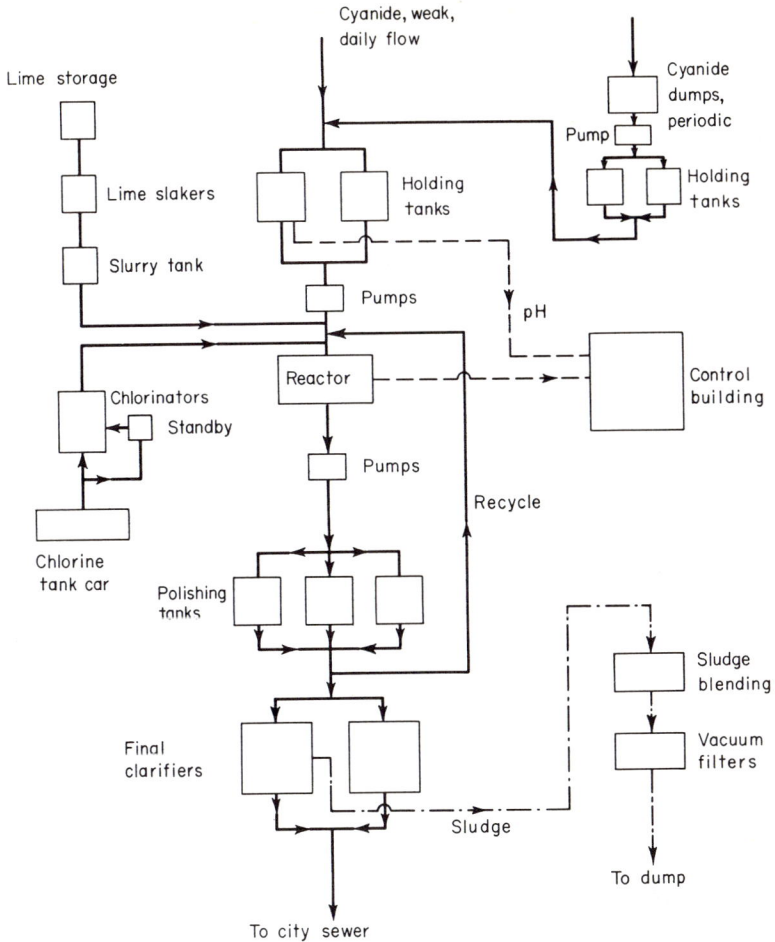

Fig. 11-2 Cyanide treatment, Pontiac Motor Division, General Motors Corporation, Pontiac, Mich.

In this case the largest car, 55 tons, was able to provide all the chlorine needed for almost 1 month. To provide for uninterrupted chlorine service while the cars were being changed, two 1-ton cylinders were installed in the chlorination room. When a car is to be changed, these cylinders are cut in to the treatment lines and supply chlorine to the chlorinators during the change interval. When the new car is connected, the cylinders are refilled from the car tank and wait unused until the next car change. They are, of course, available in case of emergency.

Large users of chlorine can find savings in tank-car delivery. The costs will vary according to the distance between the place of manufacture of the chlorine and the point of usage. The table on the following page shows the amount of

Fig. 11-3 Cyanide reaction units, waste treatment plant, Pontiac Motor Division, General Motors Corporation, Pontiac, Mich. (*K & H Engineering Company.*)

cyanide per month that can be treated on the basis of 8 lb of chlorine per pound of cyanide, for the different sizes of cars.

Size of car, tons of chlorine	Total amount of chlorine in car, lb	Pounds of cyanide that can be treated in 1 month
16	32,000	4,000
30	60,000	7,500
55	110,000	13,750

The progress of treatment of the cyanide-bearing wastes in this plant is as follows: The daily weak wastes flow by gravity from the elevated pipeline into one of the two weak-cyanide holding tanks. Into this tank is also pumped a small proportion of the strong-cyanide wastes from one of the strong-cyanide holding tanks. This filling proceeds during one entire operating day of the plating line. At the end of this day the flow to this tank is shut down, and the mixer unit in this tank is kept running continually during the night and during the next day, when the entire contents are treated over a period of 16 h. While the contents of this tank are undergoing treatment, the other weak-cyanide holding tank is being filled. The object of the mixing in each holding tank is to produce a waste uniform in character for treatment, eliminating continuous manual control of valves and chemical feeders, etc. When the liquid level in a holding tank reaches a point about 6 in above the impeller of the mixer, the mixer is automatically put into motion, and this motion continues until, on the withdrawal cycle the following day or treatment period, the level drops to the point 6 in above the mixer impeller. The point of this is to avoid running the impeller when it is not properly submerged, with the risk of causing damage to it, especially if the shaft from the top of the tank is long, and whipping of the shaft might damage the gears in the drive unit. Also, it would be a waste of electric power to run it when the tank was empty. When treatment is started on a full holding tank, the exact level of the liquid is indicated on the level indicator for that unit, on the main control panel, and the operator, by reading the gallonage chart for that same unit, knows exactly how much liquid is in the tank. Since no more liquid is admitted to that tank during treatment of the contents, he knows exactly how much waste is being treated for any period of time, and with the volume meters on the chlorination units, he has an accurate measurement of the ratio of chlorine to cyanide treated. Before starting the treatment, the operator makes one test for cyanide content, and the waste having been continually homogenized during the period of the night, the concentration of cyanide remains uniform throughout the entire treatment.

Since the weak-cyanide holding tanks are designed to be emptied during a treatment period of 16 h, the pumps provided for emptying the tanks are sized on that basis. Therefore, if the treatment is begun when the tank is less than full, the treatment period will continue only for the length of time required to empty the holding tank.

Information received from the superintendent of this plant showed that for one entire week in 1967 the total cyanide treated was, 4700 lb, or 940 lb/d for a 5-d week. For the same period the chlorine usage was 7512 lb/d, within the limit originally set up in the design. The three chlorinator units were ample for the work. The cyanide wastes were treated in a three-compartment reactor tank, each compartment being equipped with mixer units to maintain uniformity in the wastes.

Another unique provision in this plant to lessen manual control was the system employed to assure that no improperly treated waste was discharged from the plant. At the time of the design in 1955, automatic analyzers had not been sufficiently developed, and the engineers did not consider them warranted for this important plant. Instead of automatic units, a series of three tanks, called "polishing" tanks, was provided to receive wastes after chlorination. Each of these tanks had a capacity equal to 2 h flow of wastes. They were used in series. The treated wastes flowed into tank 1 until it was full. Then the flow was diverted to tank 2, and a sample of the wastes, thoroughly mixed by the tank mixer, was taken and analyzed for cyanide content. Since this test required about 2 h time, no more waste was admitted to tank 1 during that period. If the analysis showed that the contents of tank 1 were satisfactory from the cyanide standpoint, the contents of tank 1 were discharged for the final treatment phase. During this time tank 2 was filling, and when full, the flow was diverted to tank 3, and tank 1 was again available for another round. If, however, analysis showed that the cyanide content of tank 1 was too high, the wastes in that tank were recycled back to the cyanide reactor for retreatment. This system was expensive, and if the automatic analyzers had been perfected at the time of design, they could have been installed for much less. They are insurance against the possibility that a slug of cyanide might slip past while samples are being analyzed.

Since pH is important in the treatment of cyanide, it is considered best to maintain it at about 8.5 minimum. To assure this in the case of the Pontiac plant, electrodes were placed in each chamber of the two reactors, at the influent and effluent ends, and also on the exit line to the pumps from the holding tanks. The readings of these electrodes were transmitted to the indicating and recording pH meters in the operating building.

3. *In this plant the wastes containing chromium were treated separately by sulfur dioxide fed through sulfonator units, each with a capacity of 500 lb/d.* These units were fed from two 1-ton cylinders mounted on skids in the chlorination area. Figure 11-4 is an illustration of this room. The routine for the chromium treatment was the same as for cyanide wastes, from the holding tanks to treatment tanks, and then through a second set of polishing tanks, to the final treatment phase. This system for treating chrome wastes was dubbed the metallic-waste system.

4. *The treatment of wastes containing free and emulsified oil was the fourth actual treatment phase.* The elevated pipe from the plating department followed the other lines in the pipe cluster shown in Figure 11-1. Since the oily wastes contained both types of oil, another innovation in this plant was the elevation of the free-oil skimming clarifier, shown in Figure 11-5. In this installation the surface water level in the clarifier was just sufficiently below the level of the gravity feed line to permit flow through the tank and to the oily waste holding tanks. The free oil rose to the surface of this tank, and then flowed by gravity to an elevated free-oil storage tank adjacent to a paved road through the plant. This free-oil tank was at a level to permit discharge from it to an oil-tank truck, which carried the reclaimed free oil back to the factory for reuse. The oily wastes, freed of the light unemulsified fraction, flowed from the elevated skimming clarifier to one of the two holding tanks. From these the emulsified oil wastes flowed to the Sediflotor unit, where they were treated with aluminum sulfate and ferrous sulfate to break the emulsion. The sludge from the oily wastes treatment were combined with all other sludges before being pumped to the vacuum filters.

The pipelines leading from holding tanks to the treatment tanks, etc., were

Fig. 11-4 Segregated chlorination and sulfur dioxide compartments with standby chlorine cylinders, waste treatment plant, Pontiac Motor Division, General Motors Corporation, Pontiac, Mich. (*K & H Engineering Company.*)

all placed below ground in concrete trenches, with wooden covers, so that any part of any line could be inspected immediately, and leaks or breaks found and repaired promptly.

Each of the twenty-nine steel tanks in the system was completely above ground, resting on all-concrete pads or ring foundations so that any leakage would run into the trenches and would be detected at once.

Lime being required at several points in the treatment cycle, that is, in the cyanide, chromium, acid, and oily wastes, its handling and distribution required considerable thought and design innovations. Since the total amount of lime required was sufficient to warrant purchasing the material by the carload, the handling facilities were planned accordingly. At a point in the plant siding at the operating building, a chamber was constructed with a screw conveyor to move the raw lime as it was discharged from the bottom of the lime car. The conveyor carried the lime to the boot of a bucket elevator, which carried the lime to the top of the lime storage bin located on the roof of the operating building. From this storage bin the lime passed down to the second floor of

Fig. 11-5 Elevated clarifier for treatment of free and emulsified oily wastes. Tank is elevated to permit gravity flow through treatment system. Waste treatment plant, Pontiac Motor Division, General Motors Corporation, Pontiac, Mich.

the operating building and through a weightometer into one or other of the two paste-type lime slakers. Here the lime was slaked, and the resultant slurry then passed into the slurry tank in the basement of the operating building. Since milk of lime is really a suspension and will form cakes in feed lines if the flow is interrupted, special precautions were taken to prevent this. As shown in Figure 11-6, the milk of lime was circulated through a closed main delivery pipe to all areas where lime was needed. From this main line, in which the flow was kept continuously at a velocity of 7.5 ft/s, branch lines were led to the various usage points. At the point where each branch line left the main line, an electrically operated valve was placed so that the operator, when shutting down any line, could immediately stop the flow of lime to that point. At the same time, by means of a water connection to the branch line, the lime slurry remaining in the branch line was immediately flushed out, leaving the line clear. By this system of continuous recirculation the lime feed lines were kept open at all times.

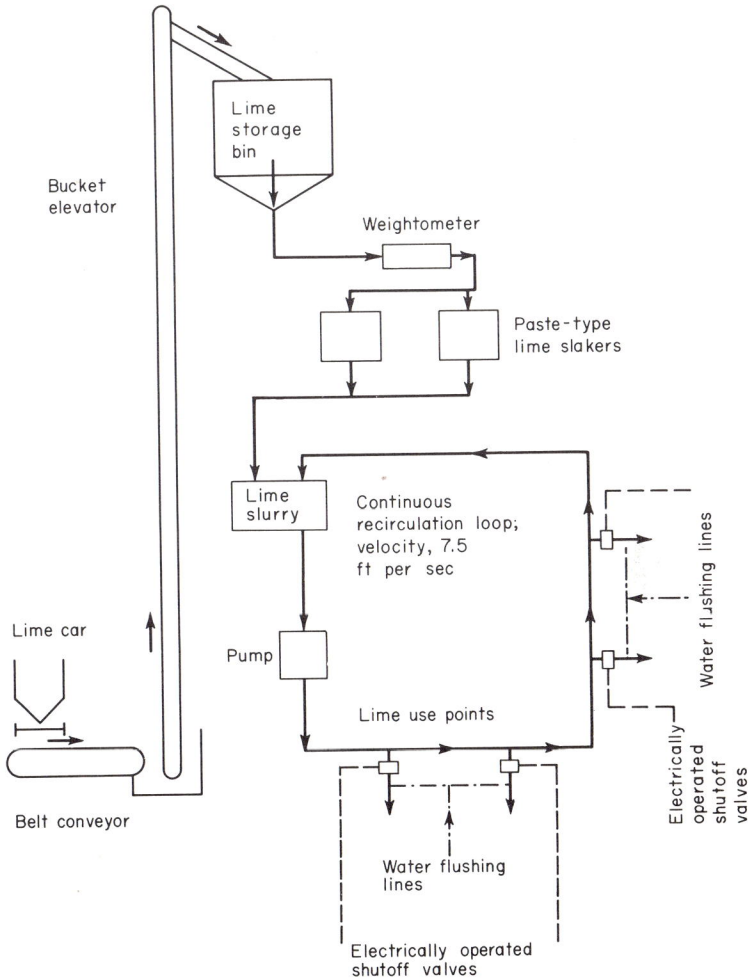

Fig. 11-6 Schematic diagram of lime-supply system, Pontiac Motor Division, General Motors Corporation, Pontiac, Mich.

No matter how good the housekeeping may be in an industrial plant, leakage will occur. Therefore all fifty-two pumps in this plant were located around the wall of the basement of the operating building. Since mixture of leakage from the different types of wastes could create hazardous conditions, each set of pumps handling a specific waste was surrounded by a low wall about 6 in in height, so that any leakage in that area would flow to a sump and be pumped back into the proper holding tanks.

Each treatment line in this plant was in duplicate in all phases—holding, treatment, final treatment, pumps, etc. Each half of each treatment line was based on handling the flow from the factory of one operating day, with the second half in reserve or for alternation of use. This proved to offer great flexibility in the operation of the plant, since it could be used in several ways:

1. The wastes of one day could be treated on the following day in any period of time desired.

2. If the flow exceeded the capacity of one holding tank in 1 day, another tank was always available to receive the excess. (*Note:* This actually did happen shortly after the plant was put into operation, but the duplicate system handled the excess and there was no difficulty.)

3. If, for any reason, such as repairs, one side had to be closed, the other side was serviceable.

4. If the flows increased, because of expansion of the plating facilities, to such an extent that the volume exceeded the capacity of one holding tank, the other one was available to handle the excess. This could continue while additional facilities were being installed, without interfering with treatment.

Each of the twenty steel and concrete tanks used for holding or treatment of wastes was equipped with level-indicating devices of the bubble type. Each tank had a corresponding indicating gauge on the main control panel which indicated the level in each tank at any moment in feet of depth of liquid. This is shown in Figure 6-32. To prevent miscalculation of the volume of a tank as shown by the depth, a table was provided for each tank, showing the exact gallonage at each foot. Each operator was provided with this table, so that any human error in calculation was eliminated.

As shown in Figure 6-32, the main control panel contained all major controls. This was a graphic panel, showing the progression of each individual waste throughout its treatment phases. In order to facilitate ease of control and prevent errors, the panel was divided into five clearly marked and identified sections. Each section contained all the control elements, switches, running lights, etc., essential in the operation of that section. These sections were, in order of progression along the panel from left to right:

1. Cyanide system

2. Alkali-acid system

3. Metallic waste system (chromium wastes)

4. Alkali-oily system

5. Final treatment

One man of the operating staff was responsible for one line only, and by this system he had only to watch the dials and lights on that section of the panel which controlled the line of which he was in charge. To facilitate his handling of his area, the operating manual provided by the designers was sectionalized, in loose-leaf form, so that the proper section could be given to each operator and thus eliminate the possibility of confusion.

The main panel board was mounted in an entirely separate room, immediately adjacent to the office of the superintendent, so that access to all sides was available. The panel faced a large, permanently fixed glass window looking into the superintendent's office, and his desk was placed facing this window, so that if any alarm sounded or a signal light flashed, showing trouble in any mechanical unit in the plant, he was immediately warned and could initiate correction of the malfunction.

The sludge from all operations in this plant was combined in an agitated tank and fed to one of the two vacuum filters located on the main floor of the operating building. In the initial design just the two filters, each 8 ft in diameter and 10 ft long, were provided. Each had 150 ft^2 of filter area. Cloth used was nylon, and string discharge of cake was used with nylon discharge strings. The filters were designed to handle 35,333 gal of wet sludge, containing an average of 6.40 percent solids. This was a rate of 7.36 gal of wet sludge per square foot per 16-h operating day. The sludge cake, with an average moisture content of 75 percent, is discharged to a belt conveyor, which passes through the wall of the operating building at such an elevation that the sludge cake is dumped directly into a dump truck and hauled away. Later, a third and

Fig. 11-7 Vacuum filters for handling combined sludges, waste treatment plant, Pontiac Motor Division, General Motors Corporation, Pontiac, Mich. Right-hand filter is running, but left-hand filter is stopped to show quality of filter cake produced. (*K & H Engineering Company.*)

**Table 11-8 Design and Actual Waste Flows,
Pontiac Motor Division**

Line	Design flow, 1955, gal/d	Flow, 1967, gal/d
Weak cyanide	392,733	255,400
Strong cyanide*	20,380	
Alkali, acid	543,338	317,600
Weak metallic	537,300	464,600
Strong metallic*	93,697	
Alkali, oily	154,981	390,000
Total	1,742,429	1,427,600

* These are the dumps from tank cleanout, and in the 1967 figures are included in the totals shown for the weak lines.

Since the actual holding capacity of all tanks in the system is 3,561,684 gal, the plant as now operating, after 11 years, has ample capacity for operational expansion without additional units of tankage.

larger vacuum filter was installed to enable the operators to handle all the sludge in one shift of 8 h.

Table 11-8 shows the design factors used and the volumes actually being treated in 1967, indicating that the design factors of the original plan were logical.

One element that required correction was in the lime storage bin on the roof of the operating building. This, when installed, was equipped with Bin-Dicator agitators placed inside the bin to prevent caking and clogging of the outlet. These agitators are soft pads blown out by compressed air. The material of the pads being semiporous, the air supply must be moisture-free, especially since, in this case, unslaked lime was used, and there was sufficient moisture in the plant air to initiate caking. Because the cost of drying the air was considered excessive, the units were removed and replaced by exterior agitation units.

The waste treatment plant described in the foregoing pages was designed and built under the "package plan" developed by the author as Manager and Chief Engineer of the Industrial Wastes Division, Kaighin & Hughes, Inc., and its subsidiary, the K & H Engineering Company of Toledo, Ohio.

Since the state of Michigan requires that any chlorination equipment be installed in an entirely separate room from which access may be had only from the outside of the building, and with a separate ventilation system, this was done, as shown in Figure 11-4, by closing off one side of the main floor of the operating building. To enable the operators to inspect operations, the wall between the chlorination room and the main floor of the operating building, in the section from 4 ft to the ceiling, was of heavy glass, sealed in place, and the chlorinators and sulfonators were placed so that their dials could be read through this window wall.

One stipulation of the Pontiac Company management which materially increased the cost of this plant was that there be no spray painting of tanks or outside structures. Immediately adjacent to the waste treatment area, new automobiles, freshly enameled, were parked as they came from the assembly lines. The management feared that if spray painting was allowed, the wind would carry the spray over the cars. This made it necessary to paint all tanks and structures by hand.

In a few of the tanks in which corrosive wastes were to be stored or treated, it was necessary to provide protective linings. It was learned that the Pontiac Company had had satisfactory service from tanks lined with Koroseal, and this was used. This is an expensive material, carefully applied by hand, but is rugged and has a long life without repair unless accidentally punctured. The cost of this lining, in place, is shown in Table 6-3.

Case 4: An Aircraft Manufacturing Plant in Ohio

This was a waste treatment system for the various uncommon wastes of a large aircraft manufacturing plant. It was a very complex operation because the functions producing wastes that required treatment were scattered all over an area of many acres in four plant buildings. Several of the points of origin of wastes were ¾ mi from the point which initially seemed logical for the placement of a treatment facility. The first survey of this plant produced an astounding array of 100 tanks at various points into which wastes requiring treatment were drained. The wastes to be treated were from a variety of processes for plating or metal finishing of small parts of the assembly of the planes. These plating, or processing, stations had been placed at the points that seemed most convenient to the employees in that area, but the practice materially complicated the treatment system that would be required. The wastes requiring treatment were

1. Cyanide from plating operations
2. Chromium-bearing wastes
3. Hydrochloric acid
4. Nitric acid
5. Sodium hydroxide

Because only small parts were handled at these various points, the total volume of wastes to be treated per day was small, averaging for the continuous wastes 158,182 gal. Dumps from tank cleanout totaled 49,311 gal, but not all of them were dumped on the same day, sometimes not for months, and in several cases only once a year. In a case where all dumps did happen to occur on the same day, and to prevent unusual control of treatment processes, a practice of holding the dumps in holding tanks and dribbling the contents into the flow of daily wastes over a long period was instituted.

This problem was complicated by the number of small flows of wastes from isolated tanks, scattered in seven major locations throughout the factory. Another complication was that due to constant and irregular changes in metal conformation, size and shape of parts, and changes in metal-treating practices; and there was no daily regular program or schedule of certain quantities of parts, pieces, etc., handled in any one day, succession of days, or in one operating period of the factory. It was found that in sixty-eight of the tanks there were continuous running rinses totaling 19,772 gal/d.

The survey of the waste characteristics showed that mutual reactions between them could be utilized to advantage. Plant data showed that the total volume of acid wastes was greater than the total of alkaline wastes, and therefore they could be combined. As stated, many small volumes of waste were produced at stations far removed from the final treatment site. Naturally, the first suggestion was to close down these isolated points and treat all the wastes at a more convenient, common spot. Surprisingly, this met with great resistance from those responsible for the operations at these outlying points. Their argument was that the parts they were turning out were used immediately adjacent to the treatment spot, and if they had to run back and forth to some distant station to take care of their work, it would cause considerable loss of time and add cost to the operation. It was finally decided, in the spirit of compromise, that all operations would be continued where they were, but all waste treatment would be consolidated in one location.

The next question was how to get these small-volume wastes to the main treatment point. The first idea was to use small tank buggies to haul the wastes to the treatment area, but this was ruled out because of the labor cost. The engineer then recommended that at each outlying station the wastes should be drained into a pneumatic ejector. Long familiarity with this much-neglected method of transport of liquids showed that in this case there were manifest advantages; namely, the ejector pot receives the wastes as they come, holds them until the pot is full, and then discharges the contents by air pressure into the sewer or pipeline. The ejector has the advantage of being an entirely closed system, and there is no chance of fumes or odors escaping into the environment.

Since the waste volumes in these distant stations were small, ejectors with a capacity of 50 gal would be sufficient and could be installed in small pits adjacent to the stations. Some of these stations being in various buildings at considerable distance from the treatment site, with roads and travelways for trucks and airplanes intervening, it was recognized that these roads and ways could not be interrupted by excavations for pipelines. Since the pipes would be small, it was finally decided to carry them along on the lower member of the roof trusses of the various buildings, and cross the roads and ways on bridge structures at the roof level of the buildings.

When the method of treatment of the wastes was decided upon, the ultimate discharge point remained to be determined. A small stream ran through the property close to the treatment area. The first thought was to discharge the final effluent into this stream, and it was discussed with the State Department of Health officials to ascertain the requirements for this method of final disposal. The author considers that this discussion resulted in one of the best examples of unbiased cooperation he has ever encountered. The Chief Engineer of the State Department of Health advised the consultants and the aircraft company that he felt that the interests of the company would be best served by treating the effluent for discharge into the sewerage system of the city instead of into the small stream mentioned. The conclusions of this discussion were recorded in a formal letter to the management of the aircraft company, stating:

By this procedure your Company is freed of the responsibility of disposal of industrial wastes and of the problem of pollution of surface waters. Also your dealings will be with the City and you will have no direct responsibility to this Department unless we should be requested by the city officials to review and comment on plans of proposed industrial waste pre-treatment facilities.

This suggestion had great and evident merit when the requirements of the state regarding discharge into open waters and of the city for discharges into its sewerage system were compared. The requirements of the city were much less strict. This made the treatment less expensive for the company. Since the aircraft organization was one of the largest employers in the area, the good relations that already existed between the company and the municipal officials were sustained by this cooperation.

In this project the question of recovery of acids and chemicals arose. A survey with the purchasing office of the company produced the following information on the amount and cost for acids in the year before the waste project came up:

Hydrochloric acid	128,368 lb	$4300
Nitric acid	13,904 lb	869
Sulfuric acid	27,385 lb	280
Chromic acid	6,460 lb	1970
Sodium hydroxide		606
Total purchase for the year		$8025

Studies showed that recovery of chromic acid by the ion-exchange process would cost from 75 to 80 cents per pound of chromium recovered, whereas the price of new acid was only $30\frac{1}{2}$ cents per pound, delivered. The reduction of chromium by sulfur dioxide can be done for about 26 cents per pound. Hydrochloric acid can be recovered by dialysis, but the installed cost of the equipment would be about $15,000, and the annual operating cost, including amortization, would be about $2400 per year. This, plus the cost of the ion-exchange equipment and other units for the other acids, did not present an attractive picture for recovery, and the recommendation was not to attempt it. That was in 1958. The author advanced the idea that, because of the more rigid requirements resulting from the federal government's entrance into the waste treatment picture, it might pay to install recovery as a treatment method regardless of the return, if any.

In this project cyanide was being treated satisfactorily by chlorine in a simple plant. Since the amount was small and the treatment effective, this was continued.

Many ways were found in which savings could be made. For instance, it was found that the operators of a certain phase, requiring a constant stream of fresh water into a series of tanks, judged the right amount when the stream from the open pipe jetted across the top and struck the other side; that was the only control. The amount of water was left to the whim of each operator. It was found that if the water was turned on full stream, the water used per year would cost around $15,000. It was recommended that rotameters be installed on each open line at a total cost of about $1000, so that the amount of water used could be controlled.

Case 5: A Steel Mill in Ohio

In this case the wastes from plating and finishing operations on strip steel manufactured at another point were as follows:

1. Plating wastes: rinse and discharges from the chromium-, nickel-, brass-, and copper-plating operations containing cyanide, chromium, copper, zinc, and other metallic components and free and emulsified oils

2. Oily wastes: coolant oils, lubricants, cleaner washes, and other solutions containing free and emulsified oils

3. Spent pickling liquors from an existing neutralizing facility

4. Pickling-liquor neutralization sludge to eliminate the existing sludge lagoons

With the exception of the neutralization effluent from waste 3, all other wastes had been discharged directly into a large drain or storm sewer which led directly to the river. The State Department of Health had instigated a cleanup campaign on this river and was mainly concerned with the toxic elements in the plating wastes and oils, which were causing heavy pollution in the stream used by downstream municipalities as a source of potable water. Stringent restrictions had been set up by the river authority for discharges into the river.

For some years this mill had also been treating the spent pickling liquors with carbide-lime residue from a nearby acetylene plant, hauled to them at a very low rate. Although the CaO content was lower than standard lime, the cost justified its use. The main objection to the use of this by-product neutralizer was that it created considerably more final sludge than a normal high-calcium or dolomitic lime would. This sludge had been accumulating in several large ponds on the company property. In recent years the property immediately adjacent to these lagoons had been developed for residential purposes, and h uses had been built and occupied and streets opened in close proximity to the lagoons. Many complaints had been registered by the nearby residents against the continued use of these basins. Since the area occupied by the ponds constituted land valuable for expansion of the steel mill, it was decided, as part of the waste treatment problem, to continue the use of the carbide lime, but to install a vacuum filter system to dewater the sludge and dispose of it elsewhere.

At the same time another purpose was served which assisted in the final lime-sludge problem. This was in the waste containing chromium. It was known that chromium could be effectively treated with ferrous sulfate, and in this case there was an excess of ferrous sulfate in the pickling liquor over the amount that would be needed to treat the chromium wastes with the spent pickling liquor, thus reducing the amount to be neutralized with lime and the amount of sludge to be filtered and disposed of, at the same time.

Only one more tank would be required for this operation.

For the oily wastes a flotation-type unit, with added air, was provided to treat the free oil and to add pickling liquor because the ferrous salts would also break oil emulsions.

For the cyanide treatment a normal gaseous alkaline chlorination system was proposed.

Out of all these considerations one salient fact emerged. This was that the state authorities were more concerned with the cyanide content of the wastes than with any other component. The studies by the engineers on treatment of all the wastes and estimates of cost for the plant to meet the requirements of the authority established a final cost of $900,000. However, the designing engineers knew that the main concern was the cyanide, and since the plant management protested that they could not afford the expenditure for the entire plant, a conference was arranged between the state engineers, the designing engineers, and a representative of the plant management to decide what should be done. On being questioned by the designing engineer, the state officials admitted that their main concern was cyanide. At the suggestion of the designing engineer, they agreed to consider a program which would permit the steel mill people to construct the section of the plant designed to treat the cyanide wastes, provided the designing engineers would submit complete plans for the entire plant to treat all the wastes. The total cost of the cyanide treatment section was estimated to cost $350,000. This program, however, permitted the engineers to propose an integrated design in which the units for cyanide treatment would be the first phase, but would also be a basic part of the whole. This permitted a sensible arrangement of units.

Case 6: The Manufacture of Automobile Wheels in Michigan

The wastes from this plant were spent pickling liquors and oily wastes. The only site available on company property was a pie-shaped piece of land between a main-line railroad right-of-way and a busy city street. In solving this problem the ordinary line of progress through a treatment plant had to be reversed in order to fit the plant units into the odd-shaped piece of land. This meant putting the large secondary sedimentation tanks first, in the widest part of the area, then carrying the flow to the smaller units in the point of the land, and then carrying the sludge back again to the treatment building. This arrangement naturally increased the cost of the plant by doubling the pipelines and pumping facilities, but it eliminated any need for purchase of very costly adjoining residential property.

In this problem, after the wastes had been tested and a treatment worked out and plans drawn for the treatment plant, another stumbling block appeared. The requirements of the city in which the industry was located specified, among other things, all of which could be met satisfactorily by the method of treatment proposed, "total solids not over 1000 ppm." In the testing it had been shown that a sparkling, water-white effluent could be produced at moderate cost, but the total solids ranged to about 4000 ppm. Realizing that the city officials might insist on adhering to the 1000-ppm stipulation, which would be very expensive to obtain, the designing engineer arranged to meet in conference the city manager, the superintendent of the sewage treatment plant, the city attorney, and a representative of the company. The engineer had a gallon jug of the effluent produced from the wastes by the treatment method proposed. This showed no signs of sedimentation, and it had been on the engineer's desk for more than 3 months. The effluent from the waste treatment plant was to be discharged into the sewerage system of the city. The 1000-ppm requirement was stated to be the stumbling block. The superintendent of the sewage treatment plant spoke up and said he would be delighted to have such an effluent come into his plant since it was superior to the normal sewage flow. Since the total-solids requirement was in a city ordinance, changes could be made by the City Council only. What was to be done? The estimated cost of removing the additional 3000 ppm of total solids was about $250,000. The company said it could not afford this, and would shut down the factory if it was compelled to adhere to the restriction. The designing engineer suggested that by the addition of one word the whole matter could be resolved: that word was "precipitable," to precede "total solids" in the ordinance. The suggestion was adopted. It took 6 months for the City Council to act, but the matter was resolved.

In a branch plant of this same industry in the same city, an acid waste was discharged, and the city had insisted on treatment. Investigation showed that there were three points of discharge of liquids from this plant. As a criterion for their demand for treatment, the city engineers had used samples of the waste from the one line which had an acid content. The other two outlets into the branch sewer were clean water. It was demonstrated to the engineers of the city that the combined wastes from these three outlets, when they finally entered the city sewerage system, had an acid content within the limits acceptable to the city. This being the crucial matter, no treatment was required.

This case points out the importance of ascertaining the exact components and characteristics of a waste at the point where it actually enters a stream or a municipal sewerage system.

Case 7: A Small Metal-finishing Plant in
Ohio Located Just outside a Large City

From this plant there was a small daily flow of wastes with a relatively high BOD. A waste treatment facility had been installed which was performing fairly well, but the final effluent had a BOD of 150 mg/l. The discharge was into a small stream which entered into a bay of one of the Great Lakes. The state health officials called for a reduction of BOD to 50 mg/l. The engineers modified the existing treatment to reach this limit, but then the state revised its demands to 20 mg/l BOD. This would have imposed a heavy expense on the industry for a very small volume of flow. The designing engineer pointed out that the large city had a limit of 250 mg/l for BOD in wastes discharged into its sewerage system. Since a main sewer of this system passed within several hundred feet of the plating plant, he suggested that the company ask the city to accept its wastes with the BOD of 150 mg/l. They did so, and found the city willing. The solution meant the construction of several hundred feet of a 6-in sewer line to the main sewer of the city and a small volume charge for discharging into this system. This proved much cheaper for the company than installing a complex treatment facility.

Many more cases could be cited where important economies were introduced into industrial waste treatment problems, but those described in this chapter should suffice to show that there are many angles to the procedure, and that they are worth exploring.

Chapter Twelve

Sludge and other solid wastes

"The deposited solids in precipitation or sedimentation tanks are probably the source of more trouble and expense than the collection of the waste liquids."*

This statement was true in 1902, when Naylor wrote it, and is still true today. A system designed to handle sludge can be anything from a simple clarifier to a multistep oxidizer. There are two general sludges we must deal with. The first is organic sludge, first in its raw, untreated form and later in its digested, inert ash form. The second type of sludge is inorganic and generally consists of hydrates of toxic heavy metals.

ORGANIC SLUDGE

The need for units to handle organic sludge efficiently, with the least unpleasantness, was recognized even before the 1900s. A mechanism had been designed that would collect raw sludge from the flat-bottomed tanks. The unit consisted of a scraper, rotated by gears turned by hand at the top of the tank.

In the endeavor to find a means of eliminating the messy job of cleaning tanks by hand, the steep-sided hopper-bottom tank was proposed. One of these was Fulner's Patent Stuff Catcher, invented before 1902. In this unit, sludge was discharged by the hydrostatic head of liquid in the upper portion of the tank. Similar to the Fulner tank was the tank developed by Karl Imhoff of Essen, Germany. The Emscher tank was developed primarily for sewage treatment and was introduced into the United States about 1912 by Rudolf Hering. It was in common use for a number of years. However, this unit developed its own troubles as organic matter was carried up by gas into the gas slots. This

* Naylor, W.: "Trades Waste: Its Treatment and Utilisation," London, 1902.

accumulation on the surface of the gas vents was frequently so dense that it sealed off the escape of the gas, which then found an outlet up through the settling compartment of the tank, disturbing gravitational settling in that area.

SLUDGE DIGESTION

As the years passed, more effective units were designed to collect the sludge. But there still remained the problem of what to do with the raw sludge. Early in the 1900s, the Dorr Company furnished the Sanitary District of Chicago with a mechanism to collect the settled sludge. This led, in 1921, to the installation at one of the sewage treatment plants at Rochester, New York, of an open-top tank designed to provide a place for the organic sludge to undergo decomposition, or digestion. The tank was equipped with a set of two arms at the top, rotated by a central shaft driven by a drive unit on the crossbeams. The arms were provided with heavy chains which hung down into the tank. Their purpose was to break down any heavy surface scums which might form because of gas buoyancy. The operation of this first separate digestion unit showed that the system was feasible, but the gas from the operation had a foul odor that was due to a high percentage of hydrogen sulfide.

Continued observations of the Rochester experiment led to a realization that a digestion tank was in reality similar to human metabolism, and that when an environment similar to that in the human system was provided—temperature and alkalinity—the result was a lack of odor and the production of a quantity of methane gas with a very small content of hydrogen sulfide. In Atlanta, Georgia, gas collected from the gas vents of the Imhoff tanks of the sewage treatment plant was used in the home of the superintendent in his kitchen and laboratory. In Plainfield, New Jersey, John Downes found that this gas could be used to produce power, and he demonstrated this fact by installing a Ford motorcar engine running on the gas from the tanks in the sewage treatment under his charge.

To maintain the environmental conditions conducive to the odorless digestion of the sludge and the production of the useful gas required the application of heat to elevate the internal temperature of the digester contents to a point between 90 and 95°F and the use of an alkaline correctant in the event that the pH of the tank contents fell below the neutral point of 7.0.

Since sludge digesters for large volumes of sludge were very large tanks, sometimes 100 ft in diameter and 25 ft deep, the provision of heat for such a large mass of material could well be a considerable expense; it was found that when the ideal conditions for digestion were maintained in the tank, the amount of gas produced, with an average BTU content of around 550 ft³ and a methane content of more than 65 percent, would be sufficient to provide the necessary heat. In the early installations this heat was provided by circulating hot water through a series of pipes fastened on the inside wall of the tank. The water was heated in heaters of the type normally used in homes and fired by the gas produced in the digestion tanks. Later, batteries of pipes suspended in the tank were used to permit cleaning and inspection. Further experiments indicated that in the development of power in internal gas engines, provided with water cooling systems, the circulating water was heated to almost boiling point. The latest development, then, was to circulate the cooling water through the digester heating system and return it to the engine. It was found essential that the contents of the primary digestion tank be thoroughly agitated to ensure contact between the digesting sludge and the incoming raw sludge to maintain the environmental conditions conducive to rapid and complete digestion. This need was met in a number of ways: mechanical agitators to stir the contents; mixers in draft tubes to move the contents up or down to create circulation; and introduction of gas recirculated through tubes to cause circulation of the contents. Mechanical agitators were used in hundreds of digestion tanks for many years, and the agitation by gas has been reasonably satisfactory, but there are inherent difficulties in the use of any of these methods. All these systems required the introduction into the digestion tanks of equipment, piping, etc., which were difficult to handle in case of trouble. In a digestion tank 100 ft in diameter and 30 ft deep, containing 1.767 million gal of liquid sludge it became a serious matter if the tank had to be emptied in order to get at a troublesome part at the bottom of the tank. The emptying of the tank was a task in itself, requiring time and money for pumping. An additional problem was finding a place to put the sludge already undergoing decomposition and liable to create an odor nuisance. If the mixing and heating units were so arranged that they could be lifted out, these deep tanks required rigging or rental of equipment for that purpose.

The possibility that such difficulties would arise led to continued search for better means of achieving the desired mixture of sludges in the digestion tank. The result was a different method of creating mixture and agitation. The system, developed by Dorr-Oliver, Inc., in 1962 was known as the Dynomix system.

Dorr-Oliver Dynomix System

All heating and agitation equipment is located outside of the digestion tank, readily accessible for inspection, repair, or maintenance at any time. This system accomplishes the objectives by a jet action which introduces the raw sludge into an intense mixing zone by jetting the sludge at a point close to the bottom of the tank. In this action the raw, already-heated sludge is mixed throughout the tank and maintains circulation for a uniform temperature and environment throughout the mass of sludge.

The light, volatile solids and greases which tend to accumulate on the surface are returned to the sludge preparation system by withdrawal from the surface of the tank. They are returned through the jets to the active digestion zone. This eliminates the characteristic blanket of surface scum common in other methods. The spiral flow within the tank ensures that the heavier solids, which digest more rapidly, remain in the lower portion of the tank. They move in a spiral motion toward the sludge drawoff at the center of the tank. The lighter solids are caught in the upflow currents which prevail around the vortex deflector tube and are recirculated into the active digestion zone.

The Dynomix system includes the following units: (1) a sludge disintegrator to assure the breakup of all solids in the raw sludge so as to avoid clogging of the jets; (2) a heating unit to heat recirculated supernatant liquid; and (3) a pumping system which produces the necessary velocity and capacity to force the heated sludge through the jets into the tank and to set in motion the circulatory currents inside the unit.

To maintain the internal temperature of the digester, covers became essential. These were of several types, such as solid concrete domes and floating metal or wooden covers buoyed up by the liquid in the tank. All these expedients served to handle the sludge, but the sludge did not disappear. It simply changed its form. It was still the most expensive phase in the treatment plant and an operating problem.

As early as 1930 difficulties with digestion tanks showed the desirability of developing ways to handle and dewater sludge other than simple digestion and drying in lagoons. The author was granted a U.S. patent, the preamble of which stated:

Therefore, the object of this invention is to devise a method or process relating to the treatment of sewage, or other trade wastes, which will hasten the disposal of those solids removed from the sewage by sedimentation, which will reduce the cost of sewage-sludge handling by reducing the number of necessary operations thereon; which will eliminate some of the construction and structures now required, thus also reducing the ground area required; and which will reduce the labor now entailed. In short, one of my objects is to render unnecessary the step of digestion. A further object is to arrange the sewage treating plant to be continuous, whereas at present the step of digestion is more or less intermittent. My aim is to devise a sewage-treating plant which will be like a modern factory; the raw materials will be fed in at one end, and the finished product will be run out at the other end, the finished product in this case being an inert and odorless material containing only fully consumed materials capable of immediate disposal on land or fill.

This statement indicates that more than forty-one years ago the troubles encountered in the digestion of sludge were present and that constructive thinking was producing a better way. Since then many companies and individuals have attacked this problem, with the result that today there are numerous developed and tried methods of dewatering and handling industrial waste sludges, or raw liquid sludge, which have eliminated the anaerobic digestion of this material. A few methods are described below.

Bartlett-Snow-Pacific

This is a multiple-hearth incinerator. Following vacuum filters, centrifuges, other types of dewatering units, the cake falls on a conveyor belt and is then introduced into the incinerator unit. The ribbon screw type of feeder also acts as a seal. Each incinerator unit consists of a vertical cylindrical shell, lined with refractory material that encloses nine circular hearths. Extending vertically through the unit is an air-cooled rotating shaft carrying rabble arms with rakes at each hearth level. Entering the unit at the top of the incinerator, the sludge is raked in a spiral path and moves in and out of each successive hearth by falling through alternate drop holes. Since all the organic matter in the sludge is burned, the ash coming out from the last (bottom) hearth at about 500°F is completely sterile. After slurrying, the ash may be discharged to a waste lagoon.

The gaseous products of combustion are exhausted from the top hearth and pass through a precooling chamber and into a wet scrubber. The fly ash is removed and final gases, said to be completely odorless, are vented through a stack. They usually contain only about 0.2 lb of fly ash per 1000 lb of gas. The system is economical in operation because the fuel used by the multiple-hearth incinerator is in some cases sludge gas from an anaerobic digestion system, or lacking such fuel, natural gas may be used. Published information indicated that a system designed to handle 500 lb/h of dry solids would have an estimated capital cost of around 180,000, including vacuum filtration, incineration unit, and essential accessories. Operating costs, per ton of dry solids, including fuel, power, and conditioning chemicals, are said to run between $2 and $16 per ton of solids handled.* These prices were true for the 1960s. For current prices see the Comparative Cost Index Chart in Chapter 11.

This system has been used on several types of industrial wastes, among them:

Metalworking

Reclaimed rubber wastes

Waste chemicals from photo processing

Waste fats from food processing

Spent catalyst from the petroleum industry

Mill scale from steel mills

Wastes from the aluminum industry

Waste solids from the pulp and paper industry

Sugar liquor from sugar processing

Beloit-Passavant System

This is a development of the Passavant-Werke of Germany, and was introduced into the United States by the Beloit Corporation. Essentially, this is a system

* Albertson, Orris E.: "Low Cost Combustion of Sewage Sludge," paper, annual meeting WPCF, Seattle, Wash., Oct. 1963.

using plate and frame presses to dewater the sludge, and normal incineration to further destroy the organic matter.

Dorr-Oliver, Inc. (F-S System)

As indicated in Figure 12-1, this system consists of the following sequence units: the sludge from primary or secondary clarifiers is fed into a hydroclone-type degritting unit, which the manufacturer claims can remove 95 percent of the 200+-mesh inorganic material. The degritted sludge is then concentrated in a sludge thickener. The sludge from the thickener is ground in a disintegrator to at least $\frac{3}{8}$-in particle size for easier handling in the following steps: After the thickening and disintegrating processes, the sludge is further dewatered by bowl centrifuge, rotating at 4000 r/min, developing 3700 gravities. The high speed of this unit, coupled with the long-bowl design, provides maximum solids capture and concentration. This unit is constructed of stainless steel, and all surfaces exposed to abrasion are hard-surfaced.

Following the centrifuge, the dewatered sludge is then screw-fed or pumped into the fluidized-sand-bed reactor. This unit, the heart of the F-S system, possesses the characteristics of boiling water, which provide the extreme mixing so vital to efficient combustion. The combustion zone of this unit, operating at from 1300 to 1400°F, is said to have a heat capacity of 16,000 BTU/ft³,

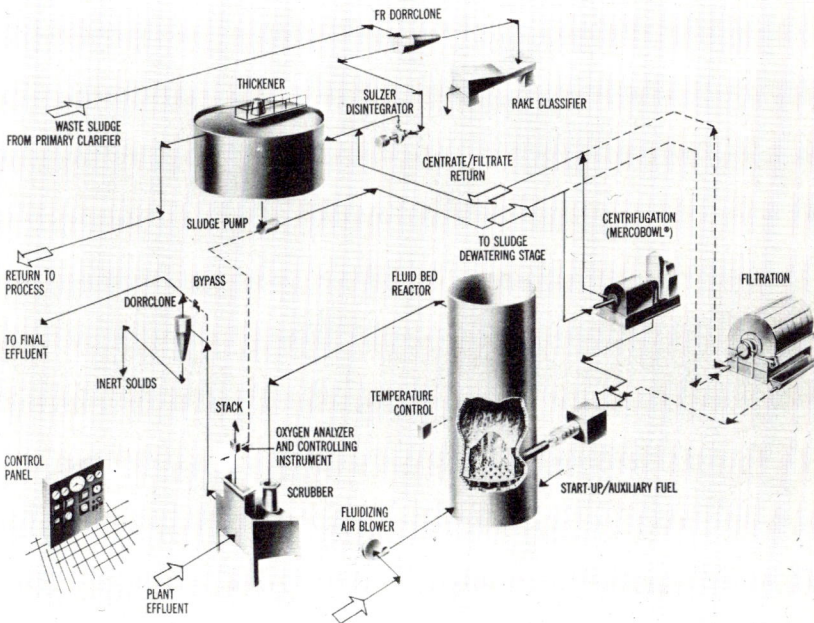

Fig. 12-1 Flow sheet of the Dorr-Oliver FS disposal system. (*Dorr-Oliver, Incorporated.*)

and this, in combination with the extreme mixing, produces complete drying and combustion in a very short period, said to be a matter of seconds. The hot sand in the reactor bed serves as a heat reservoir during periods when the unit is not operating. The minimum temperatures in the reactor ensure complete control of odors. As an economy, the hot gases leaving the reactor may be used to preheat the incoming air. Inert ash is stripped from the reactor bed by the upward-moving hot gases, and is carried out at the top of the reactor. These gases are then scrubbed to remove fly ash. The manufacturer claims that the final gases discharged to the atmosphere contain less than 0.1 gr of particulate matter per standard cubic foot of combination gas. The sterile ash which is the solid product of this system has a very high sealing rate, essential to economy of area in lagoons. If lagooning is not considered advisable (frequently the case in industrial installations where land is scarce), the inert ash may be further concentrated in a hydroclone to approximately 25 percent solids, or if still further solids concentration is desired, the underflow from the hydroclone can be fed to a rake type of classifier and thus produce a solids concentration of up to 75 percent. This system has been used in the destruction of sludges from a number of industrial waste operations. It may be operated either manually or automatically. Figure 12-2 shows the installation of a plant of this type.

Fig. 12-2 Dorr-Oliver CR FS disposal system installed at the Richardson Bay sewage treatment plant at Tiburon, Calif. (*Dorr-Oliver, Incorporated.*)

Zimmerman Process (Zimpro) of the Sterling Drug Company

This system, known as the "wet-air" oxidation system, oxidizes organic materials in liquids, of either the dissolved or the suspended type. The waste sludge and air are fed continuously to the system in proportionate quantities by a pump and an air compressor. The two streams are joined and mixed in a simple pipe tee or wye. The mixture is then heated in a heat exchanger to about 300 to 400°F, depending on the heat value of the waste and the extent of oxidation desired. When entering the reactor, the oxidation starts rapidly and the temperature rises. As the temperature rises in the reactor as well as in the heat exchangers, steam is formed directly from the wastewater. The degree of oxidation is a function of the concentration of the organic material in the sludge, the reactor outlet temperature, and the residence time in the reactor.

Operation depends upon the type of waste, the concentration of the solids in the waste, the desired degree of oxidation, and the purpose of the plant. This system has been found satisfactory on pulp and paper plant wastes and also for glue factory and textile wastes.

Traditional Methods

The cost of land in industrial areas has become so high that the use of the former open-type sludge drying bed is questionable. The much smaller area used by any modern dewatering and thermal method of final destruction is far superior. The decision as to whether to use the available land for sludge drying beds, rather than for expansion of productive factory purposes, must be given considerable thought. An interesting situation arises where an industry on a restricted site is within reasonable distance of a municipal sewage treatment plant with a modern sludge treatment installation. Providing a pipeline and pumping facilities to deliver the sludge from the industrial waste treatment center to the sludge dewatering and handling system of the municipality may warrant consideration. This, of course, would depend upon (1) the distance over which the sludge would be pumped, (2) the capacity and willingness of the municipality to handle the extra load, (3) the type of sludge from the industrial waste, and (4) the charge that would be made to the industry for this service. A feasibility and economic study would be required.

DEWATERING

Dewatering sludges has long been recognized as an important part of plant design. Any of the units for sludge disposal would run more efficiently if the sludge contained less water. Quiescent settling of liquids containing settleable solids, the oldest method of dewatering, will produce a sludge which may have a final solids content of from 2 to 4 percent solids. Sludge of the greatest density will be produced in those sedimentation tanks which are equipped with mechanical sludge collectors. In these units, especially the circular type, the motion of the rakes opens up channels in the sludge mass and permits some of the occluded water to pass out. However, a sludge with 5 percent solids is difficult to handle. If it is to be handled on vacuum filters, the size of the filter drums must be related to the moisture content.

It has always been the desire of the engineers and waste treatment plant operators to reduce the moisture in the sludge, realizing that the moisture content or, if preferred, the solids content, of the sludge has an important bearing on the amount of material to be handled. Let us consider this. A sludge with 2 percent solids is 98 percent water; that is, there is a ratio of 50:1. A sludge with 4 percent solids has a ratio of 25:1. In other words, the greater the percentage of solids in the sludge, the smaller the amount of water to be handled. Since the greater part of the water in the sludge is occluded water, that is, water that surrounds the particles of solids, any action that would open channels permitting the water to escape would be beneficial. Since the normal action of the circular type of mechanically cleaned sedimentation tanks is to open up channels in the sludge as they convey it to the central sludge discharge pit, an early type of "sludge thickener" provided riser blades. These blades, similar to a picket fence, extended upward from the arms, on the theory that the pickets would more effectively open the desired escape channels. However, experience with this operation has shown that the usual structural members of the arms act as well as the additional pickets, and they are now not used.

In practice, the thickener is a separate tank following the sedimentation unit, with a capacity of 8 to 12 h flow of the sludge from the primary unit. The sludge from the settling unit is pumped into the thickener and is usually returned to clarifiers. The action of these thickeners, when properly designed and controlled, can produce final sludge densities of 12 to 15 percent solids.

Another method of sludge thickening using the dissolved air method offers an auxiliary means of economically reducing the moisture in the sludge. This process assists materially in the reduction of moisture of the sludges from activated sludge treatment processes which normally have a solids content of about 2 percent and are difficult to dewater by normal thickening processes. This system consists of three elements: (1) the formation of minute air bubbles, (2) the attachment of air bubbles to the sludge solids, and (3) the separation of the air-entrapped solids from the water. A prime factor in the success is the production of small air bubbles. The process works successfully in a combination of textile and laundry wastes which resulted in a thickened sludge concentration of 6.1 percent solids. In brewery wastes in Colorado, combined with municipal sewage, the concentration of solids in the sludge produced by air flotation was five times that of the sludge entering the unit.

The advantage of this extra phase of treatment must be carefully considered in order to reach a proper economic balance. The sludge thickener adds another unit to the plant, the expense of which must be justified. The factors to be considered are

1. Reduction of the total volume of sludge to be handled, which materially affects the operating cost of following phases.

2. Smaller sludge pumps may be possible, resulting in

   a. Less initial cost.

   b. Less power required to operate the pumps, a reduction in operating cost.

   c. Less area required for the pumps.

3. Reduction of the size of the vacuum filters or centrifuge units, resulting in

   a. Less initial cost for the filters.

   b. Less floor area required.

   c. Less area required for the pumps.

   d. Smaller vacuum pumps (the most expensive item in vacuum filter operation).

   e. Less power for the vacuum pumps, a major item in filter operation.

   f. Less effective filter area, which means less medium area, with consequent less cost.

4. Less filter cake to handle. This reduces the cost of hauling to final disposal points or to further dewatering or destruction units.

5. Less attention is required.

6. If further dewatering by fluidized sand or other thermal methods are used, the smaller amount of sludge and less water mean smaller and less expensive units and less fuel.

7. A balancing unit is provided for further dewatering steps—vacuum filters, centrifuges, and thermal systems—which operate more satisfactorily and economically if the density and volume of sludge received are uniform in character. To take raw sludge as it accumulates in a sedimentation unit, varying in moisture and solids content during a day, means more manual control of dewatering phases. The thickener acts as a reservoir, enabling the thickened sludge to be fed into the final stage at a constant density.

Plate and Frame

There are several other ways to remove occluded water from sludges before they are further treated in thermal destruction processes. The records of industrial waste treatment installations in Great Britain before the turn of the century indicate that the unit most used was the plate and frame press. This is simply a squeezing process in which the occluded water passes into the hollow center of the leaves, separating off the solids. This unit is still used in some cases, but because of the labor involved in cleaning the plate sections when they accumulate a heavy layer of solids, other methods are more popular today.

Traveling Belt Cloth

Another device developed before 1902 consisted of a traveling belt cloth passing between rollers which squeezed out the occluded water. Water jets were used to clean the cloth. An American manufacturer came out with a device operated on somewhat the same principle (Figure 12-3). Naylor's remarks about the early device are interesting:

> The whole device is clever and would be effective if it worked with care under a constant volume of liquor having a uniform composition. But it can hardly lend itself to constant changes in liquid trade wates, and moreover, the mechanisms are too delicate

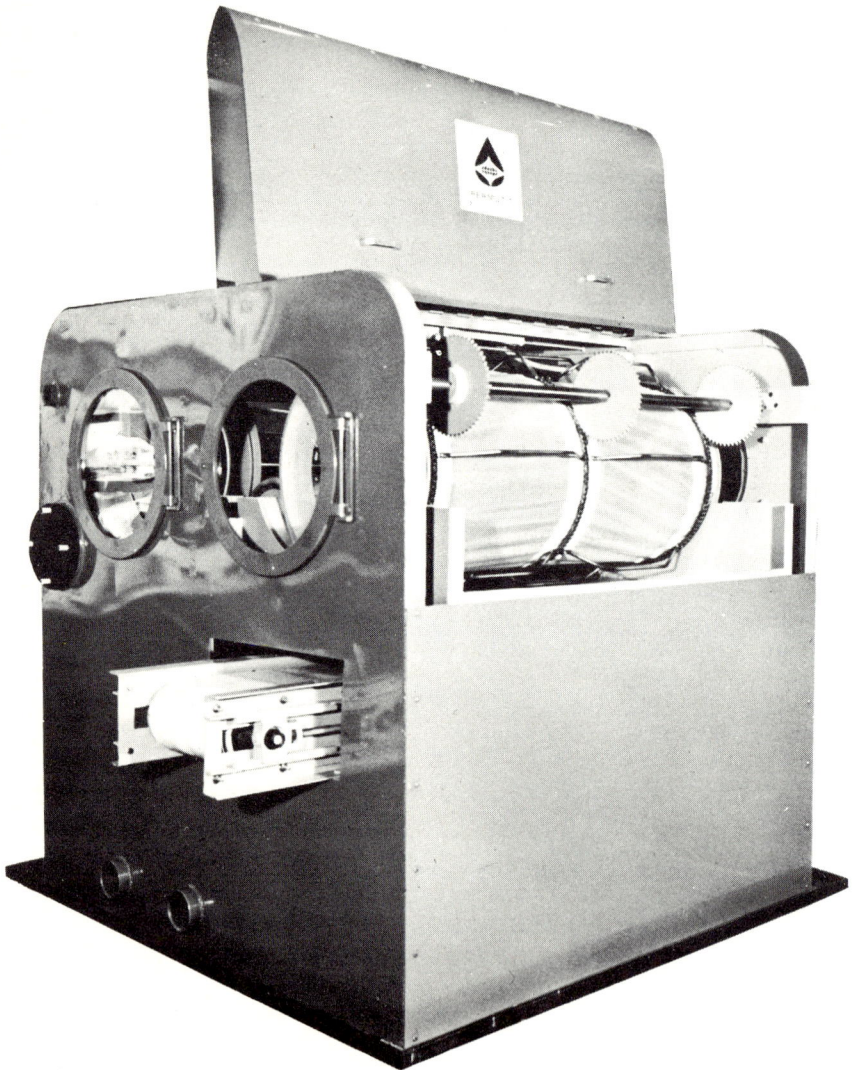

Fig. 12-3 DCG unit for dewatering of sludge. (*The Permutit Company.*)

to be handed over to the usual class of caretaker appointed to "get rid" of trade wastes, be it much or little, weak or strong, who can rarely be persuaded that the business is really worth any trouble at all, much less that involved in hurriedly refixing trippet or cam motions, repairing or withdrawing torn cloths.

Vacuum Filter

Another method of dewatering sludges is the vacuum filter, usually the round drum type. There are other types covered with appropriate media, called disk-type filters. These have wheellike segments. The usual vacuum filter is simply a round drum covered with a fine woven medium. The sludge is usually picked up as the drum rotates in a pool of sludge. After the drum emerges from the pool, a vacuum is applied from inside the drum, drawing the water through the medium and leaving the solids on the outside of the drum in the form of a flat blanket. As the drum continues to revolve, the vacuum is continued until a point on the downward travel of the rotating drum is reached. At this point the vacuum is cut off, and air pressure is exerted from the inside of the drum. The sludge blanket is loosened so that it can be picked up on the strings, or scraped off the drum by the scraper blades which rest on the surface of the drum. The "cake" then falls off onto a belt conveyor, which carries it to the final disposal point or into trucks, for hauling to a dump. An advantage of the vacuum filter is that the sludge is always visible, so that changes necessary in filter speed or sludge adjustment may be easily made.

In some types of vacuum filters there is a series of strings or cords wound around the drum which tend to lift the cake, freeing it from its hold on the filter medium so that it will fall off onto the conveyor. In one type of filter the entire filter medium consists of stainless steel springlike cords placed close to each other, through which the water is drawn. This is known as the "coil" filter. As the drum revolves and approaches the cake takeoff point, some of the cords flex in passing over a roller and release the hold of the cake so that it falls off onto the belt. Other types of filters include pressure filters.

Filter Medium. One of the most important controlling factors in the operation and cost of vacuum filters is the selection of the filter medium best suited to the type of sludge to be dewatered. In the early years this medium was usually a closely woven woolen or cotton cloth, but in later years the synthetic textiles were found to be more resistant to the acid or alkaline or other chemical action. They also provided longer life. To evaluate the true cost of certain types of filter medium, certain items other than the bare cost of the cloth must be added. One is the labor required to remove the cloth and attach another. Depending upon the construction of the filter drum and the means provided for holding the cloth, the size of the filter, the facilities for changing cloths, anl the experience of the operators are factors in this cost. The operation of changing cloths may take several hours. Experience with various types of filter cloths indicates that they may last for a few days to several months or more. It can be seen, therefore, that a cloth which has the most durable life may be the cheapest one to use in a given case, regardless of the initial cost. In selecting one cloth rather than another, an additional consideration is the possible interruption to filter service if a short-lived cloth is used and the filter is shut down every few weeks.

The information needed in the selection of the best type of filter cloth for a specific type of industrial waste sludge is as follows:

1. Chemical composition of the sludge and density of the liquid

2. Specific gravity of the solids in the sludge

3. Kind and concentration of liquid in the sludge, that is, alkaline, caustic, or acid

4. pH of the sludge

5. Temperature of the sludge at time of filtration

6. Physical characteristics of the solids in the sludge, that is, crystalline, granular, slimy, colloidal, etc.

7. Filter test information (if available), amount of solids retained, etc.

8. Maximum particle size that it will be permissible to have in the filtrate

9. Relative proportion, by weight, of solids to liquids in the sludge to be filtered

10. Filter aid; if necessary, what kind

11. Volume of sludge to be filtered, that is, gallons or cubic meters to be handled per hour or per day

The present types of filter cloths and their general effectiveness for certain sludge characteristics are as shown below.

Type of cloth	Resistance to:	
	Acid	Alkali
Acetate	Poor	Fair
Cotton	Poor	Fair
Dacron	Very good	Good
Dynel	Excellent	Excellent
Glass:		
Spun	Excellent	Fair
Continuous-filament	Excellent	Fair
Nylon	Fair	Excellent
Orlon (acrylic)	Excellent	Fair
Polyethylene	Excellent	Excellent
Polypropylene	Excellent	Excellent
Polyvinyl chloride	Good	Excellent
Rayon	Poor	Fair
Teflon	Excellent	Excellent
Wool	Very good	Fair

Economy in Filter Media. The criterion for selection of a filter cloth is its "useful life." This is the actual length of time that a filter cloth may be used before it shows signs of deterioration, breaks or cracks, or actually disintegrates. If a cloth with an estimated useful life of 1000 h were used on a filter operating 8 h/d, or 125 d, the estimated replacements would be three times per operating year, or for 16-h operation, six times per year. The total cost of this type of cloth then would be (1) the cost of the cloth delivered at the plant, and (2) the charges for the labor to remove the old cloth and replace the new cloth, probably 2 or 3 h per change, or a total of 9 to 12 h/year. On the other hand, if a cloth is selected which may have an estimated life of 11,000 h, the economy is different. On the basis of 8-h operation per day, this cloth would, theoretically, last 1375 d, or 3.76 years. In this case the total cost would be for one cloth and 3 h of labor in the period of 3.76 years versus the cost of eleven cloths and more than 33 h of labor for the cheaper cloth. The better cloth would be more expensive than the less useful one, but in the long run, over several years, the economy would be in favor of the better cloth.

Tests made in one of the largest vacuum filter installations in the world, with ninety-eight individual filter units, have developed much of this information on the life-span of filter cloths or blankets, as they are sometimes called.

The definite advantage of the plastic filter cloth is that it offers less resistance to air flow and greater surface area. On the other hand, it is reported that cloths of the polystyrene and polypropylene types are subject to attack by turpentine, an ingredient in most kraft mill effluents.

Centrifuges

Many engineers have preferred to use centrifuges for the dewatering of sludge before thermal methods of final destruction. Such units are manufactured by Bird Machine Company, De Laval Separator Company, and Dorr-Oliver, Inc. One of the advantages of a centrifuge which appeals to many users is that the sludge is contained at all times within the bowl of the unit. The total space required for a centrifuge unit is considerably less than that for a vacuum filter and its numerous accessories. On the other hand, the power requirements of a centrifuge are very high. The apparatus must maintain speeds as high as 4000 r/min.

INORGANIC SLUDGE

Inorganic wastes also produce their share of solids. Although the actual amounts of solids produced by treatment of chemical pollutants is insignificant when compared to the volume produced by organic treatments, their presence constitutes a serious problem. The quantity of suspended solids in plant effluent is limited by governmental regulations. But strict attention is given suspended solids of heavy metals. Industrial facilities remove the majority of their heavy metals—copper, gold, chromium, mercury—by precipitation reactions. Once the reaction is complete, the precipitate must be removed from the effluent. When initial concentrations range from 200 to 500 ppm heavy metal, the task of removal is often frustrating.

The first step must be to concentrate the precipitate. This is usually done in a clarifier. Conventional clarifiers can be circular or rectangular. Some recommended standards are:

1. Maximum service rate should not exceed 800 gal/d (ft²).

2. Velocity over final weir should not exceed 10,000 gal/d (linear ft) of weir.

3. Inlet should be baffled to dissipate inlet velocity to a maximum of 2 ft/s.

4. Sludge line should have a minimum of 30 in of head.

Typical clarifier underflow is 1 to 2 percent suspended solids. In a large operation, this could represent tens of thousands of gallons of unnecessary water. In the past, clarifier "muds" were sent to lagoons for "drying." Regulatory rulings forbid such discharge of industrial wastes onto the ground. The regulations are particularly severe where heavy metals are concerned.

As an alternative or subsequent dewatering step, rotary vacuum filters or pressure filters are used. Feed concentrations to the vacuum filters range from 4 to 8 percent solids. To attain this concentration, a "thickener tank" is often used after the clarifier.

Pressure filters, when properly designed, will give essentially continuous service and produce a concentrate with 20 to 25 percent dry solid. The filtrate will average 0 to 3 ppm suspended solids. The filter stations are usually designed at 0.05 to 0.06 gpm/ft² filtration surface, for service on the clarifier underflow. Properties for continuous dewatering filters for clarifier underflow are

Main flow, gal/m	Underflow, gal/m	Filter size, ft²	Cost, $
4.2	4.2	90	14,000
8.3	8.3	165	16,000
16.6	16.6	335	19,000

Automated tank-type pressure filters are designed for 0.05 gal/m (ft²) of surface. The unit will dewater the clarifier underflow of metal hydrate to 40 to 60 percent suspended solids. The filtered effluent will have a turbidity of 0 to 5 ppm suspended solids.

In batch treatment of industrial waste, filter cartridges are often best for removal of precipitates. Where treatment is infrequent or quantity of precipitate is small, a disposable cartridge may be the best solution. The main components of a filter of this type are the chamber, the fiber, and the core of the cartridge. The chambers are made of plexiglass, heat-resistant glass, polypropylene, CPVC, Teflon, aluminum, stainless steel, or vinyl-lined carbon steel. The fiber material and suitability are

Polypropylene—for strong acids and alkalies; to temperature of 200°F

Cotton—for water, portable liquids, dilute acids, alkalies, organic solvents, and gases; to temperature of 300°F

Rayon viscose—similar to cotton but more sensitive to alkaline solutions and other polar liquids; to temperature of 300°F

Rayon jute—for viscose fluids; to temperature of 250°F

Nylon—for concentrated strong alkalies and solvents to temperature of 300°F

On single-pass filtration, the cartridge covering the nominal micrometer retention range should be selected. When particle retention is critical, select a cartridge which offers one-fifth the micrometer selected for absolute removal of all particles. For example, use 5 μm when removal of all 25 μm and larger is required.

The range of micron porosity is as follows:

100 Extra coarse

75 Very coarse

30 Coarse

20 Medium

10 Fine

5 Extra fine

3 Dense

2 Medium dense

1 Extra dense

0.5 Most dense

The centrifuge is also a well-used method of dewatering inorganic sludges. It is particularly useful when the precipitate is gelatinous. A centrifuge will capture 75 to 80 percent of the metal hydrates. Oftentimes the concentrate must be sent to a clarifier or thickener tank. Centrifuges for clarifier underflow have the following approximate properties:

Underflow, gal/min	Centrifuge size, inches	Price, $	% capture	% cake moisture
4.2*	standard	5,155	75	20
4.2†	36 × 20	31,000–32,000	90	20
8.3†	36 × 20	31,000–32,000	90	20
16.6†	48 × 30	37,000–39,000	90	20

* Figures courtesy of Haviland Products Company.
† Figures courtesy of De Laval Separator Company.

Hauling of Chemical Wastes

In small facilities complete dewatering systems may not be economically justified. As a temporary expedient, many firms have their chemical wastes hauled away by commercial tank truck. The clarifier underflow can be piped into a holding tank, or if the mud contains too much water, a thickener may be used. The waste when removed is taken to an approved public dump, one in which there is no danger of contaminating the public water supply. Such dumps have limited capacity and can therefore be considered only as a temporary solution.

In order to design for disposal by commercial tank truck, one must know the maximum volumes of the truck tanks. There are two sizes of tank trucks—50 bbl and 100 bbl. A barrel is 42 gal; therefore the capacity of the trucks in gallons is 2100 and 4200 respectively.

The waste is transferred to the tank truck by means of a vacuum pump. The pump can develop a head of 28 in, enough to lift waste 10 ft with a 150-ft hose. The hose diameter for nonacid wastes is 4 in. For acid wastes, the diameter is 2 to 3 in.

The lining of the tank must be compatible with the wastes it hauls. The tank truck owners must be told of the contents of the wastes. Tanks designed for noncorrosive wastes are made of black iron interior. Corrosive wastes are carried in tanks of stainless steel or tanks lined with rubber.

The average charges for hauling away industrial wastewater are as follows:

For the 100-bbl black iron tank truck	$21.00 per hour
For the 50-bbl black iron tank truck	$18.75 per hour
For the 100-bbl rubber-lined/stainless steel	$24.55 per hour
For the 50-bbl rubber-lined/stainless steel	$22.30 per hour

All the hourly rates are based on time from and to truck owner's yard. Cyanide and acid-bearing wastes are charged an additional 2 to 3 cents per gallon at the dump. Oily wastewater is charged about $2 per ton additional. An average chemical waste dump would cost between $170 and $250 for 4200 gal.

SUMMARY

In summary, liquids-solids separation, or sludge handling, forms a critical part of any modern waste treatment plant. The regulations regarding limitations of suspended solids must be carefully examined. When found to be warranted, the equipment needed to treat the waste problem must be carefully selected so as to effectively treat the waste and conform to the regulations. It must be remembered that the limitations on discharge are not static. Many governmental agencies have set goals for water quality that are more stringent than those now on the books. With this in mind, treatment equipment must be flexible enough to change with the changing times.

Chapter Thirteen

Coagulants, reagents, and aids

In the early days of industry, the only industrial wastes that posed any problem were those consisting of suspended solids. Many of these solids were of a biological nature and, if left alone, would decompose into harmless, inoffensive ash. It was the length of time required for the natural process that eventually began to worry industrial management.

Labor as well as time became an important consideration. The efforts of scientists, engineers, researchers, inventors, and others were directed to search out, develop, and put into practice many new methods and materials which would reduce the amount of labor and time required for many operations. Automation, automatic analyzers, coagulants, coagulant aids, sludge reduction by thermal means, high-speed dewatering, and new media for filters have all been steps in this new science. The ultimate aim of all the new processes and operations has been the reduction of time to accomplish the desired result. This chapter is devoted to consideration of methods and means that have been developed to expedite the treatment of wastes by the addition of chemical compounds, both natural and synthetic, to the wastes, and the estimation of the true economy in the selection and actual use of one of these materials in preference to the others.

In the past, industrial waste treatment adopted many of the means and methods developed for the treatment of water and sewage. Many of the problems involved in the removal of solids and other ingredients in industrial wastes reacted to these same methods. Nature has always been a guide in the treatment of wastes. The natural forces of interception, gravitation, bacterial action, and atmospheric utilization were the original methods of treatment. Nature is not notoriously speedy in action, but in the early years time and space were not considered as important as they are today. Barriers will intercept solids, but

313

the barriers will clog. Sedimentation tanks will remove suspended and settleable material to a fair degree, but it takes from 2 to 4 h to produce an acceptable result. Biological operations to stabilize the remaining organic matter which did not respond to gravity take more time—in activated sludge 4 to 8 h, and more in many cases. If left to nature, the anaerobic bacterial destruction of the active organic matter in the residues (sludge from the sedimentation units) takes 13 to 50 d or more. The final drying of the inert ash takes time when exposed to the atmosphere. Much of the success of open-air drying depends on the climatic conditions of the locale. All this demands large tanks and large areas. In the early years land was not too expensive and plants covering many acres were built. As time passed, however, industrial activity and the increasing demands of the population expanded beyond their former limits. As land prices skyrocketed and labor's demands increased, the answer to the new problem was to do the job in less time, less space, and with less labor.

DEFINITIONS

To expedite treatment, chemical compounds were used in all phases of wastewater treatment. In the realm of solids removal, there is a whole science of coagulants and aids used as expediters. A few definitions are needed to begin this part of the discussion.

Coagulation

Coagulation is the "agglomeration" of colloidal or finely divided suspended matter by the addition to the liquid of an *appropriate* coagulant, by biological processes or by other means. *Appropriate* is the key word.

Chemical Coagulation

Chemical coagulation is the "process of forming flocculant particles in a liquid by the addition of a chemical coagulant; also the removal of colloidal or finely divided suspended matter from the liquid by the floc, and the agglomeration of the flocculated matter."

Coagulant

A coagulant is a chemical which, in solution, furnishes ionic charges opposite to those of the colloidal turbidity particles in the water. Coagulants neutralize repelling charges on colloidal particles of turbidity and produce jellylike spongy masses called "floc." Floc has an enormous surface area per unit of volume which entraps or adsorbs particles of turbidity, organic matter, and bacteria.

Coagulant Aid

Coagulant aids are not in themselves coagulants, but are of considerable assistance in the coagulation process. These chemicals form large floc particles that settle down through the solution and absorb the particles of coagulated turbidity. Coagulant aids act as "binders" or "bridges" in that they mechanically entrap or stick floc particles together. By creating larger and heavier floc, coagulant aids speed up the settling process, facilitating clarification at higher throughput rates.

Flocculation

Flocculation consists in the mechanical entrapment of agglomerated particles by adsorption onto the floc formed with the coagulant chemicals. It is also the molecular bridging of individual molecules of the coagulants. Although coagulation is a chemical reaction, the gathering or adsorption is a mechanical process. Flocculation causes considerable increase in the size and density of coagulated particles, resulting in a faster rate of settling of the floc particles. The rate may be further accelerated by the use of a coagulant aid.

USE OF COAGULANTS AND AIDS

As early as 1867 it was found that lime and salts of iron were very helpful in reducing the time required for solids to settle naturally. Their use, however, produced large amounts of high-moisture sludge which required expensive dewatering equipment. Lime sludge dries very slowly. Such sludge accumulating for 5 years will still show a high moisture content. In one instance, the lime sludge from a beryllium ore plant was left to dry in lagoons. After 9 years the crust was hard enough to walk on but the lower strata were still in a semi-liquid state. The science of coagulants had far to progress.

The early treatment plants added the lime directly to the waste flow. This was inefficient since it did not assure complete mixing. It also allowed deposits in the pipelines. The first step in correcting the situation was to agitate the waste solution as the time was added. This assured thorough diffusion through the liquid waste and facilitated contact with all small solids in the waste. The first mixer units used for this purpose were simple revolving blades which served to keep the solids in motion.

Liquid Alum

Aluminum sulfate has been a favorite coagulant for water and wastes treatment, both municipal sewage and industrial waste, and has been used for many years. It has generally been used in the form of a powder, purchased in bags holding about 80 lb, and is dispensed in one of the several forms of mechanical dry-feeder units. These units automatically control the amount of alum fed, and measure it at the same time. The dry alum is dropped into a solution pot, an integral part of the feeder unit, to form a solution, which is then transported to the mixing, or flocculation, basin of the plant. The handling of dry alum in bags entails certain manual operations which add cost, such as handling of the bags from the delivery vehicles; stacking them in a dry place; then handling them again from the stack to the hopper of the feeder; stacking and handling of the empty bags; and filling of the storage hoppers of the feeders at intervals. Even with the largest storage hoppers available, and extending them through an upper floor of the operating building to facilitate handling by the operators, labor is entailed.

Users of large amounts of aluminum sulfate can obtain important economies in treatment of wastes by the use of liquid alum instead of the dry powder form. Shipped as a liquid coagulant, the liquid contains 5.14 lb of aluminum sulfate per gallon. Costs of liquid alum delivered at points within 100 mi of the production plant in 1962* were $36.60 per ton, based on the weight of aluminum, or a cost of $0.01839/lb of alum, which compared favorably with the delivered cost of dry alum. The liquid alum may be delivered in tank trucks holding from 2000 to 4000 gal and single-unit tank cars holding 10,000 gal. If delivered in tank trucks, there must be provisions at the delivery or usage point for storage tanks, pumps, etc., for unloading and holding the liquid. If delivered by tank car, the car may be held at the delivery point for 30 d and serves as the reservoir from which the liquid can be dispersed to the flocculation tanks in the treatment plants by means of the available metering pumps. Since liquid alum is corrosive, the storage tanks must be lined with protective coatings and the piping must be of noncorrosive material.

To illustrate the economy to large users of alum in the use of liquid alum, the case of the Richmond, Virginia, water treatment plant may be cited. In this plant definite economies were obtained, despite the cost of providing lead-lined storage tanks and special metal in pipelines and pumps. In this case the alum was delivered in tank trucks holding 3200 gal, and transferred at the treatment plant to an old elevated water storage tank with capacity for 71,000 gal of liquid alum. The cost of conversion of the facilities to the use of liquid alum was stated to be $20,000, but since the estimated saving in total operation costs of the water treatment plant was $15,000/y, the remodeling cost would be amortized in a very short period. The savings were in the reduced labor for handling the material and in the base cost of the alum. The truck load of 3200 gal would produce the equivalent of 16,448 lb of aluminum sulfate in the dry form. At a dosage rate of 2 gr/gal, or 286 lb/million gallons of water or waste, one truckload of the liquid alum would provide the treatment for 57 million gal of waste. However, 2 gr may prove to be a very low dosage in treating industrial wastes. It may be that manufacturers have smaller tank trucks, which will eliminate the need for large storage facilities and make the use of this material economical for smaller plants. One manufacturer of this material did not recommend the use of liquid alum if the point of usage is more than 100 mi from the point of manufacture of the alum.

It was obvious that more sophisticated methods were needed. Research at the Richmond plant continued for means of expediting the performance of coagulants, reducing the amounts needed, and improving their effectiveness. This resulted in the development of synthetic electrolytes, of ionic nature, which when added to the

* See Cost Index Chart in Chapter 11 for comparative current prices.

basic coagulants, increased their adsorbent action. These "aids" tended to cause the more rapid coalescence of the colloidal particulate matter in the wastes and the formation of denser masses by the impingement of small particles upon each other. Other particles sweeping through the liquid added impact and adsorption and formed quick-settling agglomerations. The formation of particulate matter into quick-settling masses was dubbed "flocculation," and the resultant dense masses, "floc."

There is a big difference in the size and density of flow which materially influences the rate of deposition of the particles. In past years, the appearance of large flaky floc particles floating around in a basin seemed to be the criterion of excellence in floc formation. But the basic idea of flocculation is to get the floc masses to the bottom of the tank, or at least out of the path of flow currents in the tank. It was realized that the best floc was not the big flaky kind with the great surface area, but the dense pelletlike kind which offered less resistance to surface tension and settled to the bottom of the tank rapidly. The rate of settlement of a particle is in relation to the surface area presented and the decrease in velocity of transporting liquid. In a straight-line tank in which the velocity is constant throughout the length, a particle of densified solid matter will settle in a straight downward path to the bottom of the tank. In a tank in which the flow is introduced at the center of a circular unit and is discharged over a peripheral weir, the velocity of the flow gradually decreases as the liquid approaches the weir, and the solids settle to the bottom in a parabolic curve.

The equipment developed to promote the formation of dense, quick-settling floc has taken various forms. These include revolving impellers, similar to mixer units; rows of paddles running parallel with the long walls of a tank; paddle units which rise and fall in the liquid; and air to agitate the mass. The idea is to develop pinpoint floc which mixes the coagulant and waste into dense quick-settling masses by the impact of particles upon each other. In very few of these units, however, is there any motion creating multiple crosscurrents needed to create the dense, pelletlike form. There is, however, one type of floc-forming unit where multiple currents operating in a sort of "barrel rolling" travel in all directions. The idea of this unit was conceived at the water treatment plant at Richmond, Virginia, in 1930.

The idea originated with Marsden Smith, engineer of the Bureau of Water and Electricity, who thought it might be possible to obtain better water clarification in the Richmond water plant if, by some means, he could improve the diffusion of the coagulant through the mass of water and lengthen filter runs. Within his tight budget the only solution was to put four rows of paddles transversely across the 75-ft tank, 25 ft apart. It seemed to be a logical idea; but it had never been done. The four transverse paddles were installed in 1932 and immediately demonstrated that the idea was sound, as evidenced by a statement made by Mr. Smith in a publicly presented paper:

It is evident that the rotation of the paddles is such that the outer edge of the last paddle is downward while the inner edge of the first paddle is upward. Assisted by this downward motion of the last (fourth) paddle, the settled floc soon forms a pile near the last paddle and the current produced by the last paddle gently sweeps some of the settled material along the bottom of the tank to the first paddle, where it is elevated by the upward motion and mixed with the incoming treated water.*

This action induced adsorption of pinpoint floc and expedited agglomeration. The return of previously settled material resulted in a saving of from 0.25 to 0.50 gr of aluminum sulfate per gallon of water treated. The total saving ranged from 700 to 1400 lb of alum per day.

A comparison of results obtained before and after the installation of this new form of flocculation equipment testified to its advantage, as shown in Table 13-1. The saving of alum alone in 1 year following the installation of the equip-

Table 13-1   Average Quality and Operating Results before and after Installation of Flocculation

Item	Before	After
Raw-water turbidity, mg/l	53.13	70.79
Raw-water color	46.85	41.12
Raw-water oxygen consumed	13.50	7.48
Alum dosage, gr/gal	4.58	2.78
Finished-water color	5.46	3.96
Finished-water oxygen consumed	4.46	3.10
Finished-water alum content	0.26	0.06
Percent wash water required	1.31	0.59

ment was estimated at $25,000 or approximately twice the cost of the installed equipment. This saving did not take into account the saving in wash water or the labor saving in controlling filter operation.

Aids

Coagulant aids are used to assist the diffusion through the wastes. Their role is crucial to the economic operation of the facility. They have come into widespread use because of their demonstrated effectiveness in speeding up treatment processes or enabling the flocculated material to be handled with dispatch in later plant phases. The selection of a particular aid from the more than eighty varieties of coagulant aids on the market is not a simple process. All too frequently the selection is due to the persuasive effort of the sales staff of the manufacturer. Sometimes the choice is based on the effectiveness of a particular compound at a given place on a certain waste. Whatever the reason for selection, the "cost per pound" should not be the guiding factor.

These compounds cover a wide range in price and amount to be used.

* Smith, Marsden C.: Improved Coagulation at the Richmond, Va., Filter Plant, *J. Am. Water Works Assoc.*, vol. 25, no. 2, Feb. 1933.

A survey made in 1964 showed that the quoted prices on coagulant aids ranged from 0.0085 cents per pound to $2.70 per pound. Too often selection of a coagulant aid has been based on a low price per pound without considering the many other factors. It may be said that the cost of investigating all the factors given in this discussion could outweigh any economy that might be indicated. The author contests this. Admittedly, such an investigation and study might cost several hundred dollars. Each of the factors listed must be investigated. But it is assumed that the chemical selected will be used for a number of years. A demonstrated saving of a few dollars a day would soon pay off. A list of factors to be considered is as follows:

1. Will the proposed aid reduce the amount of coagulant required per unit of wastes? Usually this is expressed as 1000 gal.

2. Will it produce more effective floc formation?

3. Will it produce a dense, heavy, pelletlike floc which will settle quickly?

4. Will it enable a settling tank of given size, or one existing in a treatment plant, to handle more waste per day per unit of area, or per volume?

5. Will it reduce the size of settling basins for a new plant?

6. Will it increase the ability of filters to handle a larger volume of waste in a given time period, thus enabling an existing plant to treat an increased amount of waste, or to design a new plant with smaller filters?

7. Will it reduce the cost of treatment per unit of waste?

8. Will it reduce the by-product sludge and the problem of handling and disposing of it?

9. Will it increase the handling and safety problems in a plant?

10. Will it require additional plant personnel or highly skilled operators?

11. Will it require more operating personnel or highly skilled operators?

12. Will it enable the treatment facilities to turn out a better final effluent, at less cost to the owners, or will it reduce the overall cost of operation of the works?

The foregoing items are the initial basic considerations. In practice, it is normal for an engineer to try out a number of coagulants or aids in the laboratory on samples of waste to be treated. This work results in certain conclusions, as far as the use of specific compounds in a particular case is concerned. However, the translation of laboratory experiments into full-scale plant operations is an entirely different and much more complex proposition. A pinch of coagulant or aid on the tip of a spatula is a small matter; when this small quantity is expanded into the amounts required to treat a waste which may have a volume of millions of gallons per day, many factors requiring careful consideration are introduced.

OTHER TREATMENT CHEMICALS

The coagulation of suspended solids is not the only use made of industrial reagents. Chemicals are required in many other areas of industrial waste treatment. Cyanide-, chromium-, and copper-bearing wastes all require treatment with special reagents before discharge. Ion-exchange units require reagents to regenerate their spent resins.

Ozone

Chapter 8 lists the various treatment packages available for nonbiological contaminants. Among these is the typical treatment for cyanide-bearing wastes—chlorine. One of the recent advances has been the use of ozone instead of gaseous chlorine or one of the chlorine-containing compounds such as sodium or calcium hypochlorite. The use of ozone requires a fairly expensive plant, but if the plant is properly designed, the cost of destruction of cyanide and chromium is materially less with ozone than with chlorine or hypochlorite. A study must be made to determine the best chemical for the specific case. Figure 13-1 shows the comparison in cost of gaseous chlorine, ozone, and hypochlorites. Ozone equipment is mainly supplied by the Welsbach Company of Philadelphia, the pioneers of ozone production and use in this country. The most notable use of ozone for the treatment of industrial wastes containing cyanide is at the plant of the Boeing Company, manufacturers of aircraft, at their plant in Wichita, Kansas. This treatment plant was put into operation in 1957 and has been giving very satisfactory service. Figure 13-2 is a photograph of the ozone installation at the Boeing plant. In a letter from this company it was stated that the actual cost of producing ozone by the Welsbach system is 19.8 cents per pound of cyanide, made up of the following cost items:

Power	$0.120/lb
Compressed air	0.065/lb
Maintenance	0.013/lb

Since it requires about 2 lb of ozone to oxidize 1 lb of cyanide to hydrogen and carbon dioxide, the chemical cost for this operation is about 40 cents per pound of cyanide. The above itemized costs do not include the amortization of the ozone equipment, but this is low. At the Pontiac Motors plant at Pontiac, Michigan, cyanide is destroyed by gaseous chlorine. It was reported that the actual usage of chlorine was 8 lb/lb of cyanide. Depending on the type of chlorine container used, the cost of this method of treatment would be from 60 to 80 cents/lb of cyanide.

Hypochlorite is not suitable for large concentrations of cyanide since its price does not decrease when purchased in bulk. Its use and handling require considerable additional equipment. There is a definite rate of deterioration which makes it impractical for use in large amounts.

The cost of installation of an ozone-producing unit may be materially reduced if the ozone treatment phase can be extended over 24 h. The ozone generator provided must have a 24-h rating equal to three times the amount of ozone required in the 8-h period. In other words, if 100 lb of ozone is indicated for treatment in 8 h, the ozone generator provided must have a 24-h capacity of 300 lb. Naturally, the equipment for this amount will cost considerably more than if a 100-lb installation were adequate.

Ozone appears to have many applications in the treatment of industrial wastes containing cyanides, chromium, phenols, etc. It is a simple process, and the raw materials, electricity and air, are inexpensive or free, and available everywhere. As can be seen from the figures given for the Boeing installation, maintenance cost is also low. The Welsbach Company makes the generators and auxiliary equipment to suit all needs, including package units for small volumes. Several manufacturers such as Invex, Inc., make small assemblies also.

Chlorine

If, after a thorough study of the practical and economic factors in a given problem, the final analysis indicates that the use of gaseous chlorine alkaline treatment is warranted, a decision must be made as to the form in which the chlorine will be purchased. This decision is governed mainly by the amount of chlorine needed per day. Chlorine is sold in three forms:

1. Cylinders

    a. Small cylinders, holding 150 lb of chlorine

    b. Large cylinders, holding 1 ton of chlorine

2. Cylinder carloads

    Flat freight cars with fifteen 1-ton cylinders fastened on board

3. Single-tank cars, with capacities of

    a. 16 tons of chlorine

    b. 30 tons of chlorine

    c. 55 tons of chlorine

There are definite advantages and disadvantages in each of these forms. The price per pound of chlorine scales downward as the capacity of the container increases. In all forms except the single-unit tank cars, additional equipment and attendance are necessary in handling and using the cylinders. The proper approach is to relate the type of container to the amount of chlorine that will be needed in a given period of time.

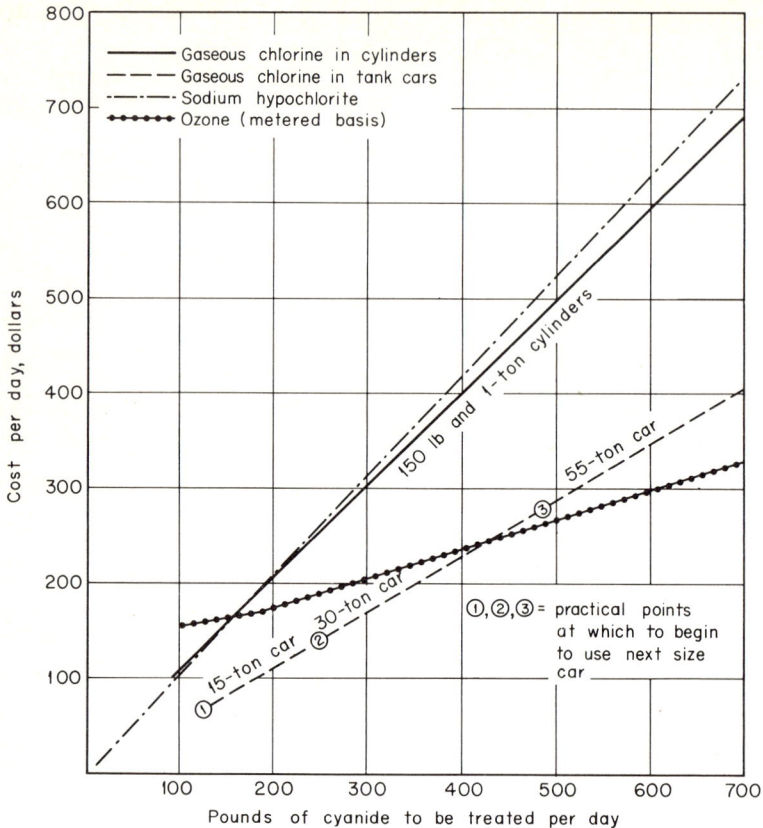

Fig. 13-1 Comparative cost of chemicals for the destruction of cyanide in wastes (1957 prices).

Chemicals for Regeneration of Ion-Exchange Resins

Chemicals are required also for the efficient operation of ion-exchange units. In the operation of ion-exchange equipment, one of the most important factors in the maintenance of exchange capacity is the quality of the regenerant chemical used in the regeneration step. Although chemically pure (CP) chemicals are not required for regeneration of ion-exchange equipment, it has been found that certain contaminants in regenerant chemicals collect on the ion-exchange resins and eventually cause difficulty in operation.

Technical grade chemicals, free of oils and other organic materials, suffice for regeneration of cation-exchange resins. For use with water softeners, evaporated salt is generally used for household units. Industrial water softeners generally use rock salt for economy. Hydrogen exchangers may be regenerated with either sulfuric or hydrochloric acid. Sulfuric acid is generally the most economical regenerant to use on large ion-exchange equipment. Technical grade 66°Bé sul-

Fig. 13-2 Ozone installation for treatment of cyanide wastes at plant of the Boeing Corpora-tion, Wichita, Kans. (*The Welsbach Company.*)

Table 13-2   Recommended Regenerants

CATION-EXCHANGE RESINS

Hydrochloric (Muriatic) Acid

Grade—muriatic acid, technical (HCl)
Color—white to light yellow
Concentration—minimum (18°Bé) 28% HCl
Sulfuric acid, as SO_3 = 0.4% maximum
Iron (Fe)   = 0.01% maximum
Freezing point   = $-40°F$
Organic contaminants = 0.01% maximum
Weight per gallon   = 9.5 lb
(NOTE—Acid should be free of turbidity.)

Sulfuric Acid

Grade—sulfuric acid, technical (H_2SO_4)
Color—water white to light brown
Concentration—minimum (66°Bé) 93.2% H_2SO_4
(NOTE—An equivalent amount of 60°Bé acid
   may be used.)
Iron (Fe)   = 50 ppm maximum
Nitrogen compounds   = 20 ppm maximum
Arsenic   = 0.2 ppm maximum
Freezing point   = $-24°F$
Organic contaminants = 0.01%
Weight per gallon   = 15.3 lb (60°F)
Acid should be free of turbidity and sediment.
   Acid containing inhibitors should not be used.

WEAK-BASE ANION-EXCHANGE RESINS

Sodium Hydroxide (Caustic Soda)

Grade—regular technical flake
Color—white

Typical analysis		Average
Na_2O		76.2%
NaOH		97.7%
Na_2CO_3		0.85%
NaCl		1.18%
$NaClO_3$	Less than	2 ppm
Fe_2O_3		0.0016%
SiO_2		0.057%
Al_2O_3		0.0015%
CaO		0.0041%
MgO		0.0035%
Cu		0.00001%
Mn		0.00011%
Pb		0.00005%
Ni		0.00011%
Na_2SO_4		0.26%
As		0.1 ppm

NOTE—A comparable amount of technical grade liquid (50%) caustic
   may be used.

Soda Ash—Sodium Carbonate

Grade—soda ash, technical, dense
   (NA_2CO_3)
Color—white

Typical analysis	Average
Na_2O	58.26%
Na_2CO_3	99.6%
$NaHCO_3$	Nil
NH_3	Nil
H_2O	0.15%
NaCl	0.21%
Na_2SO_4	0.02%
Fe_2O_3	0.0022%
Water insolubles	0.013%

Aqua Ammonia

Grade—ammonium hydroxide (NH_4OH), 26°Bé
Gravity at 60°F   = 26.0°Bé (minimum)
Total ammonia as NH_3 = 29.4%
Freezing point   = $-40°F$
Color   = water white
Pounds per gallon (at 60°F) = 7.5

Table 13-2 (continued)

STRONG-BASE ANION-EXCHANGE RESINS

Sodium Hydroxide *Liquid Caustic Soda* *Mercury Cell Grade*		*Sodium Hydroxide* *Caustic Soda, Rayon Grade*		
Typical analysis		*Average*	*Typical analysis*	*Average*
			Na_2O	76.61%
NaOH		50.6%	NaOH	98.25%
Na_2CO_3		0.02%	Na_2CO_3	0.76%
NaCl		0.002%	NaCl	0.57%
$NaClO_3$	Less than	1 ppm	$NaClO_3$	Less than 2 ppm
Na_2SO_4		10 ppm	Fe_2O_3	0.001%
SiO_2		10 ppm	SiO_2	0.0039%
Al_2O_3		3 ppm	Al_2O_3	0.001%
CaO		3 ppm	CaO	0.0027%
MgO		0.6 ppm	MgO	0.0038%
Fe		2 ppm	Cu	0.00002%
Ni		0.6 ppm	Mn	0.00005%
Cu		0.2 ppm	Pb	0.00005%
Mn	Less than	0.2 ppm	Ni	0.00009%
Hg		1.0 ppm	Na2SO4	0.20%
			As	0.01 ppm

SOURCE: Illinois Water Treatment Company, Rockford, Ill.

furic acid, free of suspended matter, and light in color, should be used. This material should mix completely with water, without formation of any precipitate.

Hydrochloric or muriatic acid, when used, should be technical grade, of a minimum of 30 percent HCl by weight (18°Bé). The hydrochloric acid should not contain excessive iron or organic materials. The use of the so-called by-product hydrochloric acid obtained from the hydrolysis of chlorinated organic chemicals is not recommended. This is especially true if the solution passing through the deionizer is to be used for human consumption. Hydrochloric acid obtained by the salt-acid process or by the hydrogen-chlorine process is satisfactory.

Hydrochloric or sulfuric acid containing inhibitors should positively not be used. When buying the chemicals for use in the regeneration of ion-exchange resins, it is a good idea to request an analysis of these materials from your chemical supplier.

Anion-exchange resins are generally regenerated with sodium hydroxide. Weak-base anion exchangers are most economically regenerated with technical grade flake sodium hydroxide, 76 percent Na_2O. Strong-base anion exchangers are best regenerated with nylon or rayon grade sodium hydroxide, 76 percent Na_2O. This grade is low in iron, silica, and chlorides. This eliminates fouling of the strong-base anion exchangers. If caustic is purchased in the liquid form as a 50 percent solution, the same grades should be purchased as mentioned above. The caustic must be low in chlorates. If you desire to regenerate a weak-base anion exchanger with soda ash or ammonium hydroxide, technical grade powdered soda ash should be purchased. Ammonium hydroxide may be purchased in 12.5-gal carboys as 28 percent NH_3 solution.

A list of the types of ion-exchange resins and an analysis of the recommended regenerant chemicals is given in Table 13-2.

SELECTING THE CORRECT REAGENTS

Experience in handling a variety of industrial wastes has convinced the author that the true economy in the use of coagulants, coagulant aids, chlorine, ozone, sodium bisulfite, hydrochloric acid, sulfuric acid, sodium hydroxide, etc., cannot be discovered without due consideration of each of the following items.

1. Amount of reagent required, based on engineer's tests.

2. Cost, either a quoted price or a price from a published list, at the manufacturer's plant or some designated supply point.

3. Whether the material is available from a local stock, and if not, the extent and cost of plant storage facilities.

4. Shipping cost, in the volume purchased at one time, from factory to point of use:

   a. Direct to user's siding.

   b. To the nearest freight station.

5. Character of the shipping containers:

   a. Bags.

   b. Barrels.

   c. Drums.

   d. Cylinders.

   e. Tank cars.

   f. Tank trucks.

   Are any of the containers returnable to the manufacturer, with some recompense to the user of the aid?

6. If bulk shipment of large quantities is indicated, what type?

   a. By railroad.

   b. By truck.

7. Usual weight of individual containers.

8. If shipped to a local freight station, cost of handling, hauling to the plant, unloading, stacking.

9. Handling of containers:

    a. Small lots, small containers, suitable for handling by one person.

    b. Large containers, requiring one or more persons to handle, mechanical lifting equipment, or the engagement of extra help in unloading and stacking.

    c. Bulk shipments, carloads, etc.

    d. Rental of lifting equipment, if needed, or chargeable to the operation of the waste treatment plant, if owned by the factory.

10. Storage facilities:

    a. Space required for the amount to be ordered or kept on hand, charges to be made against the waste treatment department for the use of this space.

    b. Handling facilities available to get the material to the plant storage area.

    c. Special precautions in the storage area to prevent damage or deterioration due to moisture, temperature, etc.; dust-prevention facilities; elevators, belt conveyors, etc.

    d. Humidity and temperature control, if the storage is in a separate building.

11. Shelf life of the raw material; the period suggested by the manufacturer as the maximum during which the material will retain its effective valency.

12. Shelf life of the solutions made from the raw material to be dispensed to the treatment, that is, the maximum length of time, as suggested by the manufacturer, that the solutions can be kept on hand, retaining their full valency.

    a. Recommended volume of the solution to be prepared at one time.

    b. Cost of preparation of the solution, labor, electric power, time, handling, etc. This is an important factor in determining the cost of use of any particular aid. If the manufacturer recommends that the solution be made in batches that will serve for a week, and it requires 2 h to prepare the solution for that period, the cost of the labor required must be added to the cost of the amount of aid used in the solution; that is, if 50 lb of aid costing $1 per pound is made into a solution to last for a week, and 2 h time of a staff member is required to prepare the batch, and his rate of pay is $4 per hour, the cost of the aid will be $1.03 per pound. Or if the solution must be made up daily and 50 lb is used at the same rate of pay, the cost of the aid will be $1.16 per pound. Preparing solutions daily may seem unlikely, but some manufacturers do recommend daily preparation of solutions.

    c. Storage of solutions; the material required to prevent attack by components of the mixture.

d. Facilities for storage of solutions in bulk, if those available are not of the right type.

e. Cost of operation of these facilities, if all waste treatment costs are charged against each operation.

13. Safety factors required by the nature of the material:

   a. Protection of handlers and adjacent workers.

   b. Protective equipment or apparatus or clothing, eye guards, etc.

   c. Dust-protection equipment for the handlers or to protect machinery in the plant.

   d. Handling of acid materials.

   e. Prevention of contact of the material with workers in the area.

   f. Provision of special wash stations or facilities, the cost to be charged against the operation.

14. Dispensing equipment:

   a. Weightometers.

   b. Metering pumps.

   c. Manual measurements.

15. Procurement time, the time for an order for material to be delivered to the plant after the order has been placed with the manufacturer or agent:

   a. Extent of storage capacity required to provide ample supply during the procurement time.

   b. Investment in the supply of material if long delivery time is required; interest and amortization charges on the storage facilities.

16. Need for corrective agents, in addition to the aid selected, to obtain the maximum efficiency of the coagulant or aid:

   a. Correction of pH of the waste treated by a certain aid to obtain the best reaction.

   b. Additional equipment for pH determination and correction.

   c. Cost of the corrective agents.

   d. Time required for correction of pH and care of the pH equipment. Electrodes require frequent inspection and cleaning, and pH meters must be calibrated periodically.

17. Comparison of recommended dosages by manufacturer; that is, 1 to 5 mg/l or more, and the cost of the aid as compared with the actual dosages found necessary by laboratory tests on the wastes to be treated.

18. Production of by-products by the use of a given aid as compared with others, which may cause an additional expense chargeable to that particular aid:

   a. Volume of sludge produced and percentage of solids in the sludge.

   b. Dewatering equipment for the sludge, type and cost, and cost of operation:

   (1) Vacuum filters; cost of cloths and replacements (see Chapter 12).

   (2) Centrifuges; maintenance and repair.

   (3) Hauling or pumping to disposal grounds; land is available for dumping or the cost of land purchased for dumping.

19. Special materials of construction needed for various elements of the handling and usage system:

   a. To prevent corrosion.

   b. Special linings for tanks.

   c. Special metals for tanks and piping.

20. Acceptance by the regulatory officials having jurisdiction over the use of the selected aid when the final effluent from the waste treatment plant is to be discharged into public waters used for potable purposes or as components of food products. Not all coagulants or coagulant aids are certified for such uses.

21. Reliability of the manufacturer or supplier for supply of the aid during the normal life of the treatment plant.

22. Comparison of cost of the selected coagulant or aid with common coagulants, lime, aluminum sulfate, ferrous sulfate, diatomaceous earth, clays, etc., which are frequently available from local sources or as by-products of other operations, such as carbide lime.

23. Study of chemical compounds used in the factory producing the wastes to be treated to determine if these will be satisfactory and effective in the proposed waste treatment; that is, if the factory uses lime, sulfuric acid, or other acids in large quantities, they will be able to purchase these at low prices, and there will also be an ample supply on hand. In such cases these chemicals can be used to economic advantage in place of the more expensive proprietary compounds.

24. Survey of local sources of supply of coagulants to reduce the cost of investing in large supplies and to reduce the size and cost of storage facilities.

25. The effect of local environmental conditions on coagulant or aid:

   a. Temperature.

   b. Humidity.

STORAGE AND HANDLING FACILITIES

If chemicals, coagulants, and/or coagulant aids are required in the treatment of the wastes, handling and storage facilities must be included in the plant design. In many cases certain chemicals are used in large quantities in the production phase of plant operations. If they can be used also in the treatment of the wastes, purchase of other chemicals or coagulants may be eliminated. This will save storage space and facilities and will ensure a constant supply for the treatment. Treatment costs may also be lower because of the lower price for large quantities.

Lime

If chemicals are to be stored in large quantities, such as carload lots, the storage facilities, space occupied, handling, environmental protection, etc., may be a considerable item of cost. Lime must be stored in bins, with agitators to prevent clogging in the discharge lines. If quicklime is used, the Bindicator type of agitator, fastened to the inside wall of the bin, must be activated by air which is freed of all moisture. If such air is not available from the factory supply, it may be necessary to install air-drying equipment at considerable cost. If conveyors are used to carry the lime from a railroad car to a bin, they must be installed in a dry pit and covered at all times. If bagged lime is used, storage space must be provided. The dimensions of this space can be determined from the dimensions of the bags to be used and the number purchased or kept on hand at one time. This, in a large plant, may require considerable space in the operating building, entailing additional expense. Elevators or lifts of some kind may be necessary to carry the bags to an upper floor, and labor will be required in the handling and stacking.

If ferrous sulfate is used in the treatment, it must be protected against wetting or moisture. Since ferric sulfate is extremely hygroscopic, a dry atmosphere must be provided for storage. Aluminum sulfate is usually purchased in bags and must be stored in a dry place. If liquid alum is used, it is usually delivered in tank trucks holding from 2000 to 4000 gal, and tank storage must be provided at the plant. Sulfuric acid in small quantities is usually delivered in glass carboys, and these must be stored in safe places to prevent accidental breakage.

Chlorine

The small 150-lb cylinders of chlorine can be handled by one man with a hand truck. The 1-ton cylinders require a hoist of some kind and a conveyor system to carry the cylinders from the truck or freight car on which they are delivered to the point of use or storage. Small numbers of 1-ton cylinders may be delivered by truck directly to the platform of the chlorine building, but flatcars containing fifteen 1-ton cylinders may not be able to approach the plant. This will require additional trucking and handling to the point of usage. The single-unit tank car may be used as the dispensing container from which the chlorine is drawn and fed to the chlorination equipment, but this requires the construction of a special railroad spur on an isolated track, with safety devices to prevent collision from other cars. This system also requires special attachments at the parked car to convey the chlorine to the point of use. However, this type of delivery, if the chlorine need justifies it, saves a lot of other expense and handling; storage space and handling of cylinders are eliminated. Where single cylinders are used, a certain number must be kept on hand for ready replacement. Those stored must be protected from the weather and the hot sun. Time is involved in disconnecting an empty cylinder, carting it away, carting in a new full cylinder, and connecting it to the dispensing system. A third handling is required when the delivery truck picks up the empties and delivers new ones. The cost of this labor must be added to the base cost of the chlorine to arrive at a true cost of chlorine treatment. In the case of the single-tank car, the railroad brings the car to the special siding and takes away the empty car. A small charge is made for this service, but in large plants having their own shunting engines, this may be reduced. Attaching the new car to the dispensing system may take an hour, but this may happen only once a month.

In using single-tank-car units—the cheapest in the delivered price per pound of chlorine—arrangements may be made with the manufacturer of the chlorine, who has his own tank-car units, to park the car at the user's plant for a period of 30 d at a low demurrage rate. Therefore if this type of service is used, the demand of chlorine should be such that the car may be retained for the full 30-d period, to reduce costs of car changing. Here again, the cost of moving the car and attaching withdrawal lines must be added to the actual cost of the chlorine delivered. However, this cost will be spread over the time the car is retained and will add a very small unit cost per pound of chlorine.

To overcome the lag in time required in changing cars, which may be an hour or two, there should be one or two 1-ton cylinders mounted on platform scales in the chlorination room to be put in use during the changeover. These cylinders are not used at any other time and after use are refilled from the tank car.

This system, as actually used in a project for a large manufacturer of automobiles, is described in Chapter 11. Figure 13-3 shows the installation of the tank car at the plant.

Without evaporators, no more than 40 lb of chlorine may be withdrawn from 100- to 150-lb cylinders during a 24-h period. When 1-ton cylinders are used, not more than 450 lb may be withdrawn. If evaporators are used, these rates can be increased considerably. When using 1-ton cylinders, the distance from the supplier's plant will govern the number of spare cylinders to keep on hand.

In very small plants using only a few pounds of chlorine per day, the demand may be satisfied with one of the small hypochlorinators which dispense HTH tablets.

Important economies result from using the proper container. To assist in this, the price of chlorine in the different types of containers, delivered at the point of use, should be obtained by the engineer from a manufacturer or supplier before completing the design of this phase.

In the chlorinator room or cubicle certain precautions must be taken to meet the requirements of the regulatory authorities. Certain safety appliances

Fig. 13-3   Installation of single-unit chlorine tank car used as supply reservoir for treatment of cyanide wastes, waste treatment plant, Pontiac Motor Division, General Motors Corporation, Pontiac, Mich. (*K & H Engineering Company.*)

must be kept on hand in usable condition at all times. Chlorine cylinders and the dispensing equipment must be installed in a room or cubicle with access only from the outside, with doors opening outward and with proper ventilation.

FEEDER SYSTEMS

Feeder units for systems requiring large amounts of lime are generally of the weightometer type. These units can be adjusted to deliver any desired quantity of lime per unit of time. They keep a record of the usage for any time period. The same is true for the feeders of aluminum sulfate, ferrous sulfate, or other dry chemicals, coagulants, or aids. Figure 13-4 illustrates an installation of such feeders for the various chemicals used in a large plant.

Chlorine is usually dispensed through standard chlorinator units which have built-in dials and recorders for regulation of the chlorine flow and the usage. Sulfur dioxide is fed through similar units. If chlorine cylinders are used, one or more are usually permanently laid on platform scales having a dial which shows the weight of chlorine remaining in the cylinder. The same is true of sulfur dioxide from cylinders. If chlorine is fed from a single-unit tank car parked at the plant, there should be a meter and recorder showing the amount drawn

Fig. 13-4 Installation of dry-feed chemical feeders, waste treatment plant, Pontiac Motor Division, General Motors Corporation, Pontiac, Mich. (*K & H Engineering Company.*)

from the car each day or period of operation. This meter should be inspected daily since it is important to know when the supply is getting low, so that a new supply may be ordered.

In specifying a feeder unit, the size of the unit selected should be one whose middle range of capacity corresponds with the average hourly anticipated demand for the chemical. The storage hoppers connected to each feeder should be sized so as to hold enough chemicals for 1 d or the better part of an operating period. In some cases the upper portion of the hopper of the feeder extends through the floor above the feeder about 18 in, so that the operators can easily dump a bag of the chemical into the hopper. The extended hopper should be provided with a hinged metal cover as a protection against dust and moisture. If the storage is on an upper floor and an elevator is not provided, pallets loaded with bags may be raised by a hoist on the outside of the building or by a chain hoist on the inside.

Many systems use liquid reagents. Primary among these are ion-exchange units (during regeneration), neutralization units, and chromium destruction units. Chemicals are frequently added by means of metering pumps. The pump may be activated manually or by some automatic mechanism. The pump is adjusted to deliver a particular volume of liquid per unit time. The length of time is regulated by an automatic or manual start-stop switch. Metering pumps have the advantage of becoming completely automated. Not only can the pump be activated when a treatment reaction is not complete, but a float mechanism can refill the holding tank when its contents run low.

In the use of all chemicals, coagulants, and aids, certain precautions must be observed by the plant operators. These are generally obtained from manufacturers. They should be posted in a prominent place. If the chemicals, coagulants, or aids are purchased dry and are to be used in solution form, the manufacturer should furnish the engineer with information on the preparation of the solutions and the time period in which they are satisfactory for use.

In discussing the use of their products, manufacturers commonly express concentrations in "parts per million," "milligrams per liter (mg/1)," or "grains per gallon." To convert these terms, use the following formulas:

Parts per million to cost per 1000 gal of waste:

$$\frac{\text{ppm} \times 8.345 \text{ lb}}{1000} = \text{pounds per 1000 gal} \times \text{cost per pound of reagent}$$

Milligrams per liter to cost per 1000 gal of waste:

$$\frac{\text{mg/1} \times 8.345 \times \text{cost per pound of reagent}}{1000} = \text{cost per 1000 gal}$$

Grains per gallon to cost per 1000 gal of waste:

$$\frac{\text{Grains per gallon} \times \text{gallons of waste} \times \text{cost per pound of reagent}}{7000}$$
$$= \text{cost per 1000 gal waste to be treated}$$

The true cost factor is very important in industrial waste treatment.

In the treatment of water and sewage, if the cost of an item of treatment, or the entire cost of treatment, increases, the municipality has the power to increase the charge for water to the user, to increase the sewer service charge, or to raise the taxes sufficiently to recoup the extra cost. The industrial plant management does not have that advantage or privilege. If by insufficient care in selection, a coagulant or aid or chemical additive is used that is found to be costly, the industry must add that extra cost to the daily production costs of the plant product. Of course, an industry may try to recover excess costs by raising the price of its product to the public, but this will only result in a loss of sales.

In summary, where the addition of chemicals, coagulants, or coagulant aids is needed, the selection of the best kind for a given case is very important. The handling and use of these products require considerable study and judgment. Omission of this study and the selection of a reagent or additive simply on the basis of advertised benefits or sales talk may result in excessive costs to the client. There are many—perhaps too many—factors to be considered and evaluated in order to reach the most practical and economical selection. In the treatment of industrial wastes, chemicals, coagulants, and/or aids will probably be the greatest single item of expense in running the plant, and therefore deserve

Table 13-3   Cost of Coagulants or Coagulant Aids Based on Minimum Dosage Recommended by Manufacturers and Effect on Ultimate Cost of Use

Dosage, min. recommended by maker, ppm	Quoted cost of material at maker's stated point of shipment, $/lb	Cost per 1000 gal waste treated, $	Cost per million gal waste treated, $
0.1	0.58	0.000484	0.484
0.1	1.75	0.00146	1.46
0.06	1.26	0.000063	0.63
0.25	0.75	0.00156	1.56
0.25	1.15	0.0024	2.40
0.5	0.30	0.00125	1.25
0.5	0.85	0.00358	3.58
0.1	1.25	0.00105	1.05
0.05	0.28	0.00012	0.12
0.2	0.45	0.00075	0.75
0.2	0.34	0.00057	0.57
0.5	0.21	0.00088	0.88
0.5	0.26	0.00109	1.09
1.0	0.12	0.00101	1.01
1.0	0.75	0.00628	6.28
1.0	2.00	0.01668	16.68
1.0	0.012	0.00010	0.10
1.5	0.16	0.00201	2.01
2.0	0.08	0.00134	1.34
2.00	0.10	0.00167	1.67
2.00	0.12	0.00201	2.01
2.00	0.26	0.00434	4.34
2.00	0.35	0.00584	5.84
3.0	0.065	0.00163	1.63
5.00	0.02	0.00084	0.84
5.00	0.13	0.00542	5.42
5.00	1.15	0.04795	47.95
5.00	1.25	0.05213	52.13
7.00	0.06	0.00351	3.51
8.00	0.08	0.00534	5.34
10.00	0.02	0.000167	1.67
10.00	0.09	0.0075	7.50
10.00	0.18	0.01501	15.01
10.00	0.50	0.0417	41.70
10.00	1.85	0.154	154.29
15.00	0.01	0.00125	1.26
20.00	0.29	0.0485	48.37
50.00	0.01	0.00417	4.17
500.00*	0.08	0.36	335.60

* This excessive dosage was applied to a certain very intricate waste only.

The minimums suggested by the manufacturers are actually seldom encountered in actual practice.

mature thought and consideration. Base cost per pound of the additive is not the sole criterion. Many other considerations must be taken into account.

In the selection of lime alone, there is considerable latitude. One grade or type of lime may be obtainable locally and may sound cheap, but if that type produces more sludge than another, it may cause a definite heavy increase in the cost of sludge handling and dewatering equipment, such as vacuum filters, thermal destructors, etc. Unslaked lime may be used instead of slaked or hydrated lime. But if the storage elements for the unslaked lime (quicklime) are not moisture-proof or are open to the atmosphere, the result may be slaking in the storage unit, causing clogging of discharge valves and pipes.

Other chemicals, such as ferric sulfate, are hygroscopic; that is, they absorb moisture from the atmosphere. In the use of such materials, a very dry environment must be provided, entailing additional cost for air drying, which may well outweigh any price advantage.

It is also wise to remember that tests made in the laboratory are made under ideal conditions, in quiescent, captive quantities of wastes. The coagulant or chemicals are of the best quality, usually highly purified, and stirred under closely controlled conditions. They will of course, produce the best results, at the lowest cost. To translate the test results into predictable full-scale plant practice requires considerable judgment and the understanding of actual plant operations. In full-scale practice the chemicals will be bought in bulk lots and be of commercial grades. These are slightly less effective than the laboratory grade. In addition there will be losses in handling, etc. In the mixing tanks in a full-scale plant, the wastes will be flowing through the unit, which will introduce a lag in the effectiveness of application. A slight miscalculation of the weightometer or feeder can mount up to a considerable amount of extra poundage of material per day in a large treatment plant. It can also cause the discharge of insufficiently treated wastes.

In the author's opinion it is advisable, after obtaining a result in a laboratory test, to add at least 25 percent to the provisions for the full-scale plant. The extra amount available will be insurance to overcome slight errors in figuring, setting dials, losses in transit, etc. Illustrating this point is the case of a large plant using gaseous chlorine for the destruction of cyanide. The stoichiometric amount of chlorine required to break down cyanide to nitrogen and carbon dioxide is 6.83 lb per pound of cyanide. The design of the plant called for a supply of 8 lb/lb of cyanide. Reports show that the actual use of chlorine is 7.99 lb/lb of cyanide. In selecting dispensing equipment for a reagent or aid, it is always safe to provide a larger capacity than the tests actually indicate.

Many of the necessary data discussed in this chapter may be obtained from the literature of the manufacturers of the coagulants, aids, or chemicals. If they are not thus available, they should be obtainable on request. If they are not, then the claims made for them are suspect. Any reputable manufacturer will provide pertinent information relative to the use of this product, even though it might influence a potential customer against the use of that particular material in a specific project.

Chapter Fourteen

Incentives to industry

Industry spends millions of dollars yearly for chemicals required in their processes. They are now required to spend many times that amount destroying any of those chemicals that enter the plant wastewater. It seems that the ideal answer to this type of water pollution problem is the reclamation of the process chemicals. If these chemicals could be reclaimed from the wastewater, both they and the cleaned water could be reused.

The problem of reclaiming spent chemicals has its origins in thermodynamics. Any system tends to go to maximum randomness. Work must be expended to re-order the system. Put in simpler terms, reclaiming chemicals from their diluted state is very costly.

But as the cost of destroying spent chemicals increases, reclamation and recycling are becoming more attractive. Expensive recovery and recycling may prove to be a cheap method of waste treatment since any money saved offsets the cost of the treatment facility.

There are other incentives for pollution control. The federal government for years has provided grants and loans to municipalities to assist them in building sewerage systems and sewage treatment plants. This aid enabled the works to proceed as a moderate cost to the citizens. Until very recently, industry was denied such aid. A treatment plant generally is an added operating expense which produces no revenue, requires expensive equipment and personnel, and is a steady drain on the resources of the industry. Industry needs financial assistance to ease the drain.

Industry tried to expedite the help. The Chamber of Commerce of the United States, the organization of American business, proposed that tax credits and accelerated amortization would be needed for industry to continue making large capital investments in pollution control. They also proposed that a study of tax incentives for industrial waste treatment be related to performance of facilities, rather than to the installation of particular methods of treatment.

In the interim between 1947 and 1974 many states have put relief systems of one sort or another into effect, and industry has taken advantage of them. These measures are various:

1. Relief from state property taxes

2. Local property tax relief or remission

3. Relief from local use or production taxes

4. Expedited write-off of the investment in waste treatment facilities

5. Establishment of the right of eminent domain—condemnation of land for waste treatment purposes

6. Loans by state agencies for structures for waste treatment

7. Purchase of local bonds of municipalities by industry to cover the cost of the facilities required for treatment of industrial wastes discharged into the common system

8. Assumption by industries of the interest charges on municipal bond issues to assist the cities to provide facilities for treating industrial wastes in their sewerage works

9. Grants by state governments to assist industry in developing new processes and methods of treatment or destruction of their wastes

In general, this chapter will help acquaint industry with the various governmental programs designed to ease the burden of pollution control. Some terms used in discussing these programs need further definition. They are as follows:

Amortize—to make special periodic payments which pay off a debt.

Ad valorem tax—a tax based on the value (or assessed value) of property.

Bond—an interest-bearing promise to pay with a specific maturity.

Revenue bond—a bond payable from revenues secured from a project which pays its way by charging rentals to users, such as toll bridges or toll highways, or from revenues from another source which are used for a public purpose.

Coupon—that part of a bond which evidences interest due. Coupons are detached from bonds by the holder, usually semiannually, and presented for payment to issuer's designated paying agent, or deposited in his own bank for collection.

Par value—the face amount of a bond.

Tax-exempt bond—a municipal bond. The interest on a municipal bond is exempt from federal income tax.

In December 1970 the United States government, displaying an increase in environmental concern, established the Environmental Protection Agency (EPA). Into this new agency were merged various administrations from the Departments of the Interior and Health, Education and Welfare, and from the Atomic Energy Commission. Under the leadership of William Ruckelshaus, the EPA sought to prevent and abate degradation of the environment. The EPA not only has set up standards for wastewater control but also has provided for grant programs which cover a broad field in research, demonstration, and training.

Information concerning the EPA grant programs can be found along with interim regulations in volume 36, number 29, of the "Federal Register," dated February 11, 1974. These grants are intended to encourage accomplishment of statutory and EPA goals with a minimum of administrative requirements necessary to protect the public interest. In the specific area of wastewater, the EPA is currently concentrating its grants programs on new and developing technology.

Individuals and industries interested in the EPA grant programs must request application forms from the U.S. Environmental Protection Agency, Grants Administrative Division, Washington, D.C. 20468. For grants involved with research and demonstration, the completed application forms along with supplementary documents must be submitted with fourteen copies to the above address. The application forms request detailed information concerning the proposed project and its related personnel, costs, goals, and schedules. One limitation which can be found in the cost-sharing section (§30.207) is that the grantee must contribute no less than 5 percent of the allowable actual project costs except as otherwise required.

Although the EPA grant programs are fundamentally involved in developing the ways and means of safeguarding our environment through federal assistance, it should be noted that they, like so many federal and state agencies, are deeply tangled in a labyrinth of paper work and red tape. On the encouraging side it can be added that congressional subcommittee hearings were held with the intention of reducing some of the incongruities which exist within the Agency, Congress, and the executive branch. The latter is responsible for having limited congressional allocation for state waste treatment facilities.

The following list describes plans available in various states to aid industry with the construction of industrial waste treatment facilities.

1. Alabama   The Alabama Water Improvement Commission governs the issuance of tax-exempt financing of pollution control facilities with Environmental Protection (or improvement) Bonds. These bonds are the most economical of the financing alternatives available to credit-worthy corporations. Industrial revenue bonds are marketable at lower rates of interest than comparable corporate bonds because interest on them is exempt from federal and some state income taxes. This results in significant savings to the using corporation.

For a pollution control facility to qualify for tax exemption:

1. The facility must be designed to meet or exceed applicable federal, state, and local requirements for the control of water pollution (in effect at the time of the obligations).

2. The proceeds are to be used to provide such facilities.

3. The expenditures for property would not have been made except for the purpose of controlling pollution.

4. The expenditure should have no significant purpose other than for the purpose of pollution control.

5. The total expenditure for such purposes satisfies the test above even though such property serves one or more functions in addition to its function as a pollution control facility.

Thus, the new production equipment which is installed simply to eliminate harmful discharges would qualify as pollution control equipment, if it did not increase the capacity or the useful life of the plant. Other parts of the regulations provide that if the capacity or useful life of the plant is increased, that part of the expenditure which could be attributed to pollution control could be financed through industrial development bonds.

For latest regulations contact the Water Improvement Commission, Montgomery, Alabama.

2. California

California's present law (1974) provides two forms of tax relief for pollution control devices. One is the fast write-off for purposes of California income tax. The other is a provision for financing new facilities with tax-exempt bonds.

Fast write-off allows 60-month amortization for tax purposes for a certified pollution control facility. An important qualification is that if the facility would normally have amortization in excess of 15 years, the portion of the cost which can be subject to a first write-off is reduced by the fraction 15/normal amortization period. It is noted that the federal law also provides for a 60-month write-off on pollution control facilities. California law does not require that the facility be located in California.

To qualify for fast write-off, the facility was required to be used for water pollution control, must have been completed or acquired after December 31, 1970, and must have been placed in service before January 1, 1975. The use of the law can begin at any time, on the election of the taxpayer. It may be stopped at any time for the current month.

The interest on bonds issued for construction of pollution control facilities is exempt from income taxes. To qualify for any of the aid plans, the industry must submit an application form. The forms can be secured from the state agency. For up-to-date information contact the California Pollution Control Financing Authority, Sacramento, California.

3. Florida

The state of Florida does not have' a formalized incentive program. For the last several years the state Legislature has been considering sales tax exemptions on pollution control devices.

The latest state of legislation can be ascertained by contacting the Department of Administration, Division of State Planning, Tallahassee, Florida.

4. Georgia

There is an extensive water pollution control program in the state of Georgia which includes industrial incentives for water pollution control facilities.

The primary aid for industries faced with multimillion-dollar expenditures is financing by industrial revenue bonds issued by a county Industrial Development Authority. The facility is built by the Industrial Development Authority. The industry leases the facility, paying a monthly charge to cover the bond's retirement. Operating expenses are paid by the industry. When the bonded indebtedness is retired, the facility is generally bought by the industry. The bonds are tax-free and result in lower interest rates. This is more advantageous than utilizing company funds for the capital expenditure. This approach also provides certain tax benefits to the company.

There are other incentives to industry:

1. Exemption from ad valorem property tax of all property used in, or part of, any facility which is installed or constructed for the purpose of eliminating or reducing water pollution

2. Exemption from sales and use tax of all machinery and equipment which is incorporated into any facility used for the primary purpose of reducing or eliminating water pollution

3. Rapid amortization of water pollution control facilities meeting certain criteria for construction and operation under section 169 of the Federal Internal Revenue Code.

All of the above incentives may be applied to privately owned industrial waste treatment facilities and are contingent upon the applicant obtaining certification from the Water Quality Control Section that the equipment and facility involved are necessary and adequate for the purposes intended.

For further information, contact the Department of Natural Resources, Environmental Protection Division, Atlanta, Georgia.

5. Illinois   Under existing Illinois state statutes, relief from sales and property taxes (both real and personal) is provided for construction of eligible pollution control facilities. The Illinois Revenue Department administers the sales-type tax relief while the Illinois Environmental Protection Agency is responsible for certifying projects for property tax relief.

In addition to the above-mentioned relief, a private industry may also procure federal income-tax-free bonding for eligible pollution control facilities through the Illinois Industrial Pollution Control Financing Act. Private industry may also rapidly amortize (60-month write-off) any eligible pollution control facility, provided the facility is certified by the state agency.

Detailed information on these laws can be obtained from the Illinois Environmental Protection Agency, Springfield, Illinois.

6. Maine   The state of Maine grants a sales and use tax exemption for qualified pollution control facilities under its regulations, Title 36, Section 1760. It also provides a property tax exemption under Title 36, Section 656. In addition, the state provides for a credit against Maine income tax for the investment in pollution control facilities equal to 7 percent of the qualified investment under Title 36, new Section 5202.

Additional information can be obtained from the Division of Industrial Services, Bureau of Water Quality Control, Augusta, Maine.

7. Nevada   The Nevada Legislature in 1973 passed legislation allowing exemption for property used for the control of water pollution. All real and personal property is exempt from property taxation to the extent that the primary use is as a facility, device, or method for the control of water pollution. The exemption applies to any land, structure, building, installation, excavation, machinery, equipment, or device or any addition to, reconstruction, replacement, or improvement of such property. The facility must have been constructed, acquired, or installed after January 1, 1965. Its primary purpose must be compliance with laws for the prevention, control, and removal of water pollution.

Tax exemptions do not apply to septic tanks, sewage facilities used to move sewage, collection facilities, facilities having value or cost less than $1000 when constructed, or any facility producing a net profit to its owner or operator. Facilities that allow for the recovery and sale or reuse of tangible products or by-products, or which result in a net reduction of operating costs, do not qualify for exemptions.

A mining operation or an electric generating facility may treat its discharge water for reentry into the public streams. If it further distills the water at a minor additional cost, and then markets the distilled water so as to offset the total cost of the treatment and distilling process, it will not qualify for exemptions.

For additional information contact the Nevada Tax Commission, Carson City, Nevada.

8. Oregon

The state of Oregon, through legislation adopted in 1967 and amended in 1969, seeks to encourage the construction and use of facilities to prevent, control, or reduce water pollution. It provides tax relief for persons who do so. In order to actually obtain the allowed tax relief, the following steps must be taken:

1. A Pollution Control Facility Certificate must be obtained from the Department of Environmental Quality.

2. An irrevocable election must be made to take the allowed credit either as a credit against income or excise taxes or as an exemption from ad valorem taxes on the certified facility.

3. The Pollution Control Facility Certificate must be filed with the appropriate taxing agency in accordance with their requirements.

If the firm chooses credit against income or excise taxes, the maximum allowable in any one tax year is the smaller of the tax liabilities. The firm may choose to be credited with the following portion of the cost of the facility: 5 percent of the cost of the facility if the portion of the cost allocated to pollution control by the certificate is 80 percent or more. The percentage allowed decreases as the portion allocated to pollution control decreases. The minimum amount is 1 percent when the portion is less than 20 percent.

Tax credit is in lieu of any depreciation or amortization deduction for the facility to which the firm would otherwise be entitled for the year which the credit is claimed.

For the latest regulations, contact the Department of Environmental Quality, Portland, Oregon.

9. Pennsylvania

At the present time (1974), Pennsylvania has not issued information regarding financial aid or other incentives for the installation of waste control facilities by private industry. Grants are awarded for the installation of sewage treatment plants by municipalities.

It is suggested that any company seeking state tax benefits in Pennsylvania contact the regional office to which they have applied for their industrial waste discharge permit.

10. Texas

In Texas, an economical alternative to financing water pollution control facilities may be available through river authorities, water districts, or other public entities. Revenue bonds are used for the capital expenditures for waste treatment facilities. There are substantial advantages in using revenue bonds. Revenue bonds are provided at lower interest rates owing to the tax-free status of municipal-type bonds. These bonds may have to be fully guaranteed by the company. Certification by the Texas Water Quality Board is required to state the purpose of the pollution control.

More information can be obtained from the Texas Water Quality Board, Austin, Texas.

11. Washington   Pollution control facilities approved by the appropriate control agency may qualify for tax credits and exemptions in the state of Washington. If the owner secures a certificate, the acquisition of the pollution control facility is exempt from sales tax and use tax. This is in effect as long as the due date of payment of such taxes is subsequent to the effective date of the certificate. The tax exemption does not apply to servicing, maintenance, repairs, and replacements of parts after a facility is complete and placed in operation.

More information is available from the Department of Ecology, Olympia, Washington.

LEASING

An important method of financing industrial projects of all kinds is leasing. While most strictly classified as an incentive, it could well be considered a very desirable means of assisting manufacturing concerns in proceeding with the construction of industrial waste treatment facilities without any outlay of major amounts of funds, or borrowing for that purpose. Leasing is a simple and well-established process in which a financing organization purchases equipment, apparatus, etc., for an industrial concern and charges the industry a certain sum per month for a predetermined term of years. This monthly sum eventually amortizes the total cost of the items provided. The monthly charge, of course, also includes the interest on the money spent by the leasing organization, plus their profit. By this method the industry using it makes only a token initial payment, and the monthly payments become an item in the deductible operating costs of the plant.

Naturally, the industry pays a total amount for whatever it buys under this plan, but if, on the other hand, the management was obliged to borrow the funds to pay for the equipment, or whatever, the final difference would not be an important item. Leasing allows an industry to install immediately the needed treatment facility without a large initial outlay of funds.

A number of companies offer this service. One of them, the Gulfstan Corporation of Miami, Florida, will not only provide the equipment and apparatus for an industrial waste operation, but also design, provide engineering services, construct, equip, staff the plant, and provide maintenance and operating services under this plan. This organization is owned by Gulf Utilities, Inc., and the American Radiator and Standard Sanitary Corporation, and thus has ample resources to call upon.

The following table illustrates the operation of one leasing plan.

Leasing Plan for Furnishing of Equipment,
Based on $50,000 and 6-Year Term as of 1969

Year	Monthly payment	Annual total
First	$1400	$16,800
Second	1215	14,580
Third	950	11,400
Fourth	720	8,640
Fifth	525	6,300
Sixth	425	5,100
Total paid for 6-yr term		$62,820
Seventh	115	1,380
Eight	115	1,380
Total paid for 8-yr		$65,580

There is no capital investment on the part of the industry. For larger or smaller amounts the monthly payments would be in proportion. Although the table shows that on the 8-year basis the industry pays 32.5 percent more for the equipment, the average annual cost has been only $8197. On the other hand, had they not used the leasing plan, but borrowed the funds for the whole amount at 6 percent, the total amount paid would have been $74,000 if equal monthly payments had been arranged.

In the case of another leasing organization, the monthly charge was the same each month, and for an equipment bill of $100,000 and a 3-year term, the monthly payment would be $3200 or a total amount of $115,200; a 4-year term, $2500 per month and a total of $120,000; and a 5-year term, $2100 per month and a total of $126,000.

Of course, before entering into any leasing agreement, the contract should be studied and thoroughly understood.

It becomes apparent that there are many alternatives open to industry. They simply require investigation. Industry must remember that treatment facilities are no longer optional, but their method of financing is.

Chapter Fifteen

The renovation of industrial waste effluents

Perhaps all this discussion of water pollution has given the reader the impression that industry destroys water. And that new laws requiring treatment are simply designed to provide a decent burial. Nothing could be further from the truth. Water cannot be destroyed, only fouled. Its usefulness has been temporarily impaired. Repair the damage and useful water is the result. New laws simply ask industry to rectify the damage done to its water. The water is then returned to the lake or stream to be reused by someone else. This might seem somewhat foolish. Why should industry renovate or repair the water so that someone else may use it? This question is now being raised more and more often. It is not common business practice to spend thousands of dollars on someone else. Why shouldn't the industry itself get the benefit of the renovated, revitalized water? The answer is obvious, it should. With this realization, industries have begun recycling their plant water. If the water is to be reused, it might as well be reused by the plant footing the bill. Besides, if the water is used in-plant, it needs only to meet plant requirements as to purity. The governing agencies no longer dictate the extent to which water must be treated; plant processes do.

As of this date, the following types of industrial wastes have been successfully renovated and reused by various industries:

1. Tailing pond waters
2. Oil refinery
3. Soap factory
4. Steel mill
5. Fiberboard mill

6. Beet-sugar plant

7. Coal mining and washery

8. Brass metal industry

9. Phosphate mining

10. Potato starch

11. Food processing

12. Metal processing (plating, etc.)

13. Cane-sugar processing

14. Wallboard manufacture

15. Chemical plant

16. Fiber glass

17. Meat processing

MUNICIPAL EFFLUENTS

Until recently, treatment of wastewater for plant purposes has been confined mainly to the renovation, or cleaning, of the effluents from municipal sewage treatment plants. There are numerous outstanding examples of this. The most discussed is that of Bethlehem Steel Company, outside of Baltimore, Maryland. For a number of years, the plant has been taking over 150 million gal/d of the treated effluent from the large Back River sewage treatment works of the city. The steel company provided the extra treatment for the sewage plant effluent to polish it to the degree needed for its use in the milling as cooling water. However, even with the extra cost for this additional treatment, the total cost of the water to the steel company, including fixed charges, is only about $3\frac{1}{2}$ cents per 1000 gal. In no other way could the mill have obtained so much usable water at such a low price. In addition to the saving for the steel mill, the city of Baltimore received revenue from the sale of the effluent. This additional revenue from an otherwise waste product ranged from $11,000 in 1942 to $40,589 in 1954.

Another outstanding example of the benefits to be derived from reuse by both parties is found in Amarillo, Texas. The Texaco, Inc., oil refinery uses over 1.5 million gal/d of municipal sewage treatment plant effluent for cooling and boiler makeup. The company pays the city about 11 cents per 1000 gal. If the same volume were to be supplied from the city water system, it would cost 18 cents per 1000 gal. Texaco saves $105 per day. On the other side of the coin, the income for the city from the sale of an otherwise-wasted by-product of a necessary service for the people currently averages $65,000 per year. It was reported that a local electric utility plans to purchase about the same amount of effluent from the same source, which would then increase the income of the city to over $100,000 per year.

One point that is not usually discussed is the extent of treatment for which an industry must pay when it purchases water from a municipal supply. A municipality usually treats all water alike, to a drinking-water standard set by the health authorities. The city makes no distinction among uses of the water, whether it is in the homes for potable purposes, by industry to cool its machinery, for washing the streets, or putting out fires. In a great many cases industries do not need water of such high quality, but they have to pay the price, which averages between 15 and 18 cents per 1000 gal. An industry purchasing sewage treatment plant effluent pays a small fraction of the normal municipal price. In numerous other cases it has been shown that industrial waste effluents can be renovated at a cost of about one-third the usual price of municipal water. Therefore, economy, as well as conservation of water and elimination of stream pollution, is a strong motivating force. Other more tangible incentives have been suggested. It has been proposed that industries could be offered a remission from taxes when water is purchased from a municipal supply. This would be in the form of a rebate of 5 to 6 cents per 1000 gal of water reused by an industry which reduced its demand from the municipal supply in the same amount. This would not be a philanthropy on the part of the city, since it would then be able to sell the released water at the going rate of 30 to 40 cents per 1000 gal. This plan would also benefit the citizens in the reduction of the amount of water which required treatment to a potable degree. This would reduce the operating costs of the treatment plant.

If an industry could reduce its demand for municipal water, the remission in its taxes in the course of a year would be an important item in its operating costs. Assuming that an industry could arrange to reuse one-half of a present demand for city water of 1 million gal, the saving, based on a normal industrial water rate of 18 cents per 1000 gal, would be $32,850 per year. Then the tax saving for the 5 cents per 1000 gal would amount to a reduction of $9125, or a total annual economy of $41,975 per year. This amount is the interest at 5 percent on $839,500, which would justify the expenditure of a considerable amount by the industry for the necessary facilities to render the water suitable for reuse. In other words, it will pay industry to investigate the possibility of reuse of its water as an economy.

REUSE OF PLANT EFFLUENT

Another point in the call for reuse of plant water is that by renovating to the extent actually needed by an industrial plant, the operation may eliminate the much greater cost of treating the waste for discharge into a stream. Many of the polluting elements in an industrial waste not permitted in a discharge to a stream, such as BOD, pH, settleable solids, toxic elements, etc., may not be of any particular importance in the return of a water for reuse in cooling and other nonproduct purposes. Cyanide, chromium, and phenol are costly to remove from a waste, but if they do not affect the reuse of the water, the cost of renovation to that degree will be much less than for total removal in an effluent for discharge to a stream. The cost of destroying 1 lb of cyanide in a waste flow may run from 37 to 76 cents, depending upon the method of destruction. If there is only 1 lb of cyanide in 1000 gal, the cost of removing that will be 37 to 76 cents for the chemicals alone. Industrial wastes may usually be renovated satisfactorily for reuse for much less.

Some estimated costs for renovating waste effluents for reuse, according to Janecek, are shown in Table 15-1.

There are many examples of successful renovation and reuse of industrial effluent and municipal effluent. The Celanese Corporation reduced the intake of freshwater from 230 to 4 million gal/d by the recirculation and reuse of cooling waters.

Many oil refineries have achieved the equivalent of a sevenfold reuse of their freshwater intake. The average steel mill reuse is now 62 percent. In 1962, the nine steel mills on the Ohio River system kept 426,000 tons of mill scale out of the river by treatment and reuse of waste effluents.

The pulp and paper industry has made noteworthy progress in its reuse of water, as shown in the figures in Table 15-2.

Water use and *reuse* by other industries are shown in Table 15-3.

Table 15-1 Estimated Costs* of Renovation of Industrial Wastes

Method of renovation	Cost per 1000 gal, cents
Coagulation	4–10
Aeration	2–4
Activated-carbon treatment of method 2	7
Breakpoint chlorination of effluent from method 2 or 3 to destroy ammonia	6–10
Weak-base anion exchange treatment of effluent from method 3	10
Distillation	75
Reverse osmosis	50–100
Electrodialysis	25

* As of 1969.

Table 15-2 Reductions in Water Usage by the Pulp and Paper Industry

	Water usage per ton of pulp, gal	
	1934	1964
Sulfite pulp	89,000	49,000
Kraft	77,350	34,000
Book paper	31,000	13,800

Table 15-3   **Water Reuse by Industry Ratios in Various Industries and Usage of Water**

Industry	Total amount freshwater mg/year	Ratio of reuse to amount of total	Usage for cooling, %	Process and other uses, %
Automobile	248.7	1.62	21	79
Beet sugar	336.0	0.48	95	5
Chemical	9,857.0*	.....	63	35
Distilling	71.0	0.51	84	16
Meat-packing	144.0	3.03	25	75
Petroleum	1,761.00	6.62	82	18
Soap and detergent	57.0	2.08	66	34
Steel	10,186.0	0.60	66	34
Sugar (cane)	180.0	0.26	69	31
Textiles	128.0	0.30	10	90
Coal preparation	....	13.91		
Food processing	....	0.19		
Machinery manufacture	....	4.50		
Natural-gas transmission	....	1.32		
Photographic	....	3.20		
Poultry processing	....	6.56		
Pulp and paper	....	2.02		
Salt processing	....	0.12		
Tannery and leather	....	0.04		

* For 1962.

SOURCE: Compiled by E. B. Besselievre from "Water in Industry," National Association of Manufacturers and Chamber of Commerce of the United States, Washington, January 1965.

In areas with a short supply of water reuse is essential. In Taylorville, Illinois, the Allied Mills Paper Company and the Hopper Paper Company once competed for water from the same source. This was resolved by Allied Mills' using the source for the water it needs for cooling and then passing along the effluent to the Hopper Company for their reuse.

In Monterrey, Mexico, the nearest freshwater source was several hundred miles away. This practically dictated reuse as an economic factor. It resulted in the treatment of both municipal sewage and industrial wastes in one large plant. The local industries such as Cellulosa and Derivados paid only 10 cents per 1000 gal for effluent water. The treatment plant served local industries, including chemical, paper, and glass, with a total waste discharge of 6.7 million gal/d. The renovated water has a BOD of 10 to 20 mg/l, which is suitable for cooling. It can also be used in many industrial processes without further treatment.

In some instances, water is obtained by an industry at a low rate from the municipal supply or from its own wells. The suggestion that the industry should construct treatment facilities may not initially sound interesting or economically practical. But the reuse should be considered for the benefit of public health rather than for the sole benefit of the industry. As water needs increase, the demands will proportionally reduce the available supply of this finite resource. The municipalities will be hard put to supply these demands, going farther and farther afield for sources of water. The increasing population will also have a primary demand on the water supply, and since the responsibility of a municipality is to its citizens, there will be a reluctance to supply industries the water they need at reasonable prices. If industries will give a long and hard look at the costs of purchasing water from a public or private water utility as compared with treating their own wastes for reuse in their own plants, they will, in most cases, find that the latter course is a saving. See Figure 15-1.

Method of Renovation

Records show that many industrial waste effluents can be renovated to a point satisfactory for many purposes in a majority of industrial plants at a cost of from 7 to 10 cents per 1000 gal. Records also show that the average cost of water treated by a municipal or private water plant is 17 to 18 cents per 1000 gal to industries. Sooner or later, reuse will be most economical, especially in light of the more severe discharge regulations.

Many methods are available for renovating plant effluent. The selection depends upon the purity of water desired and the degree of contamination in the waste. Some methods include:

1. Adsorption in sludge blanket units, carbon filters, flocculation
2. Electrodialysis
3. Evaporation
4. Solvent extraction
5. Emulsion breaking
6. Foam separation
7. Freezing
8. Hydration (spray drying)
9. Oxidation
10. Ion exchange
11. Electrochemical degradation
12. Reverse osmosis

Solar Evaporation

Examples of these methods and many others are numerous. The Celanese Corporation at Bishop, Texas, has developed a unique and reportedly very efficient method for the treatment of wastes from the production of industrial organic chemicals. The method employs solar evaporation. Records with wastes containing a BOD concentration of 10,000 mg/l indicate the suitability of the treatment in areas where the rainfall is low. The area used is normally 262 acres, with an added 75 acres used during the summer months. The waste flow is 500 gal/min. The waste contains formaldehyde, aldehyde, and ketone polymers.

Desalination

A possible new process which may have use in the renovation of effluents is one proposed for the desalination of seawater. The process was recently developed at the University of California. It is called the MEF evaporator. The 90-ft-tall prototype unit at the Scripps campus of the university in La Jolla has a rated capacity of 3000 gal/d. It is felt that commercial units of this new multiple-effect evaporation system would be able to treat 50 million gal/d of seawater at a cost of 20 cents per 1000 gal. The water would be of potable quality. The MEF system is a conical, vertical flash evaporator in which treated seawater is pumped to the top through internal heating tubes. The charge is then further heated by steam from an external boiler to which the steam condensate returns.

The heating and evaporation cycles are separate. The brine then descends via a series of flash tubes that open into eight separate flash chambers, not unlike a bubble-cap column in reverse. Brine in each stage collects on a tray beneath the flash tubes, and then goes through the flash tubes in that tray to the next tray. Product water flashes at the tube exits to the next stage. Brine and product water both exit from the last, or bottom stage. The heat of the unit is the 5-in-long plastic, vortex-level-control and flow distributor inserted in each flash tube. The tube eliminates all internal pumps and controls, making the apparatus inherently self-regulating.

Pasterilization

Another method of reclaiming waste effluents for reuse is called pasterilization. This system is akin to pasteurization, because the waste effluent is sterilized and aerated, producing high-quality water, claimed to be of potable quality, for as little as 25 to 30 cents per 1000 gal. Whether or not potable quality is needed for industrial purposes, this process may offer possibilities.

"Industrial Kidney"

An interesting development, jocularly designated as an "industrial kidney," is said to be under study by the Sulphite Pulp Manufacturers Research League at Appleton, Wisconsin. The process involves the use of a synthetic membrane that removes wastes from water in much the same way as living kidney membranes do. The League was to have been developing membrane techniques to concentrate mill wastes and eliminate stream pollution. It would also provide water for reuse in the mills. This process may have application to other types of wastes.

Experiments at the Robert A. Taft Sanitary Engineering Center at Cincinnati, Ohio, are said to have shown that the fermentation of sugar in sulfite liquor will give a 40 percent by weight yield of sugar. The process would greatly reduce the BOD content of the resultant effluent.

OIL FIELD WASTEWATER

A process for the direct generation of steam from oil field wastewater without softening has been developed by Gas Processors, Inc., of Houston, Texas. Called Thermosludge, the process is claimed to make thermal recovery possible in many oil-producing areas heretofore considered impractical. The process claims to eliminate the need for freshwater pipelines, water softening, oil separation, or other pretreatment systems. The steam generator converts dirty, hard, salty, and oily low-gravity wastewater from oil-producing formations directly into 100 percent quality steam. The capacity of the unit is said to be 20 million BTU/h at 1500 lb/in² maximum working pressure. If the claims of the developers are borne out in full-scale working installations of the system, it may prove to be very useful in the elimination and disposal of oil field wastes. It could be used by oil refineries or other plants to upgrade present effluents at moderate cost, and also provide a useful by-product. It has been demonstrated in actual practice at a petroleum refinery. Although developed originally for the oil industry, it has possibilities for use in other industries which have wastes with the following characteristics:

Materials	*Milligrams per liter*
Oil	0–1000
Total dissolved solids	1000–50,000
Hardness	100–2000 as CaCOs
Silica	0–150
Sulfate	0–1000

The generator used may be fired by crude oil, diesel fuel, fuel oil, LPG, or natural gas.

Beet-Sugar Wastes

One of the largest beet-sugar waste treatment plants in this country has been put into operation by the Holly Sugar Corporation at Hereford, Texas. This plant, known as the Merrill E. Shoup plant, is designed to handle the waste from the processing of an ultimate crop of 6,300 tons of sugar beets provided by local growers. The waste treatment plant is designed to handle an average volume of about 1000 gal/min, with a maximum of 1500 gal/min. Flume and wash water are expected to be about 725 gal/min; water from beet mud, 120 gal/min; water from lime mud, 145 gal/min; boiler blowdown, 5 gal/min; and sanitary wastes, about 15 gal/min.

As shown in the flow diagram of Figure 15-1, fluming and washing of the beets are based on a recycling system. Blowdown is provided by excess seal tank water, well water, or return water from the stabilization ponds. All water in the system is screened and settled each time it is recycled. The screening station is equipped with four vibrating 20-mesh deck screens, and the flow is manually divided between them, the maximum flow through the screens being about 10,000 gal/min. Following the screens, settling occurs in a 100-ft-diameter by 8-ft water-depth clarifier, equipped with a thickener mechanism to handle heavy mud loads. At the nominal loading of 5000 gal/min, the detention period is 1.5 h, and the overflow rate is about 1000 gal/ft^2(d). Sludge withdrawal is controlled by a density meter.

The lime pond occupies about 8 acres and has a capacity of about 70 acre-ft. It receives the reslurried lime cake from the vacuum filters at the raw-juice carbonation station. At an expected inflow of 145 gal/min, the lime pond is able to store lime mud for an entire campaign, of about 130 d. Effluent withdrawal at different depths is provided for.

A variety of process acids, caustics, and brines is routed to the chemical catch basin, then pumped to the lime pond, and after mixing, it is pumped to the stabilization ponds. The mud pond covers 5 acres and has a maximum volume of 35 acre-ft. Decanting facilities for clear water are provided in the pond. The overflow from the ponds mixed with other wastes goes to the stabilization ponds.

Cattle feedlot lagoons are provided to keep rainfall runoff from draining into the local creek. The north lagoon services 96 acres of modern cattle feedlots; the south lagoon takes the drainage from 105 acres of forage area destined to handle high cattle density. The north lagoon is designed to retain 24 acre-ft, and the south lagoon, 26.5 acre-ft. Maximum depths are about 7 ft.

Other pilot-plant studies include the air oxidation of waste sulfite liquor with cobalt chloride catalysts. This process is said to have produced a final effluent satisfactory for discharge into local streams. The effluent is also suitable for reuse.

Oftentimes, changes in plant processes themselves are needed to aid in reusing the plant water. At Hoquiam, Washington, a plant designed to use a sodium-based pulping process is reported. The process would recover over 85 percent of the pulping chemicals and leave only 15 percent sulfite liquor residue. This method, if successful, could be an important step in the reduction of stream pollution.

The Brown Paper Company, in New Hampshire, has been reported to have developed a new "hydrotropic" wood-pulping process which shows promise of important economies in production and eliminates stream pollution. This process uses sodium xylone sulfate as the primary reagent. Important reductions

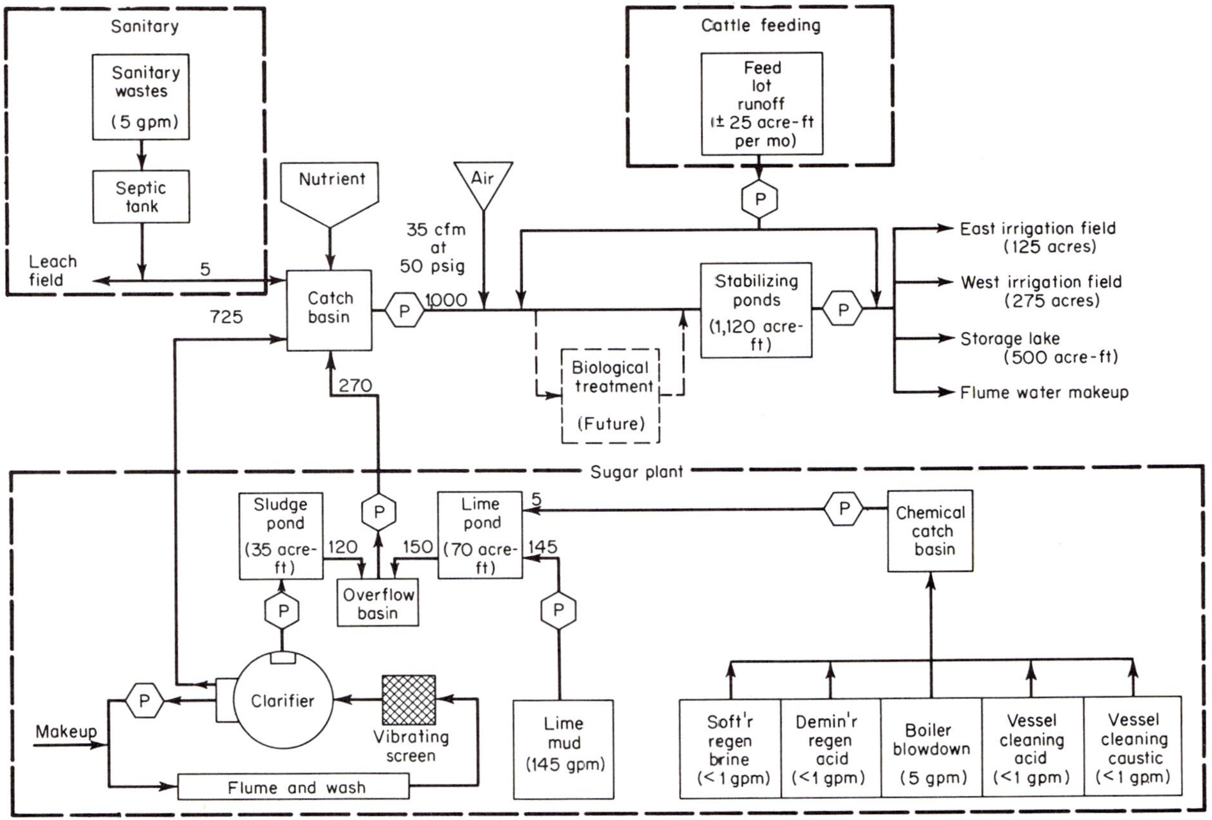

Fig. 15-1 Beet-sugar waste treatment plant, flow sheet of operations at Merrill E. Shoup plant, Hereford, Tex., of Holly Sugar Corporation. (*Forrest & Cotton, Incorporated.*)

in cost of production are reported, the chemicals used costing about 50 cents per ton compared with $5 or more for earlier methods. The process is also said to reduce water usage to about 500 gal/ton of pulp instead of the 55,000 gal reported to have been used in the older methods.

RECOVERY OF USEFUL PRODUCTS
FROM INDUSTRIAL WASTES

Potentially, there are elements of value in every industrial waste discharge. But the economy of recovery is doubtful. The values can, of course, be recovered, but if the cost of doing so makes the recovered material more expensive than new, unused material, there is no economy. Usually, materials recovered from waste flows are not in a commercially pure state, and to be effectively reused in a plant, they may require extensive and expensive purifying operations.

This factor should be very carefully evaluated before any expenditure is recommended for the reclamation of values. However, at present and in the future, reclamation may have another purpose, with a value independent of the possible reuse of the material. By removal of an element from the waste flow, for recovery purposes, a pollution problem may be solved at the same time. If such is the case, the cost of reclamation must be weighed against the cost of treatment to prevent or eliminate the pollution. In such a case, the normal price of the reclaimed material should be deducted from the cost of reclamation to obtain the true value of reclamation as a pollution prevention measure.

Thus in some cases, renovation of water may make it possible to recover elements of potential value from industrial wastes. If a waste contains chromium and phenol, these may be reclaimed at a reasonable cost by ion exchange, or dialysis, and may show a slight profit on the operation. At the same time, these chemicals have been removed from the waste stream, thus reducing the final cost of treatment. If the water resulting from the reclamation of the chemicals is satisfactory for reuse, waste treatment may be entirely eliminated.

It is reported that experiments have been carried out on the ion flotation of dichromate from aqueous solution with a cationic surfactant, using a nonionic polymer as a flocculant aid. This might find application in the treatment of plating wastes for chromium removal and recovery. In work done by Robert B. Grieves and Gregory A. Ettelt at the Illinois Institute of Technology, a dissolved-air, continuous-flow unit was used (with a surfactant-dichromate premix period of 1 hr), with a feed rate of 120 1/h of a stream containing from 25 to 100 mg/l of dichromate. Within the range of the variables studied, optimum results were obtained with a molar ratio of surfactant to dichromate feed of about 2. A feed polymer dosage of about 3 percent of the dichromate feed concentration and a recycle rate of about 200 percent of the feed rate with a column-detention time of 35 min were also used. With an air requirement (at STP) of 0.043 1 of air per liter of feed delivered at 40 lb/in^2 gauge, feed streams containing from 25 to 100 mg/l dichromate were readily reduced to

10 mg/l dichromate and 30 mg/l surfactant. The primary advantage of this new process is said to lie in the concentration of dichromate, surfactant, and polymer in a small liquid volume of collapsed foam which is less than 1 percent of the total feed throughput. Foam concentrations are of the order of 10,000 to 20,000 mg/l of dichromate.

One of the most recent processes developed by Eco Tec, Ltd., Cardiff, California, is a technique to recycle chromic acid, recovered from chromium plating rinse baths. Water being recycled is suitable for reuse in the rinsing operations. This system has a significant advantage to the plating industry since it eliminates the problem of disposal of chromic acid rinse waste while saving water.

In May 1972 a model was installed in a plating company for experimental purposes. The unit was initially operated in a variety of configurations designed to test its capabilities and limitations. The studies were completed in January 1973, and since that time the unit has been in normal plant operation.

This unit has been recovering an average of 20.9 lb of chromic acid per day. Assuming that the workload and recovery rate remain the same, a full-year projection of the cost of recovery with the cost of destruction is as follows. Recovery would result in a $9,710.22 saving.

1. 20.9 lb/d − 5434 lb/year
2. Chemical cost for destroying 1 lb chromic acid = $0.39
3. Chemical cost for recovering 1 lb chromic acid = $0.19
4. Reagent cost = $0.6325 per pound
5. Labor computed at $5 per hour

Destruction	Versus	Recovery
5434 lb @ $0.39 = $2,119.26		5434 lb @ $0.19 = $1,032.46
5434 lb @ $0.63 = $3,423.42		Labor 2 h/d = $2,600.00
Labor 6 h/d = $7,800.00		$3,632.46
$13,342.68		

Columbia-Southern Chemical Company at Barberton, Ohio, uses evaporators for the extraction of calcium chloride from its wastes. Approximately 300 tons/d is recovered and marketed. The calcium chloride is used in ice making and dust control on highways. The final wastes are later settled in large lagoons.

A new method of waste reclamation has been accomplished by the dehydration of waste materials. A process known as Carver-Greenfield method is said to achieve essentially complete water removal by adding an oil to the waste materials to render the oil fluid mixture amenable to processing through several stages of evaporation.

Pulp and paper mills using the sulfite pulping process have produced a very troublesome waste for many years. The sulfite liquor has many potentially useful ingredients, but research leading to profitable methods of recovery of some of them has not been productive of tangible results. In recent years, however, the stream pollution problem has assumed such critical aspects that more severe restrictions on the discharge of these wastes into the streams of the country have resulted. Recent laboratory and pilot-plant tests and demonstrations have developed a new and lengthy list of possible uses for the sulfite liquor itself or for the extraction of values from it. Either of these operations will reduce or eliminate pollutional constituents and possibly render the water recyclable. The uses that have been given to this waste are as follows:

1. A binder for surfacing roads

2. Stabilization of soils

3. Recovery of ammonia

4. Laminating agent in the manufacture of plywood

5. Production of cellular material to be used as insulation

6. Recovery of vanillin

7. Manufacture of detergents

8. Dispersing agents

9. A wetting agent

10. An ingredient in the manufacture of cement

11. Recovery of sulfur dioxide

12. Manufacture of industrial alcohol

13. A fodder yeast

14. A fertilizer ingredient

15. A tanning agent

Reports of recovery of vanillin from sulfite liquor may be traced back to 1904, but little or no commercial production followed until recently.

The Salvo Chemical Company started commercial production in 1936, by alkaline hydrolysis of calcium lignosulfonate. The end products of this process were vanillin and a lignin resin made into plastic sheets. This process had been patented by Howard of the Marathon Corporation in 1928. Such a process had been predicted by a German scientist in 1875. Presently, a number of plants use this process.

Another process for the extraction of vanillin from sulfite pulp liquor, developed in Germany, is a neutralization of alkaline sulfite liquor with waste CO_2 and mixed sodium, potassium, and magnesium chlorides. The solution is then oxidized with air under pressure. Vanillin is extracted from the neutral solution by ion exchange and purified with steam distillation. The process is said to yield 18.2 percent vanillin.

Container Corporation of America was developing a process to reduce black liquor from pulping mills. The liquor is evaporated to 35 percent solids in forced-circulation evaporators and burned in a fluidized-bed reactor at 1300 to 1330°F. This process produced dry pellets consisting of residual inorganic salts. The organics are completely converted to carbon dioxide and water. There is said to be no air-polluting gas formed during normal operation. Operating cost is reported as about $2.37* per ton of pellets produced.

In another process for treatment of spent sulfite liquor, the 50 percent solids left after evaporation is chemically treated and spray-dried to a powder. It is bagged and then sold for use in feed pelletizing, adhesives, gypsum board, concrete mixes, oil well drilling muds, insecticides, etc.

Another method of treating spent sulfite liquor has been developed by Dorr-Oliver, Inc., for the Green Bay Packaging Corporation. Known as the "Fluo-Solids liquor combustion system," it converts spent liquor into a dry product. The liquor is burned and converted to a dry pelletized form that is suitable for use as a chemical raw material for the kraft pulping industry.

Container-Cleveland Corporation has developed a process, which provides for burning organic material in waste liquid. This method will assist in eliminating stream pollution and will produce inorganic salts. The system is being licensed for use by others. The key unit is an especially designed fluid-bed reactor in which the organic content of concentrated black liquor is burned. The process yields the salts in pellet form ready for sale. The carbon-hydrogen content of the water is converted to CO_2 and water, and it is vented to the atmosphere. All BOD and color are destroyed.

The Copeland Process Corporation of Chicago obtained a U.S. patent on their design to eliminate their pulp mill waste liquor. In this fluid-bed process, the concentrated liquor is sprayed onto a fluidized bed of salt cake (sodium sulfate). Autogenous combustion of the noxious liquor destroys all organic matter, leaving only useful chemicals, mainly salt cake, which is available for sale or reuse. It is claimed the process will eliminate a serious pollution problem at low cost.

Centrifuging of water-softening sludge from kraft mills, followed by calcination with lime mud from the mill causticizing current is one method of lime recovery. The lime recovered is good enough for reuse in the water softening and causticizing phases.

The Mead Corporation plant at Chillicothe, Ohio, installed a process in 1960 which reclaims lime for reuse. The sludge from the soda process of manufacture of high-grade book paper; bond, coated and printing; and of magazine stock was producing a major pollution problem. The recovery of the calcium carbonate reduces purchases of lime and eliminates the waste problem.

The American Cyanamid Corporation put on stream at its Forth Worth, Texas, plant a process to recover sodium sulfate from the production catalysts. It is a unique thermocombustion and flash-drying system.

A plant on the east coast of the United States which manufactures acetone rayon studied the question of recovery of acetic acid versus activated sludge as a treatment of its wastes. The BOD of the raw wastes was 2,200 lb/d, and the local limit for discharge into a stream was 200 lb/d. The study showed that the cost of a recovery plant would be about the same as the treatment

* See Cost Index Chart in Chapter 11.

plant, estimated at $300,000. The operating cost of the recovery system was much lower. In addition the capital investment for the recovery plant could be charged off under conditions existing at the time, but the construction of a waste treatment plant could not be charged off. The operation cost of the recovery plant as a part of the manufacturing processes could also be charged off. By enlargement of two extraction columns, the BOD was reduced to less than the maximum permitted by the requirements.

In the woolen industry it has been found that grease and dirt can be recovered from wool fiber with an organic solvent instead of a detergent solution. The grease can be separated from the solvent and refined and marketed as lanolin. A profitable recovery operation has eliminated a serious pollution problem.

RECOVERY FOR FOOD PROCESSES

The Cunningham-Limp Company has developed a process for turning whiskey distillery wastes into poultry and livestock feed. The system employs normal methods to remove solids in the form of a sludge, followed by evaporation of the sludge with further drying and storage. The overflow from primary sedimentation is further diluted with water and discharged to the stream. Reuse of the effluent water as well as the solids would in the future decrease the demand for freshwater, and introduce a greater economy into the plant processes.

From Australia came a process developed by Dr. F. J. Moss, of the University of New South Wales, which converts some types of industrial wastes into food products. It is a process of continuous fermentation that is useful in converting the by-products of paper mills, sugar processing, flour milling, and milk separation into proteins, which emerge in the form of a powdery yeast. This substance, which has little taste, can be added to ordinary food to boost its protein value.

A project of Ionics, Inc., sought to convert their waste materials to food. Under a $30,000-contract awarded to the firm by the U.S. Public Health Service, they will investigate the economics of producing high-protein sources from solid waste materials. Initial work will use newspaper and agricultural waste as a raw-material source for growing protein matter, such as yeast. The process involved two stages: first cellulosic matter will be hydrolyzed and converted to sugars; second, high-protein microorganisms will ferment the sugars, producing yeast or single-cell protein.

MISCELLANEOUS RECOVERY

There are other miscellaneous materials that can be recovered from waste streams. A process of double-effect evaporation designed for recovering chemicals from plating wastes was developed. The system concentrates dilute cyanide plating solution by boiling it under vacuum and condensing the water vapor. The concentrated cyanide solution is then returned to the plating bath and the distilled water to the rinse tanks. The Pfaudler Company, which developed the system, claims that a plant could be amortized in about 2 years by the savings in chemicals. At the same time the elimination of cyanide from the wastes reduces pollution and saves on fresh water.

A process developed in 1959 by a professor at Columbia University proposes the reclamation of uranium in commercial quantities from wastes. The process uses specially treated potato starch, lime water, and other simple basic chemicals to curdle and settle the uranium-bearing solid materials out of the wastes from the manufacture of fertilizer from phosphates. The process has been offered free of charge to phosphate manufacturers.

This chapter has attempted to show that renovation, reuse, and reclamation are possible, and are becoming more and more economically feasible. One must remember that Spaceship Earth is not infinite. She possesses limited resources, which must now be safeguarded. If pollution control is still a scarce word, then thrift and frugality must become the watchwords of the future. When the resources are exhausted, the entire planet will suffer, and industry is a part of the planet. With thought, ingenuity, resourcefulness, and planning, industry will be ready for the challenges of Spaceship Earth in the twenty-first century.

Glossary

Ad valorem tax   The tax levied for the benefit of an individual district on the assessed value of property within its boundaries.

Advanced waste treatment   A process intended to remove the final 10 percent of the BOD and suspended solids as well as high levels of nitrogen or phosphorus that have not been eliminated at the primary stage. Advanced wastewater treatment techniques include filtration, carbon adsorption, ion exchange, distillation, reverse osmosis, electrodialysis, and microstraining. Full-scale physical/chemical treatment plants use such techniques as filtration and carbon adsorption in lieu of secondary treatment.

Air break   A physical separation which may be a low inlet into the indirect waste receptor from the fixture, appliance, or device indirectly connected.

Air gap   The unobstructed vertical distance through the free atmosphere between the lowest opening from any pipe or faucet conveying water or waste to a tank, plumbing fixture receptor, or other device at the flood level rim of the receptacle.

Anion   A negatively charged ion.

Assessed value   That portion of the total assessed value of the property upon which individual district taxes are levied.

Average daily flow   The number of gallons of sewage discharged into the public sewers during a 24-h period.

Backflow   The flow of water or other liquids, mixtures, or substances into the distributing pipes of a potable supply of water from any source or sources other than its intended source.

Backflow connection   Any arrangement whereby backflow may occur. Also known as a backflow condition.

362

Backflow preventer   A device or means to prevent backflow into the potable water system.

Back-siphonage   The flowing back of used, contaminated, or polluted water from a plumbing fixture or vessel into a water supply pipe owing to a negative pressure in that pipe.

Beneficial uses (of water)   The waters of the state that may be protected against quality degradation include, but are not necessarily limited to, domestic, municipal, agricultural, and industrial supply; power generation; recreation, aesthetic enjoyment; navigation; and preservation and enhancement of fish, wildlife, and other aquatic resources or preserves.

Biochemical oxygen demand (BOD)   The measure of decomposable organic material in domestic or industrial wastewaters as represented by the oxygen utilized over a period of 5 d at 20°C and as determined by the appropriate procedure in "Standard Methods."

Boiler blowoff   An outlet on a boiler to permit emptying or discharging of sediment.

Bonded sewer house connection sewer   Any house connection sewer from a lot, or part of a lot, which does not have a public sewer directly in front, at the rear, or at the side of such lot, and which has not been directly assessed for a public sewer.

Branch   Any part of the piping system other than a main, riser, or stack.

Building drain   That part of the lowest piping of a drainage system which receives the discharge from soil, waste, and other drainage pipes inside the walls of the building and conveys it to the building sewer beginning 2 ft outside the building wall.

Building sewer   That part of the horizontal piping of a drainage system which extends from the end of the building drain and receives the discharge of the building drain and conveys it to a public sewer, private sewage disposal system, or other point of disposal.

Building supply   (1) The pipe carrying potable water from the water meter or other source of water supply to a building or other point of use or distribution on a lot; (2) water service.

Cation   A positively charged ion.

Cesspool   A lined excavation in the ground which receives the discharge of a drainage system or part thereof so designed as to retain the organic matter and solids discharging therein, but permitting the liquids to seep through the bottom and sides.

Chemical oxygen demand (COD)   The measure of chemically decomposable material in domestic or industrial wastewater as represented by the oxygen utilized as determined by the appropriate procedure described in "Standard Methods."

Chlorinated polyether pipe   Pipe material with good chemical resistance and useful up to 250°F. Its price is two to three times that of PVC piping. It is recommended where chemical resistance at elevated temperatures is required and at normal temperatures where PVC and other materials lack satisfactory resistance to specific chemicals. It may be joined by threading or fusion welding.

Chlorine demand   The difference between the amount of chlorine added to a wastewater sample and the amount remaining at the end of a 30-min period as determined by the procedures given in "Standard Methods."

Coagulation   The reduction of surface charges and the instantaneous formation of hydrous oxides.

Colloids   The finely divided solids which will not settle. These are between the true solution and the true suspension.

Combination waste and vent system   A specially designed system of waste piping embodying the horizontal wet venting of one or more sinks or floor drains by means of a common waste and vent piping adequately sized to provide free movement of air above the flow line of the drain.

Contamination   An impairment of the quality of the waters by waste to a degree which creates a hazard to the public health through the spread of disease.

Continuous vent   A vertical vent that is a continuation of the drain to which it connects.

Continuous waste   A drain connecting the compartments of a set of fixtures to a trap or connecting other permitted fixtures to a common trap.

Cross-connection   Any connection or arrangement, physical or otherwise, between a potable water supply system and any plumbing fixture or any tank, receptacle, equipment, or device, through which it may be possible for nonpotable, used, unclean, polluted, or contaminated water or other substances to enter any part of such potable water system under any condition.

Developed length   The length along the centerline of the pipe and fittings.

Diameter   Unless specifically stated, the nominal diameter as designated commercially.

Discharge   Any spilling, leaking, pumping, pouring, emitting, emptying, or dumping.

Discharger   Any person that discharges or causes a discharge into a public sewer.

Dissolved solids   The solid matter in solution in the wastewater which can be obtained by evaporation of a sample from which all suspended matter has been removed by filtration as determined by the procedures in "Standard Methods."

Domestic sewage   The water-borne wastes derived from the ordinary living processes and of such character as to permit satisfactory disposal, without special treatment, into the sanitary sewer system.

Domestic wastewater   The water-carried wastes produced from noncommercial or nonindustrial activities, which result from normal human living processes.

Drain   Any pipe which carries waste or water-borne wastes in a building drainage system.

Drainage system   All the piping within public or private premises, which conveys sewage or other liquid wastes to a legal point of disposal, but not including the mains of a public sewer system or a public sewage treatment or disposal plant.

Effluent   The liquid outflow of any facility designed to treat, convey, or retain wastewater.

Estuaries   Waters at the mouths of streams which serve as mixing zones for fresh and ocean waters.

Fixture branch   A water supply pipe between the fixture supply pipe and the water distributing pipe.

Fixture drain   The drain from the trap of a fixture to the junction of that drain with any other drain pipe.

Fixture supply   A water supply pipe connecting the fixture with the fixture branch.

Fixture unit   A quantity in terms of which the load-producing effects of the plumbing system of different kinds of plumbing fixtures are expressed on some arbitrary chosen scale.

Fixture unit flow rate   The total discharge flow in gallons per minute of a single fixture divided by 7.5 which provides the flow of that particular plumbing fixture as a unit of flow. Fixtures are rated as multiples of this flow unit.

Floc   Small gelatinous masses formed in a liquid by aggregation of a number of fine suspended particles.

Flocculation   The bonding together of coagulated particles to form settleable or filterable solids.

Freshwaters   All freshwater lakes and streams downstream to the limit of tidal action.

Grade   The slope or fall of a line of pipe in reference to a horizontal plane. In drainage it is usually expressed as the fall in a fraction of an inch or percentage slope per foot length of pipe.

Groundwater   All potentially usable subsurface waters that occur in and below the saturated zone.

Horizontal branch   A drainpipe extending laterally from a soil or waste stack or building drain with or without vertical sections or branches, which receives the discharge from one or more fixture drains and conducts it to the soil, waste stack, or the building drain.

Horizontal pipe   Any pipe or fitting which is installed in a horizontal position or which makes an angle of not more than 45° with the horizontal.

House connection   The sewer connecting the building sewer or building waste drainage system to the public sewer for the purpose of conveying domestic wastewater.

Indirect waste pipe   A pipe that does not connect directly with the drainage system but conveys liquid wastes by discharging into a plumbing fixture, interceptor, or receptacle which is directly connected to the drainage system.

Industrial connection sewer   The sewer connecting the building sewer or building waste drainage system to the public sewer for the purpose of conveying industrial wastewater.

Industrial waste storm drain connection   Any storm drain connection carrying or intended to carry industrial waste from any industrial, manufacturing, processing, or servicing establishment.

Industrial wastewater   All water-carried wastes and wastewater of the community excluding domestic wastewater and uncontaminated water; and all wastewater from any producing, manufacturing, processing, institutional, commercial, agricultural, or other operation where the wastewater discharged includes significant quantities of waste of nonhuman origin.

Insanitary   Contrary to sanitary principles or injurious to health.

Interceptor (gravity separation)   Any facility designed, constructed, and operated for the purpose of removing and retaining dangerous, deleterious, or prohibited constituents from wastewater by differential gravity separation before discharge to the public sewer.

Interceptor sewer   A collecting sewer that intercepts and collects the sewage from a number of lateral or local public sewers.

Lateral sewer, collecting sewer, or main line sewer   A public sewer usually 8 in or larger in diameter and used to collect wastewater from house connection and industrial connection sewers and transport it to trunk sewers. It is normally built and maintained by the local sewerage agency.

Limitations   The restrictions on the temperature, location, or volume of a discharge; the restrictions on the concentrations of substances allowed for discharge to the appropriate discharge site.

Main   The principal artery of any system of continuous piping to which branches may be connected.

Nuisance   Anything which (1) is injurious to health, indecent or offensive to the senses, or an obstruction to the free use of the property so as to interfere with the comfortable enjoyment of life or property; (2) affects at the same time an entire community or neighborhood or any considerable number of persons, although the extent of the annoyance or damage inflicted upon individuals may be unequal; and (3) annoyance or damage that occurs during or as a result of the treatment or disposal of wastes.

Offset   A line of piping with a combination of elbows or bends which brings one section of the pipe out of line but into a parallel line with the other section.

Packaged equipment   Equipment fully assembled at the factory, checked, and shipped without disassembly (except for breakable items). It arrives as a "ready to connect" piece of equipment.

Peak flow   The instantaneous maximum of flow of sewage being discharged to the sewer.

Peak flow rate   The average rate at which wastewater is discharged to a public sewer during the highest 30-min flow period in the preceding 12 months.

Plumber (journeyman)   A person who installs, alters, or repairs plumbing as an employee, who does not furnish any materials or supplies, and who is the legal possessor of a journeyman plumber's certificate of registration.

Plumbing   The business, trade, or work having to do with the installation, removal, alteration, or repair of plumbing and drainage systems or parts thereof.

Plumbing contractor   A person who is engaged in the business of plumbing, or an individual who is in responsible charge of the installation and maintenance of plumbing for a specific employer and (1) who is not otherwise regulated by the business and professional code of the state and (2) who holds himself forth as willing to furnish materials and to do personally, or through employees or subordinates qualified and registered, any work or services in connection with the installation, alteration, or repair of plumbing or any part thereof.

Plumbing fixtures Approved-type installed receptacles, devices, and appliances which are supplied with water or which receive liquid or liquid-borne wastes and discharge such wastes into the drainage system to which they may be directly or indirectly connected. Industrial or commercial tanks, vats, and similar processing equipment are not plumbing fixtures, but may be connected to or discharged into approved traps or plumbing fixtures.

Plumbing system All potable supply and distribution pipes, all plumbing fixtures and traps, all drainage and vent pipes, and all building drains, including their individual joints and connections, devices, receptacles, and appurtenances within the property lines of the premises and potable water piping, potable water treating or using equipment, fuel gas piping, water heaters, and vents.

Pollution An alteration of the quality of the waters by waste to a degree which unreasonably affects (1) such waters for beneficial uses or (2) facilities which serve such beneficial uses.

Polypropylene (PPL) A lightweight thermoplastic piping material. It has a higher strength and better general chemical resistance than polyethylene and may be used at temperature 30 to 50°F above the recommended limits of polyethylene. Recommended uses are for laboratory and industrial drainage piping where mixtures of acids, bases, and solvents are involved; for resistance to sulfur-bearing compounds, as found in petroleum process; for use in saltwater disposal lines; for low-pressure gas gathering systems; and for crude oil flow piping. It is best joined by fusion welding.

Polyvinyl chloride pipe (PVC) A pipe with a fairly high tensile strength and modulus of elasticity. It is stronger than most other thermoplastic piping materials. The maximum service temperature is 150°F for type I (normal impact) and 130°F for type II (high impact). PVC is chemically resistant to a wide range of corrosive fluids, but may be damaged by ketones, aromatics, and some chlorinated hydrocarbons. PVC is recommended for process piping, water service, and industrial and laboratory chemical waste drainage. Joining of pipe is usually by threading or fusion welding.

Polyvinyl dichloride pipe (PVDC) A pipe useful for handling corrosive fluids at temperatures 40 to 60°F above the limits for other vinyl plastic piping materials. Its chemical resisting characteristics are comparable with those of PVC. It has low thermal conductivity and will not sustain combustion (self-extinguishing). Suggested uses include process piping for hot corrosive liquids, hot and cold water lines, and applications above the temperature range of PVC.

Potable water Water which is satisfactory for drinking, culinary, and domestic purposes and meets the requirements of the health authority having jurisdiction.

Preassembled equipment Equipment normally fully "preassembled" in manufacturer's shop. After a check to see that all necessary parts are included and fit, the equipment is disassembled. The piping is taken apart at the unions only. All parts are usually tagged and numbered with numbers corresponding to those on assembly drawings that are furnished with the equipment.

Prefabricated equipment Equipment usually made up of flanged pipe and fittings. With this flanged piping, exact dimensions can be determined from engineering drawings and preassembly is not necessary. All parts are gathered together, tagged, and checked to make sure nothing is missing, and sent out as a single shipment. At the jobsite, the assembler can identify pieces from the drawings and build up the complete system.

Pretreatment   Treatment by using screens, degritters, degreasers, and scum removal devices to eliminate settleable solids such as floating debris, sand, and grit.

Primary sewage treatment   Treatment of raw sewage to remove the largest impurities (suspended solids or liquids, such as clays, bits of organic wastes, oil droplets). In domestic wastes, BOD is reduced about 30 to 40 percent. This system is mostly physical, employing sedimentation tanks, flotation tanks, flocculation systems, and occasional screening.

Private sewage disposal system   A septic tank with the effluent discharging into a subsurface disposal field, one or more seepage pits, a combination of subsurface disposal field and seepage pits, or such other facilities as may be permitted under the local governing agency.

Private sewer   A sewer privately owned and not directly controlled by a public authority.

Public sewer   Any sewer dedicated to public use and whose use is controlled by a public corporation. A common sewer is directly controlled by a public authority.

Quality of the water (waters)   Chemical, physical, biological, bacteriological, radiological, and other properties and characteristics of water which affect its use.

Receptor   An approved plumbing fixture or device of such material, shape, and capacity as to adequately receive the discharge from indirect waste pipes, so located as to be readily cleaned.

Reclaimed water   Water which, as the result of treatment of waste, is suitable for direct beneficial use or a controlled use that would not otherwise occur.

Relief vent   A vent, the primary function of which is to provide circulation of air between drainage and vent systems or to act as an auxiliary vent on a specially designed system.

Reverse osmosis   The principle involving diffusion of water through a semipermeable membrane from a dilute to a concentrated solution. By applying pressure to the concentrated solution greater than osmotic pressure, the diffusion of water through the membrane is reversed. Pure water passes from the concentrated into dilute solution. This results in a separation of dissolved solids.

Roughing-in   The installation of those parts of the plumbing system which can be completed prior to the installation of fixtures. This includes drainage, water supply, and vent piping and the necessary fixture supports.

Sanitary sewage   (1) Domestic sewage with storm or surface water excluded; (2) sewage discharging from sanitary conveniences; (3) the water supply of a community after it has been used and discharged into a sewer.

Sanitary sewer system   All of the property involved in the operation of sewage collection, treatment, and disposal systems including land, sewers, appurtenances, pumping stations, treatment works, and equipment.

Secondary sewage treatment   Treatment operations designed to remove dissolved and colloidal organic compounds such as proteins, sugars, starches, and phenols. These organics may themselves be harmful, but additionally they consume oxygen and increase BOD. Typical processes include activated sludge, trickling filters, stabilization ponds, and aeration lagoons. All work on the same basic principle: the accelerated biological degradation or consumption of organic compounds.

Seepage pit A lined excavation in the ground which receives the discharge of a septic tank so designed as to permit effluent from the septic tank to seep through its bottom and sides.

Septic tank A watertight receptacle which receives the discharge of a drainage system or part thereof, designed and constructed so as to retain solids, digest organic matter through a period of detention, and allow the liquids to discharge into the soil outside of the tank through a system of open joint piping or a seepage pit meeting the requirements of the local governing authority.

Sewage pumping plant Any facility designed and constructed to raise wastewater in elevation or to overcome head losses due to pipeline friction.

Sewer A pipe or conduit, generally closed, but normally not flowing full, for carrying sewage and other waste liquids.

Sewerage Any and all facilities used for collection, conveying, pumping, treating, and disposing of wastewater.

Sewerage system A network of wastewater collection, conveyance, treatment, and disposal facilities interconnected by sewers.

Solid wastes The non-liquid-carried wastes normally considered to be suitable for disposal with refuse at sanitary landfill refuse disposal sites.

"Standard Methods" "Standard Methods for the Examination of Water and Wastewater" as published by the American Public Health Association.

Storm drain system All the property involved in the operation of the storm drainage collection and disposal system including conduits, natural or artificial drains, channels, and watercourses, together with appurtenances, pumping stations, and equipment.

Suspended solids The insoluble solid matter suspended in wastewater that is separable by laboratory filtration in accordance with the procedure described in "Standard Methods."

Tertiary sewage treatment Treatment operations designed to eliminate residual dissolved organic and inorganic compounds. Inorganics are usually removed in electrodialysis, reverse osmosis, or iron exchange. In large concentrations, distillation, freezing, or other desalination-type techniques are more appropriate for removing inorganics. One of the most effective means of treating organics is by absorption on activated carbon.

Tidal waters All coastal ocean waters of a state including bays and estuaries upstream to the inland limit of tidal action.

Treatment works (1) Any device or system used in the storage, treatment, recycling, and reclamation of municipal sewage or industrial wastes of a liquid nature including intercepting sewers, outfall sewers, sewage collection systems, pumping, power, and other equipment and their appurtenances; (2) any method or system for preventing, abating, reducing, storing, treating, separating, or disposing of municipal waste, storm water runoff, industrial waste, or any combination of these.

Trunk sewer   A sewer constructed, maintained, and operated to convey wastewater to treatment facilities and into which lateral and collecting sewers discharge.

Uncontaminated water   Any wasted water of the community not contaminated or polluted with wastewater, which is suitable or could be readily made suitable for discharge into the storm drainage system.

Wastewater   The water-carried wastes of the community derived from human or industrial sources including domestic wastewater and industrial wastewater. Rainwater, groundwater, or drainage of uncontaminated water is not wastewater.

Water quality control   The regulation of any activity or factor which may affect the quality of the waters, including the prevention and correction of water pollution and nuisance.

Water quality control plan   A designation or establishment for the waters within a specified area of (1) beneficial uses to be protected, (2) water quality objectives, and (3) program of implementation needed for achieving water quality objectives.

Water quality objectives   The limits or levels of water quality constituents or characteristics which are established for the reasonable protection of beneficial uses of water.

References

1. Allen, J. B., et al.: Process Design Calculations for Absorption from Liquids in Fixed Beds of Granular Activated Carbon, *J. WPFC,* vol. 39, no. 2, Feb., 1967.

2. American Standard: Scheme for the Identification of Piping Systems, ASA A13.1-1956.

3. American Standard: Graphic Symbols for Plumbing, ASA Y32.4-1955, The American Society of Mechanical Engineers, New York, 1955.

4. American Standard: Graphic Symbols for Pipe Fittings, Valves and Piping, ASA Z32.2.3-1949.

5. American National Standard: Abbreviations for Use on Drawings or in Text, ANSI Y1.1-1972, The American Society of Mechanical Engineers, New York, 1972.

6. Armco Drainage and Metal Products, Inc.: *Handbook of Water Control,* Calco Division, Berkeley, Calif., 1949.

7. Armco Drainage and Metal Products, Inc.: *Handbook of Drainage and Construction Products,* Middletown, Ohio, 1955.

8. Anon.: New Effluent Clean-up Units Polish Treated Water, *Chem. Eng.,* May 22, 1967.

9. Anon: Sample Handling: Prerequisites to Accurate Process Analyses, *Analyzer,* Oct., 1962.

10. Anon.: Complete Mixing: Something for Nothing, *Trans. ASCE,* p. 413, 1965.

11. Anon.: War Against Pollution Gets Under Way, *Chem. Eng.,* May 28, 1966.

12. Bataille, Gerald S.: Pinpoint Your Water Pollution Problems, *Plant Engineering Magazine* (reprint), Barrington, Ill., 1972.

13. Bernard, R.: Treatment of Industrial Wastes from Usina de Melle, *Tech. de l'Eau,* Brussels, no. 236, Aug., 1966.

14. Besselievre, E. B.: Design of Sewage Dosing Tanks for Trickling Filters, *Eng. News-Record,* Oct. 19, 1922.

15. Besselievre, E. B.: The Responsibility of the Community with Regard to Industry, *Public Works,* Oct., 1952.

16. Besselievre, E. B.: Pontiac Motors Treats Its Wastes, *Wastes Eng.,* Nov., 1958.

17. Besselievre, E. B.: Auto Wastes Pretreated for Discharge into City Sewers, *Wastes Eng.,* Dec., 1958.

18. Besselievre, E. B.: The Economics of Coagulants, Prts 1–4, *Water Works Waste Eng.,* Oct.–Jan., 1965.

19. Besselievre, E. B.: The Four Rs: A Solution for the Water Problem, Highlights, First International Water Quality Symposium, Washington, D.C., Aug., 1965.

20. Besselievre, E. B.: Industry Must Reuse Effluents, *Wastes Eng.,* Dec., 1960.

21. Besselievre, E. B.: Industries Recover Water and By-products from Their Wastes, *Wastes Eng.,* Dec., 1966.

22. Bishop, D. F., et al.: Studies on Activated Carbon Treatment, *J. WPCF,* vol. 39, no. 2, Feb., 1967.

23. Blendermann, Louis: *Design of Plumbing and Drainage Systems,* The Industrial Press, New York, 1959.

24. Buck, F. W.: Diatomite Filtration of Industrial Wastes, *Water Sewage Works,* June, 1966.

25. Busch, A. W.: Biochemical Oxidation of Process Waste Waters, *Chem. Eng.,* Mar. 1, 1965.

26. Calgon Corporation: *Man and Water: Water Rights,* No. 3, 1972.

27. California Regional Water Quality Control Board: "Policy Guidelines," Chapter V, Sacramento, Calif., 1972.

28. Carr, J. E.: Principles of Sampling Effluent Streams, Denver Equipment Co., Div. of Joy Manufacturing Co., Denver, Col.

29. Ceresa, Myron and Leslie E. Lancy: Wastewater Treatment, Metal Finishing Guidebook and Directory, 1972.

30. City of Los Angeles: *Plumbing Code,* Building News, Inc., Los Angeles, Calif., 1971.

31. Clarke, E. E.: Industrial Reuse of Water, *Ind. Eng. Chem.,* vol. 134, no. 2, Feb., 1962.

32. Cleary, E. F., The ORSANCO Story: Quality Management in the Ohio River Valley under an Interstate Compact, The Johns Hopkins Press, Baltimore, Md., July, 1967.

33. Day, Robert V.: Understanding Sewer Ordinances, *Plant Engineering Magazine* (reprint), Barrington, Ill.

34. Dean, B. T.: The Design and Operation of a Deep-well Disposal System, *J. WPCF*, vol. 37, no. 2, Feb., 1965.

35. Deluzio, F. C.: Issues and Problems in Water Pollution, *Water Wastes Eng.*, May, 1967.

36. Dillon, K. E.: Waste Disposal Made Profitable, *Chem. Eng.*, March 3, 1967.

37. Donaldson, E. C.: Subsurface Disposal of Industrial Wastes in the United States, U.S. Department of the Interior, Bureau of Mines Information Circ. 8212, 1964.

38. Eliassen, R., and Lauderdale, R. A.: Liquid and Solid Wastes Disposal, in H. Blatz, "Handbook of Radiation Hygiene," McGraw-Hill Book Company, New York, 1959.

39. Ewing, R. C.: Waste-treatment Plant Designed for Industry Complex, *Oil Gas J.*, March 13, 1967.

40. F. E. Myers & Bro. Co.: "ERA II, High Velocity Sewer Cleaners," Ashland, Ohio.

41. Friedlander, H. Z., and Rickles, R. N.: Membrane Separation Processes, *Chem. Eng.*, Feb. 28, 1966.

42. Genetelli, E. J.: Treatability of Industrial Wastes, *Ind. Water Eng.*, July, 1967.

43. Gordon, Culp, and Suhr, L. Gene: *Physical-Chemical Wastewater Treatment Plant Design, Environmental Protection Agency-Transfer Technology*, 1973, U.S. Government Printing Office.

44. Hach Chemical Company: *Water Analysis Equipment and Chemicals*, Ames, Ohio, 1972.

45. Hardengergh, W. A.: *Sewerage and Sewage Treatment*, International Textbook Company, Scranton, Ohio, 1946.

46. Hettig, S. B.: "Waste Farm" Takes Care of Phenolic Waste Disposal, *Chem. Eng.*, Dec. 28, 1962.

47. Hickel, Walter J.: Industry's Role in the Clean Water Battle, *Plant Engineering Magazine* (reprint), Barrington, Ill., 1972.

48. Hodges, Paul B.: Sampling and Flow Measuring Devices for Plant Surveys, *Ind. Wastes*, Dec., 1960.

49. Illinois Water Treatment Company: *IWT Mixed-Bed De-ionizers*, Bulletins SA-467 and PKA-572.

50. International Association of Plumbing and Mechanical Officials: *Uniform Plumbing Code*, Los Angeles, Calif., 1973.

51. Joint Committee of the American Society of Civil Engineers and the Water Pollution Control Federation: Design and Construction of Sanitary and Storm Drains, New York, 1966.

52. Kantawala, D., and Tomlinson, W. D.: Comparative Study of Recovery of Zinc

and Nickel by Ion Exchange Method and Chemical Precipitation, *Water Sewage Works,* April, 1964.

53. Katz, W. J., and Gernopulous, A.: Sludge Thickening by Dissolved Air Floatation, *J. WPCF,* vol. 39, no. 6, June, 1967.

54. Keefer, C. E.: Estimating the Life of Sewage Treatment Facilities, *Public Works,* July, 1962.

55. King, Horace William: "Handbook of Hydraulics," McGraw-Hill Book Company, Inc., New York, 1939.

56. Kneale, J. S.: Sludge Dewaterer Solves Space Problems, *Public Works,* March, 1967.

57. Kneese, A. V.: The Ruhr and the Delaware, *J. Sanitary Eng.,* Division ASCE, SA-5, Oct., 1966.

58. Lamb, J. C., III, et al.: A Technique for Evaluating the Biological Treatment of Industrial Wastes, *J. WPCF,* Oct., 1964.

59. Lancy, L. E.: An Economic Study of Metal-finishing Waste Treatment, *Plating,* Feb., 1967.

60. Lancy, L. E.: Neutralizing Liquid Wastes in Metal-finishing, *Metal Progr.,* April, 1967.

61. Lewis, W. L., and Martin, W. L.: Remove Phenols from Waste Water, *Hydrocarbon Processing,* Feb., 1967.

62. M. C. Nottingham Company: *Sewage and Industrial Waste Systems,* Irwindale, Calif.

63. Malina, J. F., Jr.: Don't Waste—Renovate and Reuse, *Am. City,* March, 1966.

64. Manas, Vincent T.: "National Plumbing Code Handbook," McGraw-Hill Book Company, New York, 1957.

65. Manning Environmental Corporation, *Manning Dipper Portable Sampler,* Santa Cruz, Calif.

66. Mead, Daniel W.: *Standards of Professional Relations and Conduct,* American Society of Civil Engineers, New York, 1963.

67. Means, Robert Snow: *Building Construction Cost Data 1974,* Robert Snow Means Company, Inc., Duxbury, Mass., 1974.

68. Morris, A. L.: Water Renovation, *Ind. Water Eng.,* June, 1967.

69. Moselle, Gary, and Albert S. Paxton: "National Construction Estimator," Craftsman Book Company of America, Los Angeles, Calif., 1974.

70. McGauhey, P. H.: Folflore in Water Quality Standards, *Civil Eng.,* July, 1967.

71. McKee, Jack Edward, and Wolf, Harold W., Eds.: *Water Quality Criteria,* 2nd ed., State of California, State Water Resources Control Board, Publication No. 3-A, 1971.

72. Neale, J. H.: *Advanced Waste Treatment by Distillation,* Report AWT-R-and U.S. Public Health Service, Publ. 999-WP-9, May, 1964.

73. Nobolsine, Ross and Donovan: *Treating Methods Available for Industrial Waste Water,* The Plant Engineering Co.

74. Okey, R. W., and Stavenger, P. L.: An Investigation of Potential Uses of Reverse Osmosis in Industrial Waste Treatment, *Ind. Water Eng.,* March, 1967.

75. Olsen, Alan E.: *Upgrading Metal Finishing Facilities to Reduce Pollution,* Oxy Metal Finishing Corporation, Madison Heights, Michigan.

76. Oswall, W. J., and Gotass, H. B.: Photosynthesis in Sewage Treatment, *Proc. ASCE,* vol. 81, 1955.

77. Permutit Sybron Corporation: *Permutit Continuous Destruction,* Bulletin 5605, Paramus, N.J., 1971.

78. Perry, Robert H.: "Chemical Engineers Handbook," 4th ed., McGraw-Hill Book Company, New York, 1969.

79. Porges, R., et al.: *Stabilization Ponds for Treatment of Industrial Wastes,* U.S. Department of Health, Education and Welfare Tech. Rept/W-6129, 1961. (Contains bibliography of 33 references on all types of wastes.)

80. Prenco Division, Pickands Mather and Co.: *Prenco Super E_3 Pyro-Decomposition System,* Detroit, Michigan.

81. Process Equipment-Division of Hytek International Corporation: Tech. Data #103: *Chrome Removal;* Tech. Data #100: *High Capacity Demineralizers.*

82. Ragan, J. L., et al.: Industrial Waste Disposal by Solar Evaporation, *Ind. Water and Wastes,* vol. 8, no. 4, July–Aug., 1963.

83. Recommended Procedures in the Use of Chlorine at Water and Sewage Plants, Joint Committee on Chlorine Supply, *Water Sewage Works,* 1955.

84. Richardson Engineering Services, Inc.: *Process Plant Construction and Estimating Standards,* Solana Beach, Calif., 1974.

85. Rickles, R. N.: Waste Recovery vs. Pollution Abatement, *Chem. Eng.,* Sept. 27, 1965.

86. Rock Valley Water Conditioning, Inc.: *Demineralizers,* Bulletin 71.1.1, Rockford, Ill.

87. Rudolfs, W.: "Industrial Wastes: Their Disposal and Treatment," chap. 13, Plating, Wastes, Reinhold Publishing Corporation, New York, 1953.

88. Ryckman, D. W., et al.: Evaluation and Abatement of Industrial Waste Problems, *Ind. Water Wastes,* vol. 7, no. 4, July–Aug., 1962.

89. Sanitation Districts of Los Angeles County: *Instructions for Filing an Industrial Wastewater Treatment Surcharge Statement,* 1973.

90. Sanitation Districts of Los Angeles County: *Instructions for Obtaining a Permit For Industrial Wastewater Discharge,* 1974.

91. Sanitation Districts of Los Angeles County: *An Ordinance Regulation Sewer Construction, Sewer Use and Industrial Wastewater Discharges,* 1972.

92. Saylor, Lee: Current Construction Costs 1974, Lee Saylor Inc., Walnut Creek, Calif., 1974.

93. Scharf, S.: Chrysler Attacks Pollution Problem, *Water Wastes Eng.,* Oct., 1966.

94. Schore, George: Automatic Waste Treatment Plants, Schore Automations, Inc., Westbury, Long Island, New York.

95. Schulze, K. L.: Biological Recovery of Waste Waters, *J. WPCF,* vol. 38, no. 12, December, 1966.

96. Sethco Mfg. Corp: *Bulletin WWT 5* and *Bulletin V-103,* Freeport, N.Y.

97. Shaw, G. V., and Loomis, A. W.: *Cameron Hydraulic Data,* Compressed Air Magazine Company, Inc., N.Y., 1962.

98. State of California: *Construction Safety Orders,* Building News, Inc., Los Angeles, Calif.

99. The Leon J. Barrett Company: "Liquid Clarifier," Worcester, Mass.

100. U.S. Environmental Protection Agency: *Interim Effluent Guidance for NPDES Permits,* Office of Permit Programs, Washington, D.C., 1973.

101. U.S. Environmental Protection Agency: *Federal Guidelines Pretreatment of Discharges to Publicly Owned Treatment Works,* Office of Water Programs Operations, Washington, D.C., 1973.

102. Warner, D. L.: Deepwell Disposal of Industrial Waste, *Chem. Eng.,* Jan. 4, 1965.

103. Weil Pump Co.: "Pneumatic Cast Iron Sewerage Ejectors," Bulletin LSP-2375, Chicago, Ill.

104. Western Plumbing Officials: *Uniform Plumbing Code,* Western Plumbing Officials Association, 1964.

105. Winar, R. M.: "The Disposal of Wastewater Underground," *Ind. Water Eng.,* March, 1967.

Index

Venturi tube, 32
Vibrating screens, 212
Vitrified clay pipe, 109
V-notch weir, 30
Volume measurements, 30–33
(*See also* Flow measurement; Weirs)

Washington, state of, incentives, 344
Waste products, reuse of, 2
Waste treatment plants (*see* Treatment plants)
Wastes, type of, in metal-finishing industry, 230
Wastewater characteristics, metal-finishing industry 230–231
Water logs (*see* Log forms)
Water management checklist, 18
Water Pollution Control Act, 1972 (*see* Federal Water Pollution Prevention and Control Act)

Water Quality Control Advisory Committee, 242
Water Quality Control Boards, Regional, 242
Water quality control plan for ocean waters of California, requirements, 242
Water quality control programs, state, 241–242
Weirs, 30–33
  Cipolletti, 31
  Parshall, 32
  rectangular, 31
  trapezoidal, 31
  V-notch, 30–31, 101

Zahn process, sulfuric acid recovery, 182
Zimmerman process, wet-air oxidation, 303